International Series in Operations Research & Management Science

Volume 289

More information about this series at http://www.springer.com/series/6161

Boris Sokolov • Dmitry Ivanov • Alexandre Dolgui
Editors

Scheduling in Industry 4.0 and Cloud Manufacturing

Springer

Editors
Boris Sokolov
Saint Petersburg Institute for Informatics and Automation of the RAS (SPIIRAS)
St. Petersburg, Russia

Dmitry Ivanov
Berlin School of Economics and Law
Berlin, Germany

Alexandre Dolgui
IMT Atlantique, LS2N—CNRS
La Chantrerie, Nantes, France

ISSN 0884-8289 ISSN 2214-7934 (electronic)
International Series in Operations Research & Management Science
ISBN 978-3-030-43176-1 ISBN 978-3-030-43177-8 (eBook)
https://doi.org/10.1007/978-3-030-43177-8

This Springer imprint is published by the registered company Springer Nature Switzerland AG.
The registered company address is: Gewerbestrasse 11, 6330 Cham, Switzerland

Contents

Chapter 1
Introduction to Scheduling in Industry 4.0 and Cloud Manufacturing Systems

Dmitry Ivanov, Boris Sokolov, and Alexandre Dolgui

Abstract In this chapter, we present an introduction to scheduling in Industry 4.0 and cloud manufacturing systems. We elaborate on the peculiarities of scheduling and sequencing problems in the context of Industry 4.0 and smart manufacturing. We delineate recent research streams and summarize the structure and contribution of the book.

1.1 Scope of the Research Domain

Cloud manufacturing and Industry 4.0 emerge with specific scheduling problems encompassing complex hybrid logical and terminal constraints, non-stationarity in process execution, as well as complex interrelations between dynamics in process design, capacity utilization, and machine setups (Ivanov et al. 2020). The flexibility and service orientation of digital manufacturing allow for dynamically changing process technologies (i.e., flexibly reconfigurable jobs through customer-specific sequencing of operations) to achieve the product individualization and variety. The Industry 4.0 technology and cloud manufacturing are enabling flexible production, particularly through the use of cyber-physical systems and highly customized assemblies (Hwang et al. 2017; Ahn et al. 2019; Frazzon et al. 2018; Leusin et al. 2018; Tao et al. 2018; Panetto et al. 2019; Yang et al. (2019)) in order to deliver

D. Ivanov (✉)
Berlin School of Economics and Law, Berlin, Germany
e-mail: divanov@hwr-berlin.de

B. Sokolov
Saint Petersburg Institute for Informatics and Automation of the RAS (SPIIRAS), St. Petersburg, Russia
e-mail: sokol@iias.spb.su

A. Dolgui
IMT Atlantique, LS2N—CNRS, La Chantrerie, Nantes, France
e-mail: alexandre.dolgui@imt-atlantique.fr

B. Sokolov et al. (eds.), *Scheduling in Industry 4.0 and Cloud Manufacturing*, International Series in Operations Research & Management Science 289,
https://doi.org/10.1007/978-3-030-43177-8_1

manufacturing services on-demand (Ivanov et al. 2018; Ahn et al. 2019; Liu et al. 2019; Dolgui et al. 2019b; Fragapane et al. 2020; Shukla et al. 2019; Zhou et al. 2018).

Such innovative production strategies engender new challenges and opportunities for short-term job scheduling and sequencing. One difficulty is the strong coupling when product and process are created simultaneously (Kusiak 2018). Simultaneous product and process creation direct the discussion in a class of scheduling problems that have mixed structural-temporal-logical constraints with order scheduling based on a search for free resources for free operations (Dolgui et al. 2019a). Another peculiarity of scheduling in cloud manufacturing and Industry 4.0 is its data-driven nature that, at the same time, becomes favorable for development of new decomposition methods.

Notably, the cloud manufacturing and Industry 4.0 build upon a service paradigm. Flexible processes and flexible machines allow to create a principally new entity for scheduling, that is, a service—an entity needed to fulfill a process step in a technological chain. The services are formed dynamically based on the available machines and operations. The services can be built by almost any combination of machines and operations (Ivanov et al. 2016; Ahn et al. 2019; Liu et al. 2019). As such, in digital manufacturing, the process design and job scheduling are integrated due to the flexible usage of services instead of the rigid and fixed allocations of operations and machines. In this context, the focus of scheduling shifts from assigning jobs to machines but rather emerges with a combination of process (task) composition and service composition, which are performed simultaneously to deliver manufacturing services to consumers on-demand.

Moreover, the COVID-19 outbreak—the unprecedented challenge to manufacturing industry—has clearly shown and highlighted the importance of scheduling in Industry 4.0 and cloud manufacturing. Firms with established technologies for manufacturing visibility and digital control were able to react to disruptions more flexible and responsive (Choi et al. 2020; Ivanov 2020; Ivanov and Dolgui 2020; Ivanov and Das 2020). Ni et al. (2020) point to the need of to develop "*predictive models for proactive scheduling and dynamic planning of supply demands with the consideration of uncertainties and risk factors. These predictive models will help corporate decision-makers do what-if analysis of various scenarios*".

In summary, the existing knowledge on scheduling in Industry 4.0, smart manufacturing, and cloud manufacturing is scattered over different methodologies but most of them share a certain set of attributes such as (a) flexible process design, (b) communication of products and machines, (c) flexible machines, and (d) service-oriented manufacturing.

For example, Audi smart factory in Baden-Württemberg implements a highly flexible assembly system based on the use of automated guided vehicles. Contrary to the traditional assembly systems with fixed layouts and process designs, the Audi smart factory allows for highly flexible process design and sequencing of production orders in order to achieve the highest degree of the product individualization while maintaining the manufacturing efficiency (Audi 2019). MindSphere (a manufacturing platform of Siemens) is cloud-based and represented an open Internet-of-Things (IoT) operating system where products, plants, systems, and machines are connected with each other to enable the usage of data generated by the IoT with advanced analytics in schedule optimization (Siemens 2019). Another flexibility driver are autonomous mobile robots that make it possible to develop principally new forms of assembly by flexible reallocation of the manufacturing services needed to produce a product (Nielsen et al. 2017; Fragapane et al. 2020). With the help of smart sensors and plug-and-produce cyber-physical systems, the stations in the assembly system are capable of changing the operation processing and setup sequences according to the actual order of the incoming flows and capacity utilization (Theorin et al. 2017). In the FOUP—front opening unified pods—technology in the semiconductor industry, robots are used for a real-time operation sequencing. Robots read the information about the products from the sensors and tags and decide flexibly, where to forward a wafer batch next (Mönch et al. 2012).

Despite the notable progress, scheduling in Industry 4.0 and cloud manufacturing is still at the beginning of its investigation. The results from the hybrid shop scheduling and sequencing with alternative parallel machines, resource constrained project scheduling with alternative chains, and design of reconfigurable manufacturing system and scheduling with both terminal and logical constraints can now be integrated into a unified framework of Industry 4.0 and require an extension toward models with hybrid structural-terminal-logical constraints (Dolgui et al. 2019a).

In the first published paper on scheduling in Industry 4.0, Ivanov et al. (2016) underline the role of real-time information to reduce combinatorial complexity of the NP-hard scheduling problems. Moreover, this study depicted the resemblance of Industry 4.0-based manufacturing systems to flexible multiprocessor flow shops. Rossit et al. (2019) developed a concept of smart scheduling, aiming to yield flexible and efficient production schedules that are formed dynamically in an event-oriented matter. Echoing the results in Ivanov et al. (2016), they also referred to similarity of the smart scheduling to flow shop scheduling problems. In Xu et al. (2019), online algorithms were developed to maximize the capacity of the machines in custom manufacturing environment with IoT (Internet-of-Things) capability. Despite some reported advances, doubts have been raised and considerable ambiguity remains, however, as to if and how the increased complexity of scheduling problems in the digital manufacturing can be efficiently tackled by the existing optimization techniques. Other recent advances in scheduling in cloud manufacturing have been documented for both products (Mourtzis and Vlachou 2018; Vespoli et al. 2019; Vahedi-Nouri et al. 2019; Ding et al. 2019; Helo et al. 2019; Yuan et al. 2019) and services (Chen et al. 2019) and reviewed by Liu et al. (2019) and Zhang et al. (2019).

1.2 Structure of the Book

This book has resulted from the activities of International Federation of Automatic Control Technical Committee (IFAC TC) 5.2 "Manufacturing Modelling for Management and Control." The book offers an introduction and advanced techniques of scheduling applications to cloud manufacturing and Industry 4.0 systems for larger audience. This book uncovers fundamental principles and recent developments in the theory and application of scheduling methodology to cloud manufacturing and Industry 4.0.

The purpose of this book is to comprehensively present recent developments in scheduling in cloud manufacturing and Industry 4.0 and to systemize these developments in new taxonomies and methodological principles to shape this new research domain. This book addresses the needs of both researchers and practitioners to uncover the challenges and opportunities of scheduling techniques' applications to cloud manufacturing and Industry 4.0. For the first time, it comprehensively conceptualizes scheduling in cloud manufacturing and Industry 4.0 systems as a new research domain. The chapters of the book are written by the leading international experts and utilize methods of operations research, industrial engineering, and computer science. Such a multidisciplinary combination is unique and comprehensively deciphers major problem taxonomies, methodologies, and applications to scheduling in cloud manufacturing and Industry 4.0.

Distinctive features of this book:

- Uncovering fundamental principles and recent developments in the theory and application of scheduling methodology to cloud manufacturing and Industry 4.0.
- Bridging the scheduling theory to cloud manufacturing and Industry 4.0 systems.
- Systemizing new developments and deciphering taxonomies and methodological principles to shape the new research domain scheduling in cloud manufacturing and Industry 4.0 systems.
- Innovative applications of scheduling in cloud manufacturing and Industry 4.0.
- Consideration of models with only terminal constraints, with hybrid terminal-logical constraints, and with hybrid structural-terminal-logical constraints.
- Analysis of computational algorithms.
- Data-driven scheduling models and analytics.
- Unique multidisciplinary view with utilization of operations research, industrial engineering, and computer science techniques.

The book contains 13 chapters written by the leading researchers in the field across the globe.

In the Introductory chapter "Introduction to Scheduling in Industry 4.0 and Cloud Manufacturing Systems," the book editors Dmitry Ivanov, Boris Sokolov, and Alexandre Dolgui elaborate on the peculiarities of scheduling and sequencing problems in the setting of Industry 4.0 and smart manufacturing. They delineate recent research streams and summarize the structure and contribution of the book.

Real-Time Control Dmitry Ivanov, Boris Sokolov, Frank Werner, and Alexandre Dolgui develop in their chapter, "Proactive Scheduling and Reactive Real-Time Control in Industry 4.0," a control-theoretic perspective of modeling highly flexible manufacturing systems when process and schedule need to be created simultaneously. By combining the advantages of continuous and discrete optimization, their chapter develops a decomposition method for shop floor scheduling in Industry 4.0 manufacturing systems.

Digital Twin Ícaro Agostino, Sousa Romolo, Eike Broda, Enzo M. Frazzon, and Michael Freitag focus their chapter, "Using a Digital-Twin for Production Planning and Control in Industry 4.0," on the application of simulation models in production and logistic systems to frame the digital twins. A digital-twin approach to production planning and control using current cyber-physical systems state data in real time is presented and validated.

Cloud Material Handling System Fabio Sgarbossa, Mirco Peron, and Giuseppe Fragapane propose in their chapter, "Cloud Material Handling Systems: Conceptual Model and Cloud-Based Scheduling of Handling Activities," a new material handling paradigm called Cloud Material Handling System (CMHS), developed in the Logistics 4.0 Lab at Norwegian University of Science and Technology (NTNU), Norway. With the help of CMHS, scheduling of the Material Handling Modules can be optimized, increasing the flexibility and productivity of the overall manufacturing system.

Adaptation Dimitris Mourtzis describes in his chapter, "Adaptive Scheduling in the Era of Cloud Manufacturing," the evolution of scheduling techniques during the last decade. The chapter also provides insightful and meaningful inferences from the application of innovative solutions in industrial use cases.

Robust Scheduling Pascale Marangé, David Lemoine, Alexis Aubry, Sara Himmiche, Sylvie Norre, Christelle Bloch, and Jean-François Pétin present in their chapter, "Coupling Robust Optimization and Model-Checking Techniques for Robust Scheduling in the Context of Industry 4.0," a generic methodology for robustness analysis in production systems of Industry 4.0. The authors consider this as the first milestone for coupling Operations Research models for robust optimization and Discrete Event Systems models and tools for property checking for converging toward a solution with the required robustness level defined by the decision-makers.

Information Services Dmitry Ivanov and Boris Sokolov elaborate in their chapter, "Integrated Scheduling of Information Services and Logistics Flows in the Omnichannel System," on an integration of information services and logistics flows in the scheduling of omnichannel systems. The proposed service-oriented description makes it possible to coordinate the information services and material process schedules simultaneously. It also becomes possible to optimize the use of information services needed for physical supply processes. In addition, impact of disruptions in information services on the schedule execution in the physical structure is analyzed.

Human Factors Daria Battini, Serena Finco, and Fabio Sgarbossa contribute a chapter entitled "Human-Oriented Assembly Line Balancing and Sequencing Model in the Industry 4.0 Era." Using real-time monitoring approach, a dynamic scheduling and sequencing method is proposed to guarantee the right safety level for each worker. The authors provide a general overview of smart tools for measuring and quantifying the ergonomics level. Based on the data from smartwatches, they frame a multi-objective assembly line balancing model and an ergo-sequencing model demonstrating the benefits of using smart solutions and Industry 4.0 tools.

Modularity Nathalie Klement and Cristóvão Silva propose in their chapter, "A Generic Decision Support Tool to Planning and Assignment Problems: Industrial Applications and Industry 4.0," a generic, modular decision support tool to support decision-making in planning, assignment, scheduling, and lot-sizing in Industry 4.0 systems. They show applications of their approach to several tactical and operational problems, for example, a problem of planning activities with resources assignment for hospital systems, a lot-sizing and scheduling problem taking into account the setup time for a textile application and for a plastic injection problem, and a scheduling problem with precedence constraints. At the strategic level, this tool can also be used as part of the Industry 4.0 to design reconfigurable manufacturing systems.

Cloud Paradigm Andrea Grassi, Guido Guizzi, Liberatina Carmela Santillo, and Silvestro Vespoli analyze in their chapter, "The Manufacturing Planning and Control System: A Journey Toward the New Perspectives in Industry 4.0 Architectures," an evolution of manufacturing planning and control toward Industry 4.0 and cloud manufacturing paradigms. They present different classical approaches and extend their scope of application toward the positioning of Industry 4.0 in future market scenarios. Further, the authors elaborate on the new technologies in Industry 4.0, and how they can act as enablers for bridging the gap between current and future production control approaches. Finally, they explore the state of the art of Industry 4.0 and cloud manufacturing architectural implementations, also outlining further development possibilities and strategies.

Energy Efficiency Junheng Cheng, Feng Chu, and Peng Wu elaborate in their chapter, "Multi-Criteria Single Batch Machine Scheduling Under Time-of-Use Tariffs," on the sustainability issues and energy-efficient manufacturing in Industry 4.0. They introduce a basic single machine batch scheduling problem under time-of-use (ToU) electricity tariffs. Then it is extended by further considering machine on/off switching. Finally, a single machine batch scheduling problem under ToU tariffs in a continuous-processing environment is investigated. For the three considered problems, appropriate mathematical models are established, and their problem properties and complexities are demonstrated.

Service Composition Haifeng Zhang, Yongkui Liu, Huagang Liang, Lihui Wang, and Lin Zhang devote their chapter, "Service Composition in Cloud Manufacturing: A DQN-Based Approach," to the application of deep reinforcement learning to solving cloud manufacturing service composition problems. They propose a deep Q network (DQN)-based service composition approach in which the system can learn how to find optimal service composition solutions automatically.

Rescheduling Daniel Alejandro Rossit, Fernando Tohmé, and Gonzalo Mejía introduce in their chapter, "The Tolerance Scheduling Problem in a Single Machine Case," a Tolerance Scheduling problem, which involves the decision-making issues in rescheduling processes. The solutions to this problem can be incorporated in the design of Decision Support Systems (DSS) in Industry 4.0 environments. The authors present the mathematical foundations for the solutions of the Tolerance Scheduling problem as well as the technical requirements of their embodiment in Industry 4.0's DSS.

References

Ahn, G., Park, Y.-J., & Hur, S. (2019). Performance computation methods for composition of tasks with multiple patterns in cloud manufacturing. *International Journal of Production Research, 57*(2), 517–530.

Audi (2019). Flexible Montage in der Fahrzeugproduktion Die flexible Audi R8-Manufaktur mit fahrerlosen Transportfahrzeugen. Retrieved October 4, 2019 from https://www.plattform-i40.de/PI40/Redaktion/DE/Anwendungsbeispiele/137-wandelbare-r8-manufaktur/beitrag-wandelbare-r8-manufaktur.html.

Chen, J., Huang, G. Q., Wang, J.-Q., & Yang, C. (2019). A cooperative approach to service booking and scheduling in cloud manufacturing. *European Journal of Operational Research, 273*(3), 861–873.

Choi TY, Rogers D, Vakil B. (2020). Coronavirus is a wake-up call for supply chain management. Harvard business review, march 27th.

Ding, K., Chan, F. T. S., Zhang, X., Zhou, G., & Zhang, F. (2019). Defining a Digital Twin-based Cyber-Physical Production System for autonomous manufacturing in smart shop floors. *International Journal of Production Research, 57*(20), 6315–6334.

Dolgui, A., Ivanov, D., Sethi, S. P., & Sokolov, B. (2019a). Scheduling in production, supply chain and Industry 4.0 systems by optimal control. *International Journal of Production Research, 57*(2), 411–432.

Dolgui, A., Ivanov, D., Potryasaev, S., Sokolov, B., Ivanova, M., & Werner, F. (2019b). Blockchain-oriented dynamic modelling of smart contract design and execution control in the supply chain. *International Journal of Production Research.* https://doi.org/10.1080/00207543.2019.1627439.

Fragapane, G., Ivanov, D., Peron, M., Sgarbossa, F., & Strandhagen, J. O. (2020). Increasing flexibility and productivity in industry 4.0 production networks with autonomous mobile robots and smart intralogistics. *Annals of Operations Research.* https://doi.org/10.1007/s10479-020-03526-7.

Frazzon, E. M., Kück, M., & Freitag, M. (2018). Data-driven production control for complex and dynamic manufacturing systems. *CIRP Annals, 67,* 515–518. https://doi.org/10.1016/j.cirp.2018.04.033.

Helo, P., Phuong, D., & Hao, Y. (2019). Cloud manufacturing—Scheduling as a service for sheet metal manufacturing. *Computers and Operations Research, 110,* 208–219.

Hwang, G., Lee, J., Park, J., & Chang, T.-W. (2017). Developing performance measurement system for Internet of Things and smart factory environment. *International Journal of Production Research, 55*(9), 2590–2602.

Ivanov D. (2020) Predicting the impact of epidemic outbreaks on the global supply chains: A simulation-based analysis on the example of coronavirus (COVID-19 / SARS-CoV-2) case. Transportation research – Part E, 136., https://doi.org/10.1016/j.tre.2020.101922.

Ivanov D., Das A. (2020). Coronavirus (COVID-19 / SARS-CoV-2) and supply chain resilience: A research note. International journal of integrated supply management, forthcoming.

Ivanov, D., & Dolgui, A. (2020). Viability of intertwined supply networks: Extending the supply chain resilience angles towards survivability. In *A position paper motivated by COVID-19 outbreak.* International Journal of Production: Research. https://doi.org/10.1080/00207543.2020.1750727.

Ivanov, D., Sokolov, B., Dolgui, A., Werner, F., & Ivanova, M. (2016). A dynamic model and an algorithm for short-term supply chain scheduling in the smart factory Industry 4.0. *International Journal of Production Research, 54*(2), 386–402.

Ivanov, D., Sethi, S., Dolgui, A., & Sokolov, B. (2018). A survey on the control theory applications to operational systems, supply chain management and Industry 4.0. *Annual Reviews in Control, 46*, 134–147.

Ivanov, D., Sokolov, B., Chen, W., Dolgui, A., Werner, F., & Potryasaev, S. (2020). A control approach to scheduling flexibly configurable jobs with dynamic structural-logical constraints. *IISE Transactions*. https://doi.org/10.1080/24725854.2020.1739787.

Kusiak, A. (2018). Smart manufacturing. *International Journal of Production Research, 56*(1–2), 508–517.

Leusin, M., Frazzon, E., Uriona Maldonado, M., Kück, M., & Freitag, M. (2018). Solving the job-shop scheduling problem in the Industry 4.0 era. *Technologies, 6*(4), 107.

Liu, Y., Wang, L., Wang, X. V., Xu, X., & Zhang, L. (2019). Scheduling in cloud manufacturing: State-of-the-art and research challenges. *International Journal of Production Research, 57*(15–16), 4854–4879.

Mönch, L., Fowler, J. W., & Mason, S. (2012). *Production planning and control for semiconductor wafer fabrication facilities: Modeling, analysis, and systems*. Dordrecht: Springer Science & Business Media.

Mourtzis, D., & Vlachou, E. (2018). A cloud-based cyber-physical system for adaptive shop-floor scheduling and condition-based maintenance. *Journal of Manufacturing Systems, 47*, 179–198.

Ni, J. (2020). How China can rebuild global supply chain resilience after COVID-19. https://www.weforum.org/agenda/2020/03/coronavirus-and-global-supply-chains/, accessed on April 4, 2020.

Nielsen, I., Dang, Q.-V., Bocewicz, G., & Banaszak, Z. (2017). A methodology for implementation of Mobile robot in adaptive manufacturing environments. *Journal of Intelligent Manufacturing, 28*(5), 1171–1188.

Panetto, H., Iung, B., Ivanov, D., Weichhart, G., & Wang, X. (2019). Challenges for the cyber-physical manufacturing enterprises of the future. *Annual Reviews in Control, 47*, 200–213.

Rossit, D. A., Tohmé, F., & Frutos, M. (2019). Industry 4.0: Smart scheduling. *International Journal of Production Research, 57*(12), 3802–3813.

Shukla, N., Tiwari, M. K., & Beydoun, G. (2019). Next generation smart manufacturing and service systems using big data analytics. *Computers & Industrial Engineering, 128*, 905–910.

Siemens (2019). What doesn't happen keeps our world running smoothly - the power of MindSphere. https://www.plm.automation.siemens.com/global/en/topic/mindsphere-whitepaper/28842 (accessed on November 18, 2019).

Tao, F., Qi, Q., Liu, A., & Kusiak, A. (2018). Data-driven smart manufacturing. *Journal of Manufacturing Systems, 48*(C), 157–169.

Theorin, A., Bengtsson, K., Provost, J., Lieder, M., Johnsson, C., Lundholm, T., & Lennartson, B. (2017). An event-driven manufacturing information system architecture for Industry 4.0. *International Journal of Production Research, 55*(5), 1297–1311.

Vahedi-Nouri, B., Tavakkoli-Moghaddam, R., & Rohaninezhad, R. (2019). A multi-objective scheduling model for a cloud manufacturing system with pricing, equity, and order rejection. In D. Ivanov, A. Dolgui, & F. Yalaoui (Eds.), *Proceedings of the 9th IFAC Conference on Manufacturing Modeling, Management and Control MIM 2019, IFAC PapersOnLine* (Vol. 52(13), pp. 2177–2182). Amsterdam: Elsevier.

Vespoli, S., Grassi, A., Guizzi, G., & Santillo, L. C. (2019). Evaluating the advantages of a novel decentralised scheduling approach in the Industry 4.0 and cloud manufacturing era. In D. Ivanov, A. Dolgui, & F. Yalaoui (Eds.), *Proceedings of the 9th IFAC Conference on Manufacturing Modeling, Management and Control MIM 2019, IFAC PapersOnLine* (Vol. 52(13), pp. 2170–2176). Amsterdam: Elsevier.

Xu, J., Tran, H. M., Gautam, N., & Bukkapatnam, S. T. S. (2019). Joint production and maintenance operations in smart custom-manufacturing systems. *IISE Transactions, 51*(4), 406–421.

Yang, H., Kumara, S., Bukkapatnam, S. T. S., & Tsung, F. (2019). The internet of things for smart manufacturing: A review. *IISE Transactions, 51*(11), 1190–1216.

Yuan, M., Cai, X., Zhou, Z., Sun, C., Gu, W., & Huang, J. (2019). Dynamic service resources scheduling method in cloud manufacturing environment. *International Journal of Production Research*. https://doi.org/10.1080/00207543.2019.1697000.

Zhang, J., Ding, G., Zou, Y., Qin, S., & Fu, J. (2019). Review of job shop scheduling research and its new perspectives under Industry 4.0. *Journal of Intelligent Manufacturing, 30*(4), 1809–1830.

Zhou, L., Zhang, L., Sarker, B. R., Laili, Y., & Ren, L. (2018). An event-triggered dynamic scheduling method for randomly arriving tasks in cloud manufacturing. *International Journal of Computer Integrated Manufacturing, 31*(3), 318–333.

Chapter 2
Proactive Scheduling and Reactive Real-Time Control in Industry 4.0

Dmitry Ivanov, Boris Sokolov, Frank Werner, and Alexandre Dolgui

Abstract Scheduling in Industry 4.0 systems belongs to a class of problems that have mixed structural-temporal-logical constraints. In other words, a strong coupling is considered when product and process are created simultaneously. As a result of the proven NP-hardness of such problems, solution methods have extensively utilized different decomposition principles. The known decomposition methods in discrete optimization are founded on the difficulties in deriving analytical properties. The existing solutions in continuous optimization are based on the maximum principle and yield a dynamic process decomposition using the natural logic of time. By combining the advantages of continuous and discrete optimization, this chapter develops a decomposition method for shop floor scheduling in Industry 4.0 manufacturing systems. Technically, this study proposes to decompose dynamically the large-scale assignment matrix according to the precedence relations between the operations of the jobs and considers only the operations that satisfy these precedence relations at a given time point in small-dimensional, discrete optimization models. Continuous optimization is used to generate a schedule from the assignments found in the discrete optimization models at each time point by extremizing the Hamiltonian function at this time point subject to scheduling objective(s). In addition, the execution of the operations in time can be accurately modeled in continuous time as

D. Ivanov (✉)
Berlin School of Economics and Law, Berlin, Germany
e-mail: divanov@hwr-berlin.de

B. Sokolov
Saint Petersburg Institute for Informatics and Automation of the RAS (SPIIRAS), St. Petersburg, Russia
e-mail: sokol@iias.spb.su

F. Werner
Otto-von-Guericke University Magdeburg, Magdeburg, Germany
e-mail: frank.werner@ovgu.de

A. Dolgui
IMT Atlantique, LS2N—CNRS, La Chantrerie, Nantes, France
e-mail: alexandre.dolgui@imt-atlantique.fr

B. Sokolov et al. (eds.), *Scheduling in Industry 4.0 and Cloud Manufacturing*, International Series in Operations Research & Management Science 289,
https://doi.org/10.1007/978-3-030-43177-8_2

a continuous state variable; the machine availability and capacity disturbances at the machines are also considered. The method developed provides further insights into decomposition methods for scheduling and is supported by an analytical analysis and an algorithmic realization.

Keywords Scheduling · Industry 4.0 · Flexible flow shop · Manufacturing · Optimal control · Algorithm · Real-time scheduling · Dynamic scheduling

2.1 Introduction

Individualization of products is a critical business capability and requires flexible and customized production systems. Because of the increased complexity of flexible, small batch manufacturing, the costs of individualized production are typically higher than in mass production systems. The Industry 4.0 technology has enabled new production strategies, particularly through the use of cyber-physical systems that require highly customized assemblies (Erol et al. 2016; Oesterreich and Teuteberg 2016; Kumar et al. 2016; Nayak et al. 2016; Battaïa et al. 2017a; Hwang et al. 2017). The ultimate objective of these systems is to facilitate a flexible customized manufacturing at the lower cost of mass production.

Such innovative production strategies engender new challenges and opportunities for short-term job scheduling and sequencing. In particular, Kusiak (2018) points out the issue of strong coupling in smart manufacturing when product and process are created at the same time. Simultaneous product and process creation results in a class of scheduling problems that have mixed structural-temporal-logical constraints with order scheduling based on a search for free resources for free operations (Dolgui et al. 2019b; Fragapane et al. 2020; Ivanov et al. 2020). Manufacturing processes for different customer orders may have individual structures of the stations such that the flexible stations are able to execute different functions subject to individual sets of operations within the jobs (Weyer et al. 2015; Ivanov et al. 2016a, b; Nayak et al. 2016; Battaïa et al. 2017b; Zhong et al. 2017). Therefore, an integrated problem of simultaneous, structural-functional synthesis of the Industry 4.0 customized assembly systems and job scheduling in these systems arises and becomes a visible research avenue (Chen et al. 2019; Dolgui et al. 2019a, b; Ivanov et al. 2018b; Leusin et al. 2018; Liu et al. 2019; Mourtzis and Vlachou 2018; Panetto et al. 2019; Rossit et al. 2019; Zhang et al. 2019).

In the given problem class, multi-stage, flexible, job-flow shop scheduling problems with flexible machines have been studied (Ivanov and Sokolov 2012a, 2013; Ivanov et al. 2016a, b; Božek and Werner 2018). Kyparisis and Koulamas (2006) considered a multi-stage, flexible flow shop scheduling problem with uniform parallel machines at each stage and makespan minimization. This study proposed a heuristic algorithm for this strongly NP-hard problem. Tahar et al. (2006) considered a scheduling problem for a set of independent jobs with sequence-dependent setup times and job splitting on a set of identical parallel machines such that the maximum completion time (i.e., the makespan) is minimized. For this NP-hard problem, the study developed a heuristic algorithm using linear programming (LP). Furthering these insights, Božek and Werner (2018) developed an optimization method for flexible job shop scheduling with lot streaming and sublot size optimization. It can be noted that a review of solution techniques for flexible (or hybrid) flow shop problems has been given, for example, in the paper by Ruiz and Vazquez-Rodriguez (2010), and a review on flexible job shop scheduling problems has been given by Chaudry and Khan (2016).

In light of the proven NP-hardness of such problems, solution methods for the simultaneous structural-functional synthesis of customized Industry 4.0 assembly systems and job scheduling in these systems need to be developed using different decomposition principles. This is because large-scale mixed integer linear programming (MILP) models would be negatively influenced by the computational complexity of these problems. The known solutions in discrete optimization based on decomposition, such as data-driven or clustering approaches (Chen et al. 2013a, b), are founded on the difficulties in deriving analytical properties.

The existing solutions in continuous optimization based on the maximum principle and the control of a dynamic system use a decomposition of a dynamic process and the natural logic of time (Ivanov and Sokolov 2012b). These solutions were primarily applied to technical systems (e.g., space shuttle movement control) and rely on a proven analytical axiomatic of optimal control. In the 1990s, optimal control models based on the maximum principle were applied to master production scheduling, but they did not consider the precedence relations within the jobs (Dolgui et al. 2019b) and focused mostly on small-dimensional problems (Kogan and Khmelnitsky 2000).

The major intention of this chapter is to provide further insights into scheduling in smart manufacturing with simultaneous product and process creation using decomposition methods by combining the advantages of continuous and discrete optimization. In particular, we focus on shop floor scheduling in very flexible manufacturing systems such as Industry 4.0. Technically, this study proposes a dynamical decomposition of a large-scale assignment matrix according to the precedence relations between the operations of the jobs and considers only the operations that satisfy the precedence relations at the given time point in small-dimensional LP models. Discrete optimization algorithms (B&B, Hungarian method) are used for scheduling in these matrices of small dimension at each time point. Continuous optimization algorithms (e.g., the method of successive approximations, or the method of Krylov-Chernousko; see Ivanov et al. 2016a and Dolgui et al. 2019b, and the references in these papers) are used to create a schedule from the LP results generated at each time point by extremizing the Hamiltonian function at this time point subject to some criteria (e.g., tardiness). In addition, the execution of the operations in time can be accurately modeled in continuous time as a continuous state variable, while considering machine availability and capacity disturbances at the machines.

The remainder of this chapter is as follows. Section 2.2 is devoted to a verbal problem statement. Section 2.3 develops generalized models of selecting the design of the manufacturing process and operation sequencing. It also considers a model for a simultaneous process design and sequencing. In Sect. 2.4, the generalized model from Sect. 2.4 is modified regarding a flexible dynamic scheduling in Industry 4.0 systems. Subsequently, a computational algorithm is presented. Section 2.5 concludes the chapter by summarizing the most important results of this study and outlining some issues for future research.

2.2 State of the Art

Practical environments for applying scheduling and sequencing models and algorithms to a simultaneous structural-functional synthesis of the customized assembly system are multi-faceted. With the help of smart sensors and plug-and-produce cyber-physical systems, the stations in the assembly system are capable of changing the operation processing and setup sequences according to the actual order of the incoming flows and capacity utilization (Otto et al. 2014; Theorin et al. 2017). In the FOUP—front opening unified pods—technology in the semiconductor industry, robots are used for a real-time operation sequencing. Robots read the information about the products from the sensors and tags and decide flexibly, where to forward a wafer batch next (Mönch et al. 2012).

The recent literature has included a variety of principles and approaches to the design and scheduling of flexible and reconfigurable assembly systems with a focus on balancing, scheduling, and sequencing (Boysen et al. 2007; Chaube et al. 2012; Delorme et al. 2012; Battaïa and Dolgui 2013; Battaïa et al. 2017a). In these studies, models and methods have been presented for solving problems related to the optimization of the performance intensity of an assembly system for sets of flexibly intersecting operations.

For systems that consider both the machine selection for each part of the manufacturing process and the loading sequences of the parts to the machines, Blazewicz et al. (2001) showed that these problems are NP-hard. In particular, scheduling with *alternative parallel machines* addresses the practical challenge that at each stage of the manufacturing process, alternative machines may execute the operations. This creates flexibility in the process plan and requires both a machine assignment and sequencing of tasks (Yu et al. 2011; Janiak et al. 2013; Blazewicz et al. 2015). In practice, the optimization objectives consider throughput maximization, lateness minimization, and equal machine utilization.

Józefowska et al. (2002) presented a heuristic approach to allocate a continuous resource in discrete-continuous scheduling problems to minimize the makespan. Kyparisis and Koulamas (2006) considered a multi-stage, flexible flow shop scheduling problem with uniform parallel machines at each stage and the minimization of makespan. Tahar et al. (2006) considered the problem of scheduling a set of independent jobs with sequence-dependent setup times and job splitting on a set of identical parallel machines such that the maximum completion time (makespan) is minimized.

During the last three decades, a variety of papers presented results and algorithms for flexible flow shop and job shop scheduling. The paper by Ribas et al. (2010) first classifies the papers for flexible (hybrid) flow shop scheduling according to the production characteristics and limitations and then according to the solution approaches proposed. The reader can find 164 references to papers dealing with hybrid flow shop problems. In parallel, another review on solution approaches with 225 references has been presented by Ruiz and Vazquez-Rodriguez (2010). For flexible job shop scheduling problems, a recent review has been given by Chaudry and Khan (2016). The authors found 404 papers dealing with flexible job shop problems from the period 1990 to 2014. The interested reader can also find 212 cited references in this paper. In a very recent paper, Shen et al. (2018) presented a mathematical model and a tabu search algorithm for the flexible job shop problem with sequence-dependent setup times and minimizing the makespan. The results show that their algorithm outperforms most existing approaches for the classical flexible job shop problem.

Lauff and Werner (2004); Jungwattanakit et al. (2009); Sotskov et al. (2013); and Harjunkoski et al. (2014) have pointed out that specific scheduling problems require further investigation and the application of a broad range of methodical approaches. Control approaches to scheduling are of vital importance for addressing the flow dynamics in the assembly line. The studies by Sarimveis et al. (2008); Ivanov et al. (2013b); and Harjunkoski et al. (2014) showed a wide range of advantages regarding the application of control-theoretic models in combination with other techniques for scheduling in manufacturing. These advantages include, but are not limited to, a non-stationary process view and the accuracy of continuous time.

Optimal control approaches provide a different perspective than mathematical programming methods. Optimal control approaches represent schedules as trajectories. The first studies in this area were devoted to inventory control. One of these studies was published in the first volume of the *International Journal of Production Research* (*IJPR*). Later, Hwang et al. (1967) were among the first to apply optimal control and the maximum principle to multi-level and multi-period master production scheduling, which determined an optimal control (i.e., production) for the corresponding state (i.e., the inventory trajectory). Developed a Bayesian approach to the optimal control of continuous industrial processes. Developed a dynamic model for the planning of a manufacturing system. The maximum principle was used to formulate the problem and to obtain a solution. Flexible manufacturing systems and their dynamics have been examined in numerous studies. The stream of production scheduling was continued by Khmelnitsky et al. (1997), who applied the maximum principle in discrete form to the planning of continuous-time flows in flexible manufacturing systems.

Applications of optimal control to scheduling problems are encountered in production systems with single machines, job sequencing in two-stage production systems, and multi-stage machine structures with alternatives in job assignments and intensity-dependent processing rates, such as those in flexible manufacturing systems (Ivanov and Sokolov 2013); supply chains as multi-stage networks (Ivanov et al. 2004, 2007, 2013a; Ivanov and Sokolov 2012a), and Industry 4.0 systems based on a data interchange between the product and stations, flexible stations dedicated to various technological operations, and real-time capacity utilization control (Ivanov et al. 2016a).

In previous studies, the selection of the process structure and the respective station functionality for the execution of the operations were considered in isolation. In many real-life problems, such an integration can have a significant impact on the process efficiency (Bukchin and Rubinovitz 2003). The problem of a simultaneous structural-functional synthesis of the customized assembly system is still at the beginning of its investigation (Levin et al. 2016).

Optimal control scheduling models with only terminal constraints typically address the level of master production scheduling (Hwang et al. 1967; Khmelnitsky et al. 1997; Kogan and Khmelnitsky 2000). Scheduling models with both terminal and logical constraints can also be applied to flow shop and job shop scheduling (Kalinin and Sokolov 1985; Ivanov and Sokolov 2013; Ivanov et al. 2016a, b) as well as to supply chain scheduling (Ivanov and Sokolov 2012b; Ivanov et al. 2013a, 2015).

Previously isolated insights gained in hybrid shop scheduling, scheduling and sequencing with alternative parallel machines, and optimal control scheduling models with both terminal and logical constraints can now be integrated into a unified framework of Industry 4.0 and must be extended toward models with hybrid structural-terminal-logical constraints (Dolgui et al. 2019a, b; Ivanov et al. 2018a, b). The three most important prerequisites for such an integration, that is, the data interchange between the product and stations, flexible stations dedicated to various technological operations, and a real-time capacity utilization control, are enabled by the Industry 4.0 technology (Ivanov et al. 2019).

This study extends previous publications of the authors (Ivanov et al. 2016a, b): the problem statement and the scheduling model consider the structural synthesis and sequencing decisions of the manufacturing process. In the studies (Ivanov et al. 2016a, b), only scheduling decisions were considered and the design of the process structure was assumed to be fixed, that is, the design of a flow shop process was considered.

This study develops an optimal control model for the simultaneous structural-functional design of a customized manufacturing process and the sequencing of the operations within the jobs. The developed theoretical framework presents a contribution to flexible scheduling in the emerging field of Industry 4.0-based innovative production systems.

2.3 Model

2.3.1 Problem Statement

In generalized terms, the problem considered captures the following features. The processing speed of each machine is described as a time function and is modeled by material flow functions (integrals of processing speed functions) and the resulting processing time of the operation is, in general, dependent on the characteristics of the processing channel. The processing and routing capacities are constrained. Setups are included into the analysis. The lot sizes are fixed and known in advance. The temporary unavailability of capacity as a consequence of possible disruptions is included. Material supply and consumption dynamics are considered. The optimization is performed subject to the following performance indicators (control functionals): throughput, lead time, makespan, total lateness, equal utilization of the stations in the assembly line, and waiting time.

In terms of scheduling theory, we study a multi-objective, multi-stage hybrid shop scheduling problem with alternative machines at each stage with different time-dependent processing speeds, time-dependent machine availability, and ordered jobs, where job splitting is allowed. Examples of such problems can be found in the studies by Kyparisis and Koulamas (2006) and Tahar et al. (2006). The peculiarity of the problem under consideration is the simultaneous consideration of both the selection of the process design structure and the operation assignment. On the one hand, an assignment problem is discrete by nature and requires the introduction of binary variables, that is, in this case discrete optimization techniques are appropriate. On the other hand, the execution of a non-stationary operation can be accurately described in terms of continuous optimization. An additional peculiarity of such a simultaneous consideration is that both the machine structures and the flow parameters may be uncertain and changes in dynamics are therefore non-stationary.

According to the problem statement, we integrate and synchronize the following processes:

- Customer order fulfillment dynamics in regard to the process design and operation sequencing
- Processing and transportation channel utilization dynamics
- Material supply and consumption dynamics in the assembly system
- Processing and movement dynamics in regard to processing and transportation channels in the assembly system

2.3.2 Model

Let us introduce the following basic sets and structures (indices [o], [k], [r], and [f] describe the relations of the sets to the operations [o], channels [k], resources [r], and material flows [f], respectively):

$A = \{A_\upsilon, \upsilon \in 1, \ldots n\}$ is the set of customer orders (jobs).

$M = \{M_i, i \in 1, \ldots m\}$ is the set of stations in the assembly line.

$\tilde{B} = A \cup M$ is the union of the assembly line stations and the customer orders, that is, the customized assembly system.

$C = \left\{C_\lambda^{(i)}, \lambda \in 1, \ldots l_i\right\}$ is the set of channels at the stations and in-between the stations i and j in the assembly line, where index j is used for stations that receive products from an i-station.

$D = \left\{\left\{D_{æ'}^{(i)}\right\} \cup \left\{D_{æ}^{(i,j)}\right\}, æ, æ' \in 1, \ldots s_i\right\}$ is the set of operations that can be executed in the assembly system.

$\Phi = \left\{ \left\{ \Phi S_{\pi}^{(i)} \right\} \cup \left\{ \Phi N_{\mu}^{(i)} \right\}, \pi \in K_i^{(r,1)}, \mu \in K_i^{(r,2)} \right\}$ is the set of resources at the stations in the assembly line, where K is the set of numbers.

$\Phi S^{(i)} = \left\{ \Phi S_{\pi}^{(i)}, \pi \in K_i^{(r,1)} \right\}$ is the set of storable resources at $M^{(i)}$ and $\Phi N^{(i)} = \left\{ \Phi N_{\mu}^{(i)}, \mu \in K_i^{(r,2)} \right\}$ is the set of non-storable resources at $M^{(i)}$.

$P = \left\{ \left\{ P_{<æ',\rho>}^{(i)} \right\} \cup \left\{ P_{<æ,\rho>}^{(i,j)} \right\}, \rho \in K_i^{(f)} \right\}$ is the set of material flows in the manufacturing process.

$P^{(i)} = \left\{ P_{<æ',\rho>}^{(i)} \right\}$ is the set of material flows for the ρ-types of materials subject to $M^{(j)}$.

$P^{(i,j)} = \left\{ P_{<æ,\rho>}^{(i,j)} \right\}$ is the set of material flows for the ρ-types of materials subject to $M^{(i)}$ and $M^{(j)}$.

The sets $\Gamma_{\nu 1}$, $\Gamma_{\nu 2}$ define "and" and "or" precedence relations for different jobs and the sets $\Gamma_{iæ1}$, $\Gamma_{iæ2}$ define "and" and "or" precedence relations for the operations $D_{æ}^{(i,j)}$ and $D_{æ'}^{(i)}$, respectively.

Assume that the manufacturing and transportation capacities may be disrupted, and:

- The availability of a station can be described by a given preset matrix time function $\varepsilon_{ij}(t)$ of time-spatial constraints: we have $\varepsilon_{ij}(t) = 1$, if the station is available and $\varepsilon_{ij}(t) = 0$, otherwise.
- The availability of a channel at a station can be described by the function $\Theta_{\iota\varphi\lambda}(t)$ (or $\Theta_{\nu\varphi\lambda}(t)$), which is equal to 1 if there are available channels, and equals 0, otherwise.
- The capacity degradation and recovery dynamics can be described by a continuous function of the perturbation impacts $\xi_{ij}(t)$; $\xi_{ij}(t) = 1$ if the channel is 100% available and $\xi_{ij}(t) = 0$ if the channel is fully disrupted. All other values for $\xi_{ij}(t)$ in the interval [0;1] are possible.

The formal statement of the scheduling problem is based on a dynamic interpretation of the execution processes of the operations. Let us introduce some new notations.

Parameters

$a_{\alpha}^{(o,1)}, a_{\beta}^{(o,1)}, a_{i\tilde{\alpha}}^{(o,2,\nu)}, a_{i\tilde{\beta}}^{(o,2,\nu)}, a_{is_i}^{(o,2,\nu)}, a_{iæ}^{(o,1,\nu)}$ are the planned manufacturing and transportation quantities for each operation. The values of these parameters are related to the end conditions that need to be reached in $x_{\alpha}^{(o,1)}(t), x_{\beta}^{(o,1)}(t), x_{i\tilde{\alpha}}^{(o,2,\nu)}(t), x_{i\tilde{\beta}}^{(o,2,\nu)}(t), x_{is_i}^{(o,2,\nu)}(t), x_{iæ}^{(o,1,\nu)}(t)$ at $t = T_f$.

$\mathbf{h}_0^{(o)}$, $\mathbf{h}_1^{(o)}$ are known differential functions for the start and end conditions subject to the state variables $\boldsymbol{x}^{(o)}$ at $t = T_0$ and $t = T_f$.

$\mathbf{h}_0^{(k)}$, $\mathbf{h}_1^{(k)}$ are known differential functions for the start and end conditions in regard to the state vector $\boldsymbol{x}^{(\kappa)}$.

$b_{i'æ'iæ}^{(j,\lambda)}$ is the channel setup time.

$d_{iæj\lambda}^{(\pi)}, g_{iæj\lambda}^{(\mu)}$ are given consumption rates of the resources $\Phi S_{\pi}^{(j)}$ and $\Phi N_{\mu}^{(j)}$ for $D_{æ}^{(i,j)}$ and $C_{\lambda}^{(j)}$.

$\tilde{H}_{j}^{(\pi)}(t)$, $\tilde{\tilde{H}}_{j}^{(\mu)}(t)$ are known rates for the replenishment of the resources $\Phi S_{\pi}^{(j)}$ and $\Phi N_{\mu}^{(j)}$, respectively.

$a_{j\lambda\pi(\eta-1)}^{(p,3)}, a_{j\lambda\mu(\eta'-1)}^{(p,4)}$ are known volumes (quantities) of the resource replenishment at the $(\eta - 1)$th recovery cycle; $\tilde{\rho}_{\lambda}, \tilde{\tilde{\rho}}_{\lambda}$ are the numbers of the replenishment cycles.

$a_{iæ\rho}^{(f,1)}$ is the known lot size of a product type ρ for each operation $D_{æ}^{(i,j)}$.

$\tilde{P}_{j}^{(1)}, \tilde{P}_{j\rho}^{(2)}, \tilde{P}_{ij}^{(3)}$ are known values for the maximum storage capacities at M_j, handling (throughput) at M_j for ρ, and the transportation between M_i and M_j.

$c_{iæj\lambda\rho}^{(f,1)}$ is the maximum processing rate for the operation $D_{æ}^{(i,j)}$ at the λ-channel; it determines the maximum possible value for the production rate $u_{iæj\lambda\rho}^{(f,1)}$.

Decision Variables

$x_{\nu}^{(0,1)}(t)$ is a state variable characterizing the lead time for job A_{ν} at each moment t.

$x_{iæ}^{(0,2,\nu)}(t)$ is a state variable characterizing the flow time of the operation $D_{æ}^{(i)}$ or $D_{æ}^{(i,j)}$.

$x_{\nu}^{(0,3)}(t)$ is a state variable characterizing the gap between the planned completion time for all jobs and the actual completion time of the job A_{ν}.

$u_{\nu j}^{(0,1)}(t), u_{iæj}^{(0,2,\nu)}(t), u_{\nu j}^{(0,3)}(t)$ are control variables; if $u_{\nu j}^{(0,1)}(t) = 1$, then we have a transportation of job A_{ν} to B_j, and $u_{\nu j}^{(0,1)}(t) = 0$ otherwise; $u_{iæj\lambda}^{(0,2,\nu)}(t) = 1$ if operation $D_{æ}^{(i)}$ or $D_{æ}^{(i,j)}$ is assigned to a λ-channel, and $u_{iæj\lambda}^{(0,2,\nu)}(t) = 0$ otherwise; $u_{\nu j}^{(0,3)}(t) = 1$ at the moment when A_{ν} is completed at time point t until $t = \mathrm{T_f}$, and $u_{\nu j}^{(0,3)}(t) = 0$ otherwise.

$x_{iæj\lambda}^{(\kappa,1)}(t)$ is the state variable for the channel $C_{\lambda}^{(i)}$ at M_j during the setup to prepare the channel for processing $D_{æ'}^{(i,j)}$ after completing the operation $D_{æ}^{(i,j)}$.

$u_{iæj\lambda}^{(\kappa,1)}(t)$ is a control variable; $u_{iæj\lambda}^{(\kappa,1)}(t) = 1$ if $C_{\lambda}^{(i)}$ is in the setup process and $u_{iæj\lambda}^{(\kappa,1)}(t) = 0$ otherwise.

$x_{j\lambda}^{(\kappa,2)}(t)$ is a state variable characterizing the process (run) time of a channel.

$x_{j\lambda\pi}^{(p,1)}(t), x_{j\lambda\mu}^{(p,2)}(t), x_{j\lambda\pi\eta}^{(p,1)}(t), x_{j\lambda\mu\eta'}^{(p,2)}(t)$ are state variables that characterize the current quantity (volume) of non-storable resources $\Phi S_{\pi}^{(j)}$, storable resources $\Phi N_{\mu}^{(j)}$, non-storable and recoverable (at stages η and η′) resources, and storable and recoverable (at stages η and η′) resources subject to channel $C_{\lambda}^{(j)}$, respectively. These state variables characterize a π-resource consumption and replenishment.

$x^{(\mathrm{p},3)}_{j\lambda\pi\eta}(t)$, $x^{(\mathrm{p},4)}_{j\lambda\mu\eta'}(t)$ are auxiliary state variables that are needed to define the sequence of the resource replenishments and the ends of the replenishment intervals, respectively.

$u^{(\mathrm{p},1)}_{j\lambda\pi\eta}$, $u^{(\mathrm{p},2)}_{j\lambda\mu\eta}$ are control variables characterizing the replenishment process for the non-storable and storable resources respectively; $u^{(\mathrm{p},1)}_{j\lambda\pi\eta}, u^{(\mathrm{p},2)}_{j\lambda\mu\eta} = 1$ if a π-resource is under replenishment at time point t, and equal 0 otherwise.

$x^{(\mathrm{f},1)}_{i æ j\lambda\rho}(t)$ is a state variable characterizing the quantity (volume) of the product «ρ» that is being delivered at M_j from M_i during the execution of $D^{(i,j)}_{æ}$ (or the processed quantity at M_j, if $i = j$).

$x^{(\mathrm{f},2)}_{i æ j\lambda\rho}(t)$ is an auxiliary state variable characterizing the total processing time (including the waiting time) of a product flow ρ resulting from the interaction of M_i and M_j for $D^{(i,j)}_{æ}$ at $C^{(i)}_{\lambda}$, $C^{(j)}_{\lambda}$.

$u^{(\mathrm{f},1)}_{i æ j\lambda\rho}$ is the shipment rate for the transportation from M_i to M_j (or the processing rate at M_j if $i = j$); $u^{(\mathrm{f},2)}_{i æ j\lambda\rho}(t)$ is an auxiliary control variable; $u^{(\mathrm{f},2)}_{i æ j\lambda\rho}(t) = 1$ if processing at M_j is completed, and $u^{(\mathrm{f},2)}_{i æ j\lambda\rho}(t) = 0$ otherwise, or if, after the completion of $D^{(i,j)}_{æ}$ (or $D^{(i)}_{æ}$, if $i = j$), the next operation in the technological process $D^{(i,j)}_{\tilde{æ}}$ (or $D^{(i)}_{\tilde{æ}}$, if $i = j$) begins.

2.3.2.1 Process Model of the Operation Execution

$$x^{(0,1)}_{\nu} = \sum_{j=1}^{m} u^{(0,1)}_{\nu j};\quad x^{(0,2,\nu)}_{i æ} = \sum_{j=1}^{m}\sum_{\lambda=1}^{l_j} \varepsilon_{ij}(t)\Theta_{i æ j\lambda}(t)u^{(0,2,\nu)}_{i æ j\lambda};\quad x^{(0,3)}_{\nu j} = u^{(0,3)}_{\nu j}, \tag{2.1}$$

$$x^{(\kappa,1)}_{i æ j\lambda} = \sum_{j'=1}^{m}\sum_{æ'=1}^{s_i} \Theta_{i' æ' j æ} u^{(\kappa,1)}_{i' æ' j\lambda} \frac{b^{(j,\lambda)}_{i' æ' i æ} - x^{(\kappa,1)}_{i æ j\lambda}}{x^{(\kappa,1)}_{i' æ' j\lambda}}, \tag{2.2}$$

$$x^{(\kappa,2)}_{j\lambda} = \sum_{i=1}^{m}\sum_{æ=1}^{s_i}\left(u^{(o,2)}_{i æ j\lambda} + u^{(\kappa,1)}_{i æ j\lambda}\right), \tag{2.3}$$

$$x^{(\mathrm{p},1)}_{j\lambda\pi} = -\sum_{i=1}^{m}\sum_{æ=1}^{s_i} d^{(\pi)}_{i æ j\lambda}\left(u^{(\mathrm{o},2)}_{i æ j\lambda} + u^{(\kappa,1)}_{i æ j\lambda}\right), \tag{2.4}$$

$$x_{j\lambda\mu}^{(\mathrm{p},2)} = -\sum_{i=1}^{m}\sum_{æ=1}^{s_i} g_{iæj\lambda}^{(\mu)}\left(u_{iæj\lambda}^{(\mathrm{o},2)} + u_{iæj\lambda}^{(\kappa,1)}\right), \tag{2.5}$$

$$x_{j\lambda\pi\eta}^{(\mathrm{p},1)} = -\sum_{i=1}^{m}\sum_{æ=1}^{s_i} d_{iæj\lambda}^{(\pi)}\left(u_{iæj\lambda}^{(\mathrm{o},2)} + u_{iæj\lambda}^{(\kappa,1)}\right) + u_{j\lambda\pi(\eta-1)}^{(\mathrm{p},1)}, \tag{2.6}$$

$$x_{j\lambda\mu\eta'}^{(\mathrm{p},2)} = -\sum_{i=1}^{m}\sum_{æ=1}^{s_i} g_{iæj\lambda}^{(\mu)}\left(u_{iæj\lambda}^{(\mathrm{o},2)} + u_{iæj\lambda}^{(\kappa,1)}\right) + u_{j\lambda\mu(\eta'-1)}^{(\mathrm{p},2)}, \tag{2.7}$$

$$x_{j\lambda\pi\eta}^{(\mathrm{p},3)} = u_{j\lambda\pi\eta}^{(\mathrm{p},1)}; \quad x_{j\lambda\mu\eta'}^{(\mathrm{p},4)} = u_{j\lambda\mu\eta'}^{(\mathrm{p},2)}, \tag{2.8}$$

$$x_{iæj\lambda\rho}^{(\mathrm{f},1)} = u_{iæj\lambda\rho}^{(\mathrm{f},1)}; \quad x_{iæj\lambda\rho}^{(\mathrm{f},2)} = u_{iæj\lambda\rho}^{(\mathrm{f},2)}. \tag{2.9}$$

Equation (2.1) describes the dynamics of the operation execution for the job A_ν. If $x_\nu^{(0,1)} = \sum_{j=1}^{m} u_{\nu j}^{(0,1)}$, then at each time point where, if $u_{\nu j}^{(0,1)}(t) = 1$ (i.e., the volume of the state variable x_ν is increasing), the job processing is in progress at the j-station in the assembly system. If $x_{iæ}^{(0,2,\nu)} = \sum_{j=1}^{m} \varepsilon_{ij}(t) u_{iæj}^{(0,2,\nu)}$, then the operation processing can start subject to the time windows of feasible capacity. If $x_{\nu j}^{(0,3)} = u_{\nu j}^{(0,3)}$, then the job is completed earlier than the due date (i.e., the earliness of the job completion subject to the slack time).

Equations (2.2) and (2.3) describe the state dynamics of the channel $C_\lambda^{(i)}$ at M_i and characterize the availability of the channel for the processing of operation $D_æ^{(i,j)}$. Equation (2.2) describes the setup dynamics, and Eq. (2.3) reflects the occupation time of each channel subject to the dynamics of the operation execution (i.e., variable $u_{iæj\lambda}^{(o,2)} = 1$) in Eq. (2.1).

Equations (2.4)–(2.8) describe the resource consumption dynamics (Eqs. 2.4 and 2.5) and the resource replenishment dynamics (Eqs. 2.6–2.8) subject to the assignment and setup decisions in Eqs. (2.1)–(2.3). Finally, Eq. (2.9) describes the material flow dynamics in the assembly system subject to the operation assignments to the stations, setups, and resource management decisions.

2.3.2.2 Constraints

$$\sum_{j=1}^{m} u_{\nu j}^{(0,1)} \left[\sum_{\alpha \in \Gamma_{\nu 1}} \left(a_{\alpha}^{(0,1)} - x_{\alpha}^{(0,1)}(t) \right) + \prod_{\beta \in \Gamma_{\nu 2}} \left(a_{\beta}^{(0,1)} - x_{\beta}^{(0,1)}(t) \right) \right] = 0, \tag{2.10}$$

$$\sum_{\lambda=1}^{l_j} u_{i æ j \lambda}^{(0,2,\nu)} \left[\sum_{\tilde{\alpha} \in \Gamma_{i æ 1}} \left(a_{i\tilde{\alpha}}^{(0,2,\nu)} - x_{i\tilde{\alpha}}^{(0,2,\nu)}(t) \right) + \prod_{\tilde{\beta} \in \Gamma_{i æ 2}} \left(a_{i\tilde{\beta}}^{(0,2,\nu)} - x_{i\tilde{\beta}}^{(0,2,\nu)}(t) \right) \right] = 0, \tag{2.11}$$

$$\sum_{\nu=1}^{u} u_{\nu j}^{(0,1)}(t) \leq 1, \ \ \forall j; \quad \sum_{j=1}^{m} u_{\nu j}^{(0,1)}(t) \leq 1, \ \ \forall j; \ \ u_{\nu j}^{(0,1)}(t) \in \{0, 1\} \tag{2.12}$$

$$u_{i æ j \lambda}^{(0,2,\nu)}(t) \in \left\{ 0, u_{\nu j}^{(0,1)} \right\}; \ \ u_{\nu j}^{(0,3)}(t) \in \{0, 1\}; \ \ u_{\nu j}^{(0,3)} \left(a_{j s_i}^{(0,2,\nu)} - x_{j s_j}^{(0,2,\nu)}(t) \right) = 0 \tag{2.13}$$

$$u_{i æ j \lambda}^{(0,2)} x_{i æ j \lambda}^{(\kappa,1)} = 0; \quad x_{i æ j \lambda}^{(\kappa,1)}(t) \in \{0, 1\}, \tag{2.14}$$

$$\sum_{i=1}^{n} \sum_{æ=1}^{s_i} u_{i æ j \lambda}^{(\kappa,1)}(t) \leq 1, \quad \forall j, \ \ \forall \lambda, \tag{2.15}$$

$$\sum_{i,æ,\lambda} d_{i æ j \lambda}^{(\pi)} \left(u_{i æ j \lambda}^{(o,2)} + u_{i æ j \lambda}^{(\kappa,1)} \right) \leq \tilde{H}_j^{(\pi)}(t), \tag{2.16}$$

$$\sum_{i,æ,\lambda} \int_{T_0}^{T_f} g_{i æ j \lambda}^{(\mu)} \left(u_{i æ j \lambda}^{(o,2)}(\tau) + u_{i æ j \lambda}^{(\kappa,1)}(\tau) \right) d\tau \leq \int_{T_0}^{T_f} \tilde{\tilde{H}}_j^{(\mu)}(\tau)\, d\tau, \tag{2.17}$$

$$u_{j \lambda \pi \eta}^{(p,1)} \left(a_{j \lambda \pi (\eta-1)}^{(p,3)} - x_{j \lambda \pi (\eta-1)}^{(p,3)} \right) = 0, \qquad u_{j \lambda \pi \eta}^{(p,1)} x_{j \lambda \pi \eta}^{(p,1)} = 0, \tag{2.18}$$

$$u_{j\lambda\mu\eta}^{(\mathrm{p},2)}\left(a_{j\lambda\pi(\eta'-1)}^{(\mathrm{p},4)}-x_{j\lambda\pi(\eta'-1)}^{(\mathrm{p},4)}\right)=0,\qquad u_{j\lambda\mu\eta'}^{(\mathrm{p},2)}x_{j\lambda\mu\eta'}^{(\mathrm{p},2)}=0, \tag{2.19}$$

$$u_{j\lambda\pi\eta}^{(\mathrm{p},1)}(t),\ u_{j\lambda\mu\eta}^{(\mathrm{p},2)}(t)\in\{0,1\},\quad \eta=1,\ldots,\tilde{\rho}_{\lambda};\quad \eta'=1,\ldots,\tilde{\tilde{\rho}}_{\lambda}, \tag{2.20}$$

$$0\le u_{i\text{æ}j\lambda\rho}^{(\mathrm{f},1)}\le c_{i\text{æ}j\lambda\rho}^{(\mathrm{f},1)}u_{i\text{æ}j\lambda}^{(\mathrm{o},2)}, \tag{2.21}$$

$$u_{i\text{æ}j\lambda\rho}^{(\mathrm{f},2)}\left(a_{i\text{æ}\rho}^{(\mathrm{f},1)}-x_{i\text{æ}j\lambda\rho}^{(\mathrm{f},1)}\right)=0;\quad u_{i\text{æ}j\lambda\rho}^{(\mathrm{f},2)}x_{i\tilde{\text{æ}}}^{(\mathrm{o},2)}=0;\quad u_{i\text{æ}j\lambda\rho}^{(\mathrm{f},2)}(t)\in\{0,1\}, \tag{2.22}$$

$$\sum_{i=1}^{m}\sum_{\lambda=1}^{l_i}\sum_{\text{æ}=1}^{s_i}\sum_{\rho=1}^{k_i}x_{i\text{æ}j\lambda\rho}^{(\mathrm{f},1)}\left(u_{i\text{æ}j\lambda\rho}^{(\mathrm{o},2)}+_{i\text{æ}j\lambda\rho}^{(\mathrm{f},2)}\right)\le\tilde{P}_j^{(1)}, \tag{2.23}$$

$$\sum_{i=1}^{m}\sum_{\lambda=1}^{l_i}\sum_{\text{æ}=1}^{s_i}u_{i\text{æ}j\lambda\rho}^{(\mathrm{f},1)}\le\tilde{P}_{j\rho}^{(2)}, \tag{2.24}$$

$$\sum_{\lambda=1}^{l_i}\sum_{\text{æ}=1}^{s_i}\sum_{\rho=1}^{k_i}u_{i\text{æ}j\lambda\rho}^{(\mathrm{f},1)}\le\tilde{P}_{ij}^{(3)}. \tag{2.25}$$

Constraints (2.10) and (2.11) describe the technological precedence relations in regard to the jobs and operations of the jobs. Constraint (2.12) defines the rules of operation splitting and overlapping. Equation (2.13) is a binary constraint on the control variables.

Constraints (2.14) and (2.15) determine the setup sequence at the channels and the conditions for setups at the channel $C_{\lambda}^{(i)}$. According to constraints (2.16) and (2.17), the intensities of the maximum resource consumption at each time point t are constrained subject to $\tilde{H}_j^{(\pi)}$, $\tilde{\tilde{H}}_j^{(\mu)}$. Constraints (2.18)–(2.20) determine the sequence of the replenishment actions. Equations (2.21)–(2.25) constrain the maximum processing rates subject to the operation assignments.

2.3.2.3 Boundary Conditions

$$\mathbf{h}_0^{(\mathrm{o})}\left(\mathbf{x}^{(\mathrm{o})}\left(\mathrm{T}_0\right)\right) \le 0\ ;\quad \mathbf{h}_1^{(\mathrm{o})}\left(\mathbf{x}^{(\mathrm{o})}\left(\mathrm{T}_\mathrm{f}\right)\right) \le 0, \tag{2.26}$$

$$\mathbf{h}_0^{(\mathrm{k})}\left(\mathbf{x}^{(\mathrm{k})}\left(\mathrm{T}_0\right)\right) \le 0\ ;\quad \mathbf{h}_1^{(\mathrm{k})}\left(\mathbf{x}^{(\mathrm{k})}\left(\mathrm{T}_\mathrm{f}\right)\right) \le 0, \tag{2.27}$$

$$\mathbf{h}_0^{(\mathrm{r})}\left(\mathbf{x}^{(\mathrm{r})}\left(\mathrm{T}_0\right)\right) \le 0\ ;\quad \mathbf{h}_1^{(\mathrm{r})}\left(\mathbf{x}^{(\mathrm{r})}\left(\mathrm{T}_\mathrm{f}\right)\right) \le 0, \tag{2.28}$$

$$\mathbf{h}_0^{(\mathrm{f})}\left(\mathbf{x}^{(\mathrm{f})}\left(\mathrm{T}_0\right)\right) \le 0\ ;\quad \mathbf{h}_1^{(\mathrm{f})}\left(\mathbf{x}^{(\mathrm{f})}\left(\mathrm{T}_\mathrm{f}\right)\right) \le 0. \tag{2.29}$$

Equations (2.26)–(2.29) determine the initial and final values for the state variables in regard to the operations, channel, resource, and flow dynamics.

2.3.2.4 Control Functionals

$$J_1^{(\mathrm{o})} = \sum_{\nu=1}^{n}\sum_{j=1}^{m} u_{\nu j}^{(o,3)}\left(T_f\right), \tag{2.30}$$

$$J_{<2,\alpha,\nu>}^{(\mathrm{o})} = \sum_{i=1}^{m}\sum_{j=1}^{m}\left(x_{\alpha i}^{(o,3)}\left(T_f\right) - x_{\nu j}^{(o,3)}\left(T_f\right)\right), \tag{2.31}$$

$$J_3^{(\mathrm{o})} = T_f - \sum_{j=1} x_{nj}^{(o,1)}\left(T_f\right), \tag{2.32}$$

$$J_{<4,i,\nu>}^{(\mathrm{o})} = \sum_{\nu,j,\lambda,æ}\int_{\mathrm{T}_0}^{T_f} \varepsilon_{ij}\left(\tau\right)\Theta_{iæj\lambda}\left(\tau\right)u_{iæj\lambda}^{(0,2,\nu)}\left(\tau\right)d\tau, \tag{2.33}$$

$$J_{<5,i>}^{(\mathrm{o})} = \sum_{\nu,j,æ}\int_{\mathrm{T}_0}^{T_f}\left[\varepsilon_{ij}\left(\tau\right) - \varepsilon_{ij}\left(\tau\right)u_{iæj\lambda}^{(0,2,\nu)}\left(\tau\right)\right]d\tau, \tag{2.34}$$

$$J_6^{(\mathrm{o})} = \sum_{i=1}^{m} \sum_{æ=1}^{s_i} \left(a_{iæ}^{(o,2,\nu)} - x_{iæ}^{(o,2,\nu)} \left(T_f \right) \right)^2, \tag{2.35}$$

$$J_7^{(\mathrm{o})} = \sum_{\nu=1}^{n} \sum_{i=1}^{m} \sum_{æ=1}^{s_i} \sum_{j=1}^{m} \sum_{\lambda=1}^{l_j} \int_{T_0}^{T_f} \tilde{\tilde{\beta}}_{iæ}^{(\nu)} (\tau) \, u_{iæj\lambda}^{(0,2,\nu)} (\tau) \, d\tau, \tag{2.36}$$

$$J_1^{(\mathrm{\kappa})} = \sum_{\Delta_1=1}^{m-1} \sum_{\Delta_2=\Delta_1+1}^{m} \sum_{\lambda=1}^{l} \sum_{\zeta=1}^{l} \int_{T_0}^{T_f} \left(x_{\Delta_1 \lambda}^{(\mathrm{\kappa},2)} (\tau) - x_{\Delta_2 \zeta}^{(k,2)} (\tau) \right) d\tau, \tag{2.37}$$

$$J_2^{(\mathrm{\kappa})} = \sum_{\Delta_1=1}^{m-1} \sum_{\Delta_2=\Delta_1+1}^{m} \sum_{\lambda=1}^{l} \sum_{\zeta=1}^{l} \left(x_{\Delta_1 \lambda}^{(\mathrm{\kappa},2)} \left(T_f \right) - x_{\Delta_2 \zeta}^{(\mathrm{\kappa},2)} \left(T_f \right) \right), \tag{2.38}$$

$$J_{1j\pi}^{(\mathrm{p})} = \sum_{\lambda=1}^{l_j} \sum_{\eta=1}^{\tilde{\rho}_\lambda} x_{j\lambda\pi\eta}^{(\mathrm{p},3)}, \tag{2.39}$$

$$J_{2j\mu}^{(\mathrm{p})} = \sum_{\lambda=1}^{l_j} \sum_{\eta'=1}^{\tilde{\tilde{\rho}}_\lambda} x_{j\lambda\mu\eta}^{(\mathrm{p},4)}, \tag{2.40}$$

$$J_1^{(\mathrm{f})} = \left. \sum_{i=1}^{m} \sum_{æ=1}^{s_i} \sum_{\substack{j=1 \\ i \neq j}}^{m} \sum_{\lambda=1}^{l_i} \sum_{\rho=1}^{k_i} \left(a_{iæ\rho}^{(\mathrm{f},1)} - x_{iæj\lambda\rho}^{(\mathrm{f},1)} \right)^2 \right|_{t=T_f}, \tag{2.41}$$

$$J_2^{(\mathrm{f})} = \sum_{i=1}^{m} \sum_{æ=1}^{s_i} \sum_{\substack{j=1 \\ i \neq j}}^{m} \sum_{\lambda=1}^{l_i} \sum_{\rho=1}^{k_i} \int_{T_0}^{T_f} x_{iæj\lambda}^{(\mathrm{f},2)} (\tau) \, d\tau. \tag{2.42}$$

We refer to the studies (Ivanov et al. 2010; Ivanov and Sokolov 2012a) for a multi-objective resolution of optimal control scheduling models.

The control functional $J_1^{(\mathrm{o})}$ (Eq. 2.30) characterizes the overall number of completed jobs in the assembly system at $t = T_f$. This is the performance indicator for the assembly system *throughput*. $J_{<2,\alpha,\nu>}^{(\mathrm{o})}$ (Eq. 2.31) reflects the lead time for the job A_ν. $J_3^{(o)}$ (Eq. 2.32) characterizes the *makespan* for all jobs A_ν. $J^{(o)}_{<4,i,\nu>}$ (Eq. 2.33) characterizes the *processing time* of job A_ν. $J^{(o)}_{<5,i>}$ (Eq. 2.34) is the *waiting time* of job A_ν. $J_6^{(o)}$ (Eq. 2.35) depicts the degree of the job completion at the end of the planning interval. $J_7^{(o)}$ (Eq. 2.36) expresses the *total tardiness* for all operations subject to the penalty functions $\tilde{\tilde{\beta}}^{(\nu)}_{iæ}$, that is, an *on-time-delivery*.

The control functionals (2.37) and (2.38) estimate the equality of the channel utilization at the stations in the assembly system at each time point $t \in (T_0, T_f\,]$ and at the end of the planning interval. The control functionals (2.39) and (2.40) estimate the degree of the resource replenishment and the timeliness of the resource replenishment, respectively. The control functional (2.41) characterizes the gap between the planned and actually processed operation volume and is interconnected with the control functional (2.35). The control functional (2.42) depicts the waiting time for the operations and is interconnected with the control functional (2.34).

2.3.2.5 Integration Principle

To obtain a constructive solution to the problem considered, we propose to use a functorial transition from the category of digraphs (*Cat*Φ) that specifies the manufacturing technology to the category of dynamic models (*Cat*D), which describes the operation execution. The covariant functor G: Φ → D sets the state relations in-between the nodes in the manufacturing technology plan and the operation execution schedule. The simplified mathematical model of manufacturing technology plan and the operation execution schedule integration can be presented as shown in Eq. (2.43):

$$\begin{aligned}\Delta = \Bigg\{ & \mathbf{u} | \tfrac{dx_i}{dt} = \sum_{\nu=1}^{n} u_{\nu j};\ \sum_{j=1}^{m} u_{\nu j} \le 1;\ \sum_{\nu=1}^{n} u_{\nu j} \le 1;\ u_{\nu j}(t) \in \{0, 1\}; \\ & \sum_{\nu=1}^{n} u_{\nu j} \left[\sum_{\alpha \in \Gamma_1^-} (a_\alpha - x_\alpha(t)) + \prod_{\beta \in \Gamma_2^-} \left(a_\beta - x_\beta(t)\right) \right] = 0; \\ & t \in \left(T_0, T_f\right] = T;\quad x_\nu\,(T_0) = 0;\ \ x_\nu\left(T_f\right) = a_\nu \Big\} \end{aligned} \tag{2.43}$$

where x_ν is a variable characterizing the state of the job A_ν, $u_{\nu j} = 1$ is a control action ($u_{\nu j} = 1$, if the station $M^{(i)}$ is used for job $A^{(\nu)}$), a_ν, a_α, a_β are given quantities (end conditions), the values of which should have the corresponding variables $x_\nu(t)$, $x_\alpha(t)$, $x_\beta(t)$ at the end of the planning interval at the time point $t = T_f$, t is the running time point, T_0 is the start time point of the planning horizon, T_f is the end time point of the planning horizon, T is the planning horizon, $\sum_{\nu=1}^{n} u_{\nu j} \sum_{\alpha \in \Gamma_1^-} (a_\alpha - x_\alpha(t)) = 0$ are constraints "and" that relate the condition of the total processing of all predecessor operations, $\sum_{\nu=1}^{n} u_{\nu j} \prod_{\beta \in \Gamma_2^-} \left(a_\beta - x_\beta(t)\right) = 0$

are constraints "or" that relate the condition of the processing of at least one of the predecessor operations, and Γ_1^-, Γ_2^- are the sets of processes that immediately precede the job $A^{(\nu)}$.

In the proposed approach, the assignment and flow control are considered simultaneously. Since the task times may differ subject to a varying processing speed $c_{i\ae j\lambda\rho}^{(\mathrm{f},1)}$ and the channel availability $\varepsilon_{ij}(t)$ and $\Theta_{i\ae j\lambda}(t)$, the assignments made on the basis of the planned processing volumes $a_{\nu j}$ are forwarded to the resource and flow dynamics control models and further optimized in regard to resource consumption, replenishment, and usage over time. In the flow control model, the assignment of an operation to a channel and the execution start of the operation at the channel cause dynamic flows of the processed products.

2.3.2.6 Formulation of the Scheduling Problem

The task is to find a feasible control $\mathbf{u}(t)$, $[T_0, T_f)$, which ensures that the dynamic control model meets the constraint functions and guides the dynamic system (i.e., the schedule) $\dot{\mathbf{x}} = \mathbf{f}(t, \mathbf{x}, \mathbf{u})$ from the initial state to the specified final state subject to given end conditions and the uncertainty area under the disturbances $\boldsymbol{\xi}(t)$. If there are several feasible controls (schedules), then the best one (optimal) should be selected in order to maximize (minimize) the control functionals (2.30)–(2.42).

2.4 The Role of Industry 4.0 and Digital Technology in the Implementation of the Dynamic Schedule Computation

The control model proposed can be enriched by the integration of Industry 4.0 elements such as sensors and data analytics that may have two impacts on the system. The first is the tuning: changing uncertain coefficients in the structure of the differential equations of the system, taking into account that a larger number of these coefficients implies a more accurate system response to a changing environment. The second is the learning: imposing new restrictions on the system behavior. The number of arbitrary coefficients in the structure of the differential equations changes in the process of learning, imposing dynamic restriction adjustments on the behavior of the system.

The possibility to use real-time data of the machine availability and the operation processing status on the basis of digital technologies allows the realization of different dynamic decomposition principles in the optimal control algorithm. The selection of the time points at which the Hamiltonian is extremized and small-dimensional assignment problems are solved can be implemented not only as a fixed t-step procedure but also in the event-oriented form subject to such events as "a machine becomes available" or "a new job enters the system" (Fig. 2.1).

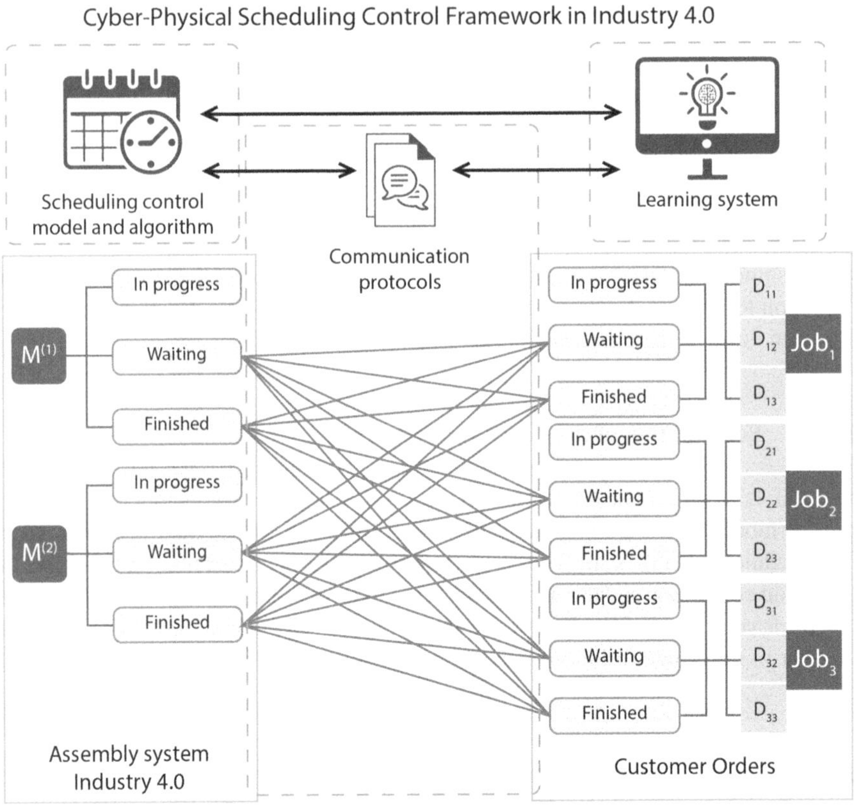

Fig. 2.1 Cyber-physical scheduling control framework in Industry 4.0

Consider the example illustrated in Fig. 2.1 with two machines M_1 and M_2 and three products. The manufacturing of the products is organized in three jobs each of which is characterized by the sequence of operations D_1-D_2-D_3 that can be processed on both M_1 and M_2. Both machines and products are equipped with sensors and a communication protocol is established between the sensors. The sensors observe the machine utilization and operation processing subject to three states, that is, "in process," "waiting," and "finished." The scheduling algorithm is activated in the case of the following events in the communication protocol: (1) a machine signals the completion of an operation processing and there is at least one operation either in the state "waiting" or "finished," (2) a product signals the completion of the processing of an operation on a machine and there is at least one machine either in the state "waiting" or "finished," and (3) there are at least one machine and one operation in the state "waiting."

At the proactive stage, our method can be used for a descriptive and diagnostic analysis and a predictive modeling to analyze the possible performances of the production system. At the reactive stage, we contribute to the control of the real-time schedule and adaptive learning. The Industry 4.0 and cloud manufacturing typically have data-driven, sensor-based environments. Sensors and data analytics may have two impacts on the system. The first is the tuning: changing uncertain coefficients in the structure of the differential equations of the system, taking into account that a larger number of these coefficients implies a more accurate system response to a changing environment. The second is the learning: imposing new restrictions on the system behavior. The number of arbitrary coefficients in the structure of the differential equations changes in the process of learning, imposing dynamic restriction adjustments on the behavior of the system. The use of real-time data allows the realization of different dynamic decomposition principles in the optimal control algorithm for task and service compositions. The selection of the time points at which the Hamiltonian is extremized and small-dimensional assignment problems are solved can be implemented not only as a fixed *t*-step procedure but also in the event-oriented form subject to events for a service composition such as "a machine becomes available" or events for task composition such as "a new job enters the system."

Moreover, the first analyses of COVID-19 pandemic impacts on the supply chains and production systems show the importance of Industry 4.0 and digital manufacturing from the perspectives of supply chain resilience and ripple effect control (Ivanov and Dolgui 2019; Hosseini et al. 2020). Firms that have visibility and digital control in manufacturing networks seem to be better positioned at the crisis time and for the future recovery coordination (Choi et al. 2020; Ivanov et al. 2019; Panetto et al. 2019; Ivanov 2020; Ivanov and Dolgui 2020a, b; Ivanov and Das 2020; Ni et al. 2020).

2.5 Conclusions

The Industry 4.0 technology enables new production strategies that require highly customized assembly systems. The ultimate objective of these systems is to facilitate a flexible customized manufacturing at the lower costs of mass production. Such innovative production strategies create a number of new challenges and opportunities for short-term job scheduling. In particular, the manufacturing processes for different customer orders may have individual machine structures such that the flexible stations are able to execute different functions subject to individual sets of operations within the jobs. Therefore, the problem of a simultaneous structural-functional synthesis of the customized assembly system arises. A flexible distributed scheduling as required by the Industry 4.0 paradigm has been addressed in this study using optimal program control theory.

The major contribution of this chapter is the development of an optimal control model for the simultaneous structural-functional design of a customized manufacturing process and sequencing of the operations of the jobs in an Industry 4.0 system. For the first time, a multi-objective, multi-stage job shop scheduling problem with alternative and flexible machines at each stage and different time-dependent processing speeds and time-dependent machine availability, without job splitting, was solved by means of optimal control and the maximum principle using the maximization of the Hamiltonian.

The method and the algorithm developed present a contribution to flexible, distributed scheduling in the emerging field of Industry 4.0-based innovative production systems. In contrast to previous studies that assumed a fixed process design, our approach is capable of designing simultaneously the manufacturing process in regard to the available alternative stations, their current capacity utilization and the processing time, and the sequencing of the jobs at the stations.

The basic *computational idea* of the computational approach developed is that the operation execution and machine availability are dynamically distributed in time over the planning horizon. As such, not all operations and machines are relevant to decision-making at the same time. Therefore, the solution at each time point for a small-dimensional system is calculated by mathematical programming. The multi-dimensionality and the combinatorial explosion of the problem faces a decreasing connectivity under the network diagram of the operations. The analysis of the manufacturing process paths for the execution of different jobs can help to reveal a real utilization of the channels at different stations as well as the stations all together. This may be helpful for estimating the requirements on the multi-functionality of the stations. Such an analysis may reveal, for example, that some channels are utilized fully while other channels are used occasionally. This analysis may be used for the capacity design and investment decisions.

The formulation of the scheduling model in the dynamic control form makes it possible to apply it both to proactive and reactive real-time scheduling. As such, integration of the planning and real-time control stages can be realized using the unified methodical and technical principles. Moreover, the formulation of the scheduling model as an optimal program control allows the consideration of a non-stationary process view and the use of the accuracy of continuous time. In addition, a wide range of analysis tools from control theory regarding stability, controllability, adaptability, etc. may be used if a schedule is described in terms of control.

In the future, a robustness analysis of the overall system, that is, both of the process design and the schedule, can extend the results of this study. In addition, computational examples may help to reveal new insights. A more detailed analysis of the Industry 4.0 technology may illuminate a taxonomy of structural-functional problems in this emerging research field. Finally, Industry 4.0 and digital technologies open new possibilities to the implementation of dynamic scheduling techniques using real-time data about the machine utilization, entering of new jobs, and the operation processing status. This makes it possible to extend the algorithmic dynamic decomposition principles of the control scheduling models, for example, by incorporating an event-oriented decomposition based on such events as "a machine becomes available," "an operation is completed," or "a new job enters the system." These extensions can be considered in light of future research topics.

Acknowledgments The research described in this chapter is partially supported by the Russian Foundation for Basic Research grants: 17-29-07073-ofi-i (Sect. 2.2), state research 0073-2019-0004 (Sect. 2.3). This research has received funding from the European Union's Horizon 2020 research and innovation program under the Marie Sklodowska-Curie grant agreement No. 873077 (MAIA-H2020-MSCA-RISE 2019) (Sect. 2.4).

2.6 Appendix. Notations

2.6.1 Sets, Maps, and Constants

Notation	Meaning
$A = \{A_\upsilon, \upsilon \in 1, \ldots n\}$	Set of jobs
$D = \left\{ \begin{array}{l} \left\{D_{æ'}^{(i)}\right\} \cup \left\{D_{æ}^{(i,j)}\right\}, \\ i, j \in 1, ..m, æ', æ \in 1, \ldots s_i \end{array} \right\}$	Set of operations
$M = \{M_i, i \in 1, \ldots m\}$	Set of stations
$C = \left\{C_\lambda^{(i)}, \lambda \in 1, l_i\right\}$	Set of channels
$\Phi = \left\{ \begin{array}{l} \left\{\Phi S_\pi^{(i)}\right\} \cup \left\{\Phi N_\mu^{(i)}\right\}, \\ \pi \in K_i^{(r,1)}, \mu \in K_i^{(r,2)} \end{array} \right\}$	Set of resources
$\Phi S^{(i)} = \left\{ \Phi S_\pi^{(i)}, \pi \in K_i^{(r,1)} \right\}$	Set of storable resources at $M^{(i)}$
$\Phi N^{(i)} = \left\{ \Phi N_\mu^{(i)}, \mu \in K_i^{(r,2)} \right\}$	Set of non-storable resources at $M^{(i)}$
K	Set of numbers
$P = \left\{ \left\{P_{<æ',\rho>}^{(i)}\right\} \cup \left\{P_{<æ,\rho>}^{(i,j)}\right\}, \rho \in K_i^{(f)} \right\}$	Set of material flows subject to $M^{(i)}$
$P^{(i,j)} = \left\{ P_{<æ,\rho>}^{(i,j)}, \rho \in K_i^{(f)} \right\}$	Set of material flows for the ρ-types of materials subject to $M^{(i)}$ and $M^{(j)}$.
$\Gamma_{\upsilon 1}, \Gamma_{\upsilon 2}$	Sets of “and” and “or” precedence relations for the jobs
$\Gamma_{iæ1}, \Gamma_{iæ2}$	Sets of “and” and “or” precedence relations for the operations
$\mathbf{U}$	Set of feasible control inputs
J	Set of performance indicators
$\Pi^t_{<\delta,\delta'>}$	Map describing the allowable transitions from one multi-structural macro-state to another one
Δ	Set of dynamic and static alternatives of the manufacturing process
$R_{\tilde{r}}$	Set of process constraints
$\tilde{\tilde{R}}_{\tilde{r}}$	Constants, which are known, and $T = (T_0, T_f]$ is the time interval for the manufacturing process design synthesis
$\varsigma \in \{1, \ldots, \Im\}$	Set of the numbers of the performance indicator
$\mathbf{X}(\mathfrak{x}(t), t)$	Area of the allowable states of the assembly line structural dynamics

2.6.2 Parameters and Functions

Notation	Meaning
a	Planned processing volume (lot size)
$\tilde{P}_j^{(1)}, \tilde{P}_{j\rho}^{(2)}, \tilde{P}_{ij}^{(3)}$	Known values for the maximal storage capacity at M_j, handling capacity (throughput) at M_j for ρ, and transportation capacity between M_i and M_j, respectively
T_0	Start time of the scheduling horizon
T_f	End time of the scheduling horizon
b	Setup time of a channel
$d_{iæj\lambda}^{(\pi)}, g_{iæj\lambda}^{(\mu)}$	Given consumption intensities of $\Phi S_\pi^{(j)}$ and $\Phi N_\mu^{(j)}$ for $D_æ^{(i,j)}$ and $C_\lambda^{(j)}$
$\tilde{H}_j^{(\pi)}(t), \tilde{\tilde{H}}_j^{(\mu)}(t)$	Intensities for the replenishment of the resources of $\Phi S_\pi^{(j)}$ and $\Phi N_\mu^{(j)}$, respectively
$\xi(t)$	Vector of perturbation impacts
$\mathbf{h}_0^{(o)}, \mathbf{h}_1^{(o)}$	Differentiable functions that determine the end conditions of the vector
σ	Duration of the planning interval
$\varepsilon(t)$	Preset matrix time function of the time-spatial constraints for the stations
$\Theta_{iæj\lambda}(t)$	Preset matrix time function of the time-spatial constraints for the channels
$\tilde{\beta}(\tau)$	Penalty function for the completion delay of an operation
$\mathbf{q}^{(1)}$ and $\mathbf{q}^{(2)}$	Vector-functions, defining the main spatio-temporal, economic, technical, and technological conditions for the machine functioning process

2.6.3 Indices

Notation	Meaning
v	Job index
n	Running numbers of a job
æ	Operation index
s	Running numbers of an operation
i	Station index Job index from Sect. 3.6.2 ongoing
m	Running numbers of a station
λ	Channel index
l	Running numbers of a channel
r	Number of the iteration of the algorithm
ρ	Product flow index
π	Storable resource index
μ	Non-storable resource index Operation index from Sect. 3.6.2 ongoing
η	Replenishment cycle
$\tilde{\rho}, \tilde{\tilde{\rho}}$	Running numbers of the replenishment cycles

(continued)

Notation	Meaning
α, β	Indices of the precedence relations "and" and "or" for the jobs
$\tilde{\alpha}, \tilde{\beta}$	Indices of the precedence relations "and" and "or" for the operations
l'	Running numbers of a structure element of the manufacturing process
δ	Running number of a multi-structural macro-state of the manufacturing process
χ	Running number of the design structure of an alternative manufacturing process
(o), (k), (r), (f)	Indexes to describe the relationships of the respective sets to the operations (o), channels (k), machines (r), and material flows (f)
t	Current time point

2.6.4 Decision Control and State Variables

Notation	Meaning
$\mathbf{u}(t)$	Control variable
$\mathbf{x}(t)$	State variables
$z_{ij\mu(o)}$	Auxiliary variable that characterizes the execution of the μ-operation
$h_{ij\mu(o)}$	The square under the integral curve $z_{ij\mu(o)}$
$g_{ij\mu(o)}$	Auxiliary variable that is equal to the time $t'_{ij\mu}$ between the completion time of the μ-operation and T_f
$w^{(o)}_{i\mu\ j}$	Auxiliary control variable that equals 1 if $x^{(f)}_{i\mu}(t) = a^{(f)}_{i\mu}$ at time t and $x^{(o)}_{i\mu} \neq a^{(o)}_{i\mu}$
$\sigma^{(and)}_{i\overline{\alpha}}; \sigma^{(or)}_{i\overline{\beta}}; \sigma^{(or)}_{i\overline{\overline{\beta}}}; \sigma^{(and)}_{i\overline{\overline{\alpha}}}; \sigma^{(5,2)}_{i\mu}; \sigma^{(2,1)}_{i\mu}$	Coefficients of the adjoint system
$\psi(t)$	Adjoint variable

References

Battaïa, O., & Dolgui, A. (2013). A taxonomy of line balancing problems and their solution approaches. *International Journal of Production Economics, 142*(2), 259–277.

Battaïa, O., Dolgui, A., & Guschinsky, N. (2017a). Integrated process planning and system configuration for mixed-model machining on rotary transfer machine. *International Journal of Computer Integrated Manufacturing, 30*(9), 910–925.

Battaïa, O., Dolgui, A., & Guschinsky, N. (2017b). Decision support for design of reconfigurable rotary machining systems for family part production. *International Journal of Production Research, 55*(5), 1368–1385.

Blazewicz, J., Ecker, K., Pesch, E., Schmidt, G., & Weglarz, J. (2001). *Scheduling computer and manufacturing processes* (2nd ed.). Berlin: Springer Verlag.

Blazewicz, J., Pesch, E., Trystram, D., & Zhang, G. (2015). New perspectives in scheduling theory. *Journal of Scheduling, 18*(4), 333–334.

Boysen, N. A., Fliedner, M., & Scholl, A. (2007). A classification of assembly line balancing problems. *European Journal of Operational Research, 183*(2), 674–693.

Božek, A., & Werner, F. (2018). Flexible job shop scheduling with lot streaming and sublot size optimisation. *International Journal of Production Research, 56*, 6391–6411. https://doi.org/10.1080/00207543.2017.1346322.

Bukchin, J., & Rubinovitz, J. (2003). A weighted approach for assembly line design with station paralleling and equipment selection. *IIE Transactions, 35*, 73–85.

Chaube, A., Benyoucef, L., & Tiwari, M. K. (2012). An adapted NSGA-2 algorithm based dynamic process plan generation for reconfigurable manufacturing system. *Journal of Intelligent Manufacturing, 23*(4), 1141–1155.

Chaudry, I. A., & Khan, A. A. (2016). A research survey: Review of flexible job shop scheduling techniques. *International Transactions in Operational Research, 23*(3), 551–591.

Chen, W., Song, J., Shi, L., Pi, L., & Sun, P. (2013a). Data mining-based dispatching system for solving the local pickup and delivery problem. *Annals of Operations Research, 203*(1), 351–370.

Chen, W., Wang, X., Petersen, J., Tyagi, R., & Black, J. (2013b). Optimal scheduling of demand response events for electric utilities. *IEEE Transactions on Smart Grid, 23*(4), 2309–2319.

Chen, J., Huang, G. Q., Wang, J.-Q., & Yang, C. (2019). A cooperative approach to service booking and scheduling in cloud manufacturing. *European Journal of Operational Research, 273*(3), 861–873.

Choi TY, Rogers D, & Vakil B. (2020b). Coronavirus is a wake-up call for supply chain management. Harvard business review, march 27th.

Delorme, X., Dolgui, A., & Kovalyov, M. Y. (2012). Combinatorial design of a minimum cost transfer line. *Omega—International Journal of Management Science, 40*(1), 31–41.

Dolgui, A., Ivanov, D., Potryasaev, S., Sokolov, B., Ivanova, M., & Werner, F. (2019a). Blockchain-oriented dynamic modelling of smart contract design and execution control in the supply chain. *International Journal of Production Research*. https://doi.org/10.1080/00207543.2019.1627439.

Dolgui, A., Ivanov, D., Sethi, S. P., & Sokolov, B. (2019b). Scheduling in production, supply chain and Industry 4.0 systems by optimal control. *International Journal of Production Research, 57*(2), 411–432.

Erol, S., Jäger, A., Hold, P., Ott, K., & Sihn, W. (2016). Tangible Industry 4.0: A scenario-based approach to learning for the future of production. *Procedia CIRP, 54*, 13–18.

Fragapane, G., Ivanov, D., Peron, M., Sgarbossa, F., & Strandhagen, J. O. (2020). Increasing flexibility and productivity in industry 4.0 production networks with autonomous mobile robots and smart intralogistics. *Annals of Operations Research*. https://doi.org/10.1007/s10479-020-03526-7.

Harjunkoski, I., Maravelias, C. T., Bongers, P., Castro, P. M., Engell, S., Grossmann, I. E., Hooker, J., Méndez, C., Sand, G., & Wassick, J. (2014). Scope for industrial applications of production scheduling models and solution methods. *Computers and Chemical Engineering, 62*, 161–193.

Hosseini, S., Ivanov, D., & Dolgui, A. (2019a). Review of quantitative methods for supply chain resilience analysis. *Transportation Research: Part E, 125*, 285–307.

Hwang, C. L., Fan, L. T., & Erikson, L. E. (1967). Optimum production planning by the maximum principle. *Management Science, 13*(9), 751–755.

Hwang, G., Lee, J., Park, J., & Chang, T.-W. (2017). Developing performance measurement system for Internet of Things and smart factory environment. *International Journal of Production Research, 55*(9), 2590–2602.

Ivanov D. (2020b) Predicting the impact of epidemic outbreaks on the global supply chains: A simulation-based analysis on the example of coronavirus (COVID-19 / SARS-CoV-2) case. Transportation research – Part E, 136., https://doi.org/10.1016/j.tre.2020.101922.

Ivanov D., & Das A. (2020b). Coronavirus (COVID-19 / SARS-CoV-2) and supply chain resilience: A research note. International journal of integrated supply management, forthcoming.

Ivanov, D., & Dolgui, A. (2019a). New disruption risk management perspectives in supply chains: Digital twins, the ripple effect, and resileanness. 9th IFAC conference on manufacturing modelling, management and control (MIM 2019), edited by Dmitry Ivanov, Alexandre Dolgui, Farouk Yalaoui. *IFAC-PapersOnLine, 52*(13), 337–342.

Ivanov, D., & Dolgui, A. (2020b). Viability of intertwined supply networks: Extending the supply chain resilience angles towards survivability. In *A position paper motivated by COVID-19 outbreak*. International Journal of Production: Research. https://doi.org/10.1080/00207543.2020.1750727.

Ivanov D., Dolgui A. (2020d). A digital supply chain twin for managing the disruptions risks and resilience in the era of industry 4.0. Production planning and control, forthcoming.

Ivanov, D., & Sokolov, B. (2012a). Structure dynamics control approach to supply chain planning and adaptation. *International Journal of Production Research, 50*(21), 6133–6149.

Ivanov, D., & Sokolov, B. (2012b). Dynamic supply chain scheduling. *Journal of Scheduling, 15*(2), 201–216.

Ivanov, D., & Sokolov, B. (2013). Dynamic coordinated scheduling in the supply chain under a process modernization. *International Journal of Production Research., 51*(9), 2680–2697.

Ivanov, D., Arkhipov, A., & Sokolov, B. (2004). Intelligent supply chain planning in virtual enterprises. In L. Camarihna-Matos (Ed.), *Virtual Enterprises and Collaborative Networks, Proceedings of the IFIP Conference on Virtual Enterprises PRO-VE 2004* (pp. 215–223). Boston, MA: Kluwer Academic Publishers.

Ivanov, D., Kaeschel, J., & Sokolov, B. (2007). Integrated modeling of agile enterprise networks. *International Journal of Agile Systems and Management, 2*(1), 23–49.

Ivanov, D., Sokolov, B., & Kaeschel, J. (2010). A multi-structural framework for adaptive supply chain planning and operations with structure dynamics considerations. *European Journal of Operational Research, 200*, 409–420.

Ivanov, D., Sokolov, B., & Dolgui, A. (2013a). Multi-stage supply chain scheduling in petrochemistry with non-preemptive operations and execution control. *International Journal of Production Research, 52*(13), 4059–4077.

Ivanov, D., Dolgui, A., & Sokolov, B. (2013b). Multi-disciplinary analysis of interfaces "supply chain event management – RFID – Control theory". *International Journal of Integrated Supply Management, 8*, 52–66.

Ivanov, D., Sokolov, B., Hartl, R., Dolgui, A., Pavlov, A., & Solovyeva, I. (2015). Integration of aggregate distribution and dynamic transportation planning in a supply chain with capacity disruptions and ripple effect considerations. *International Journal of Production Research, 53*(23), 6963–6979.

Ivanov, D., Sokolov, B., Dolgui, A., Werner, F., & Ivanova, M. (2016a). A dynamic model and an algorithm for short-term supply chain scheduling in the smart factory Industry 4.0. *International Journal of Production Research, 54*(2), 386–402.

Ivanov, D., Dolgui, A., & Sokolov, B. (2016b). Robust dynamic schedule coordination control in the supply chain. *Computers and Industrial Engineering, 94*, 18–31.

Ivanov, D., Dolgui, A., & Sokolov, B. (2018a). Scheduling of recovery actions in the supply chain with resilience analysis considerations. *International Journal of Production Research, 56*(19), 6473–6490.

Ivanov, D., Sethi, S., Dolgui, A., & Sokolov, B. (2018b). A survey on the control theory applications to operational systems, supply chain management and Industry 4.0. *Annual Reviews in Control, 46*, 134–147.

Ivanov, D., Tsipoulanidis, A., & Schönberger, J. (2019). *Global supply chain and operations management: A decision-oriented introduction into the creation of value*. Cham: Springer International.

Ivanov, D., Dolgui, A., & Sokolov, B. (2019a). The impact of digital technology and industry 4.0 on the ripple effect and supply chain risk analytics. *International Journal of Production Research, 57*(3), 829–846.

Ivanov, D., Sokolov, B., Chen, W., Dolgui, A., Werner, F., & Potryasaev, S. (2020). A control approach to scheduling flexibly configurable jobs with dynamic structural-logical constraints. *IISE Transactions*. https://doi.org/10.1080/24725854.2020.1739787.

Janiak, A., Janiak, W., Kovalyov, M., Kozan, E., & Pesch, E. (2013). Parallel machine scheduling and common due window assignment with job independent earliness and tardiness costs. *Information Science, 224*, 109–117.
Józefowska, J., Mika, M. K., Różycki, R., Waligóra, G., & Węglarz, J. (2002). A heuristic approach to allocating the continuous resource in discrete–continuous scheduling problems to minimize the makespan. *Journal of Scheduling, 5*(6), 487–499.
Jungwattanakit, J., Reodecha, M., Chaovalitwongse, P., & Werner, F. (2009). A comparison of scheduling algorithms for flexible flow shop problems with unrelated parallel machines, setup times and dual criteria. *Computers & Operations Research, 36*(2), 358–378.
Kalinin, V. N., & Sokolov, B. V. (1985). A dynamic model and optimal scheduling algorithm for activities with bans of interraps. *Automation Remote Control, 489*(1–2), 88–94.
Khmelnitsky, E., Kogan, K., & Maimom, O. (1997). Maximum principle-based methods for production scheduling with partially sequence-dependent setups. *International Journal of Production Research, 35*(10), 2701–2712.
Kogan, K., & Khmelnitsky, E. (2000). *Scheduling: Control-based theory and polynomial-time algorithms*. Dordrecht: Kluwer.
Kumar, M., Graham, G., Hennelly, P., & Srai, J. (2016). How will smart city production systems transform supply chain design: A product-level investigation. *International Journal of Production Research, 54*(23), 7181–7192.
Kusiak, A. (2018). Smart manufacturing. *International Journal of Production Research., 56*(1 2), 508–517.
Kyparisis, G. J., & Koulamas, C. P. (2006). Flexible flow shop scheduling with uniform parallel machines. *European Journal of Operational Research, 168*, 985–997.
Lauff, V., & Werner, F. (2004). On the complexity and some properties of multi-stage scheduling problems with earliness and tardiness penalties. *Computers & Operations Research, 31*, 317–345.
Leusin, M., Frazzon, E., Uriona Maldonado, M., Kück, M., & Freitag, M. (2018). Solving the job-shop scheduling problem in the Industry 4.0 era. *Technologies, 6*(4), 107.
Levin, G., Rozin, B., & Dolgui, A. (2016). Optimization of the structure and execution modes of intersecting operation sets. *IFAC-PapersOnLine, 49*(12), 105–110.
Liu, Y., Wang, L., Wang, X. V., Xu, X., & Zhang, L. (2019). Scheduling in cloud manufacturing: State-of-the-art and research challenges. *International Journal of Production Research, 57*(15–16), 4854–4879.
Mönch, L., Fowler, J. W., & Mason, S. (2012). *Production planning and control for semiconductor wafer fabrication facilities: Modeling, analysis, and systems*. Dordrecht: Springer Science & Business Media.
Mourtzis, D., & Vlachou, E. (2018). A cloud-based cyber-physical system for adaptive shop-floor scheduling and condition-based maintenance. *Journal of Manufacturing Systems, 47*, 179–198.
Nayak, A., Reyes Levalle, R., Lee, S., & Nof, S. Y. (2016). Resource sharing in cyber-physical systems: Modelling framework and case studies. *International Journal of Production Research, 54*(23), 6969–6983.
Ni, J. (2020b). How China can rebuild global supply chain resilience after COVID-19. https://www.weforum.org/agenda/2020/03/coronavirus-and-global-supply-chains/, accessed on April 4, 2020.
Oesterreich, T. D., & Teuteberg, F. (2016). Understanding the implications of digitisation and automation in the context of Industry 4.0: A triangulation approach and elements of a research agenda for the construction industry. *Computers in Industry, 83*, 121–139.
Otto, J., Henning, S., & Niggemann, O. (2014). Why cyber-physical production systems need a descriptive engineering approach—A case study in plug & produce. *Procedia Technology, 15*, 295–302.
Panetto, H., Iung, B., Ivanov, D., Weichhart, G., & Wang, X. (2019). Challenges for the cyber-physical manufacturing enterprises of the future. *Annual Reviews in Control, 47*, 200–213.
Panetto, H., Iung, B., Ivanov, D., Weichhart, G., & Wang, X. (2019a). Challenges for the cyber-physical manufacturing enterprises of the future. *Annual Reviews in Control, 47*, 200–213.

Ribas, I., Leisten, R., & Framinan, J. (2010). Review and classification of hybrid flow shop problem from asolutions procedure perspective. *Computers & Operations Research, 37*, 1439–1454.

Rossit, D. A., Tohmé, F., & Frutos, M. (2019). Industry 4.0: Smart scheduling. *International Journal of Production Research, 57*(12), 3802–3813.

Ruiz, R., & Vazquez-Rodriguez, J. A. (2010). The hybrid flow shop problem. *European Journal of Operational Research, 205*, 1–18.

Sarimveis, H., Patrinos, P., Tarantilis, C. D., & Kiranoudis, C. T. (2008). Dynamic modeling and control of supply chain systems: A review. *Computers & Operations Research, 35*, 3530–3561.

Shen, L., Dauzere-Peres, S., & Neufeld, J. (2018). Solving the flexible job shop scheduling problem with sequence-dependent setup times. *European Journal of Operational Research, 265*, 503–516.

Sotskov, Y. N., Lai, T.-C., & Werner, F. (2013). Measures of problem uncertainty for scheduling with interval processing times. *OR Spectrum, 35*(3), 659–689.

Tahar, D. N., Yalaoui, F., Chu, C., & Amodeo, L. (2006). A linear programming approach for identical parallel machine scheduling with job splitting and sequence-dependent setup times. *International Journal of Production Economics, 99*(1–2), 63–73.

Theorin, A., Bengtsson, K., Provost, J., Lieder, M., Johnsson, C., Lundholm, T., & Lennartson, B. (2017). An event-driven manufacturing information system architecture for Industry 4.0. *International Journal of Production Research, 55*(5), 1297–1311.

Weyer, S., Schmitt, M., Ohmer, M., & Gorecky, D. (2015). Towards Industry 4.0—Standardization as the crucial challenge for highly modular, multi-vendor production systems. *IFAC-PapersOnLine, 48*(3), 579–584.

Yu, W., Liu, Z., Wang, L., & Fan, T. (2011). Routing open shop and flow shop scheduling problems. *European Journal of Operational Research, 213*(1), 24–36.

Zhang, J., Ding, G., Zou, Y., Qin, S., & Fu, J. (2019). Review of job shop scheduling research and its new perspectives under Industry 4.0. *Journal of Intelligent Manufacturing, 30*(4), 1809–1830.

Zhong, R. Y., Xu, C., Chen, C., & Huang, G. Q. (2017). Big data analytics for physical internet-based intelligent manufacturing shop floors. *International Journal of Production Research, 55*(9), 2610–2621.

Chapter 3
Using a Digital Twin for Production Planning and Control in Industry 4.0

Ícaro Romolo Sousa Agostino, Eike Broda, Enzo M. Frazzon, and Michael Freitag

Abstract Simulation models are one of the most used quantitative approaches for modeling and decision-making in production and logistic systems. In the Industry 4.0 context, new paradigms arise from the possibility of collecting and storing large amounts of data in real-time and throughout productive and logistical operations, enabling the development of Digital Twins concept and related approaches. In this context, this chapter discusses the application of simulation models in productive and logistic systems. A bibliometric analysis was conducted, reviewing main concepts and applications illustrated in the literature. On the sequence, a digital twin approach for production planning and control using current cyber-physical systems state data in real-time is presented. The approach is evaluated by means of a real-world scenario involving a manufacturer supplying mechanical parts to the automotive industry. This evaluation shows that the approach is able to improve the performance of the production system for three different key performance indicators.

3.1 Introduction

The advent of the "Industry 4.0" concept addressing new paradigms involving productive and logistic systems has allowed a reformulation of industrial processes

Í. R. S. Agostino · E. M. Frazzon (✉)
Industrial and Systems Engineering Department, Federal University of Santa Catarina, Florianópolis, SC, Brazil
e-mail: enzo.frazzon@ufsc.br

E. Broda
University of Bremen, Faculty of Production Engineering, Bremen, Germany
e-mail: brd@biba.uni-bremen.de

M. Freitag
BIBA - Bremer Institut für Produktion und Logistik GmbH at the University of Bremen, Bremen, Germany
e-mail: fre@biba.uni-bremen.de

B. Sokolov et al. (eds.), *Scheduling in Industry 4.0 and Cloud Manufacturing*, International Series in Operations Research & Management Science 289,
https://doi.org/10.1007/978-3-030-43177-8_3

by considering new perspectives for integration and control of dynamic processes in real-time (Thoben et al. 2017; Zhong et al. 2017; Panetto et al. 2019). In this context, the industrial digitalization resulting from the recent changes, supported by new technologies, such as the Internet of Things, big data, blockchain, and artificial intelligence allowed the emergence of the concept of cyber-physical systems. In these systems, data and process can be connected in a dynamic and integrated way, bringing the industry to a new perspective in which intelligent systems present themselves as a feasible and promising solution for planning and control of production and logistics systems (Wang et al. 2015; Kusiak 2018; Ivanov et al. 2019; Dolgui et al. 2019; Kück et al. 2016, 2017).

Technological advances have increased data availability and volume in production systems (Tao et al. 2018), reducing the cost of collecting and storing large sets of information (Peres and Fogliatto 2018; Megahed and Jones-Farmer 2015). In this direction, the concept of "Smart Manufacturing" arises, referring to productive systems integrated to sensors and computational platforms with the intensive use of data modeling and predictive engineering (Kusiak 2018). Such approaches based on real-time information can provide high quality platforms for decision-makers (Heger et al. 2017). Simulation models are one of the most widely used quantitative approaches in the modelling of production and logistics systems, allowing to simulate the operation and decide on aspects involving various resources and contexts (Borshchev 2013; Agostino et al. 2019), and in addition enabling the integration of optimization approaches and data analysis (Lee et al. 2015; Frazzon et al. 2013; Ivanov et al. 2019).

The integration of analytical tools with real-time data becomes a potential field of research involving industrial systems (Heger et al. 2017). In this context, the evolution of information technologies and the increasing digitalization of production and operations connect physical and information flows into productive systems (Lee et al. 2015). This cyber-physical view allows the acquisition of system state data that can be used to support better decisions along production networks, with great potential to change paradigms in relation to the management of processes with a high degree of accuracy and productivity, supported mainly by Internet of Things technologies (Monostori et al. 2016; Tu et al. 2018).

New paradigms associated with technological development, involving remote sensing of machines and devices, as well as real-time connectivity allowed the development of the concepts of "online simulation" (Cardin and Castagna 2009, 2011), "coupling of simulation and production" (Bergeron et al. 2009; Zülch et al. 2002), and more recently "Digital Twin" (Kritzinger et al. 2018; Weyer et al. 2016). These approaches address the integration of sensors and quantitative models of simulation and optimization in industrial operations.

This emerging concept has been discussed both in the practical and academic environment. Digital Twin can be defined as a simulation model that reflects, in a timely manner, the state of a corresponding twin based on the historical data, real-time sensor data, and physical model (Glaessgen and Stargel 2012). Some theoretical studies such as the one developed by Kritzinger et al. (2018) and Weyer et al. (2016) point out the potential of the application of Digital Twins in the industrial environment involving production processes and logistics. Recent applications such as Zheng et al. (2019) corroborate this potential, but present the need for studies that broadly and generically systematize methods, tools, and concepts related to the development of Digital Twin applications.

This gap in the literature guides the development of this research. There is no consolidated reference model in the literature that deals with the application of simulation and optimization models for real-time data treatment for synchronous and data-oriented decision-making, which allows the connection of the shop floor and logistics operations with the management and dynamic control of processes.

In this way, this chapter presents a conceptual model for a data exchange framework in cyber-physical systems allowing the development of real-time simulation applications in industry. For this purpose, a bibliometric analysis was conducted to analyze the current state of the art and the practice of the application of simulation models to control and scheduling in dynamic production and logistics systems. This setup and the results of this analysis are presented in the next section. The reminder of this article is structured as follows: Sect. 3.3 describes the results of the analysis with a focus on the concepts and applications of simulation and Digital Twins. Section 3.4 describes the Digital Twin approach for production control and scheduling that is afterwards evaluated in the fifth section. The paper closes with conclusion and outlook.

3.2 A Bibliometric Analysis on Simulation in Production and Logistic Systems

To analyze the dynamics of research evolution considering simulation in productive and logistic systems in the context of Industry 4.0, a bibliometric analysis was performed. The final search was realized in August 2019 on the Scopus databases using the terms "Simulation," "Digital Twin," "Production," and "Logistics" in combination with "Industry 4.0," "Cyber-Physical Systems," and "Smart Manufacturing" (see Fig. 3.1) applied in the titles, abstracts, and keywords of the papers. For the portfolio only publications in journals and in English were considered. A total of 249 papers were found. The final search string is presented as follows:

Fig. 3.1 Search strategy

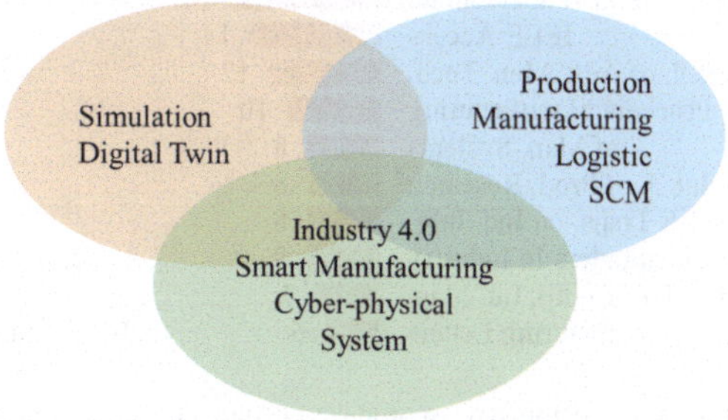

Search string

TITLE-ABS-KEY(((simulatio* OR "digital twin") AND (productio* OR manufactur* OR logistic* OR scm) AND ("industry 4.0" OR "smart manufacturing" OR "cyber-physical system*"))) AND (LIMIT-TO (DOCTYPE, "ar")) AND (LIMIT-TO (LANGUAGE, "English"))

Figure 3.2 shows the temporal evolution of publications in the selected portfolio. 2001 was the first year that a publication appeared in a journal indexed in the considered databases. Qiu et al. (2001) developed a discrete simulation system to control a flexible manufacturing system considering real-time data. In the following years, since 2011, the number of publications has grown consistently. Between 2016 and 2017, publications increased by 246%. This analysis shows the growing interest in simulation model applications in production and logistics systems in the context of Industry 4.0.

Figure 3.3a shows the ten journals with the highest concentration of publications in the analyzed group. The IFAC-PapersOnLine, IEEE access, and International Journal of Advanced Manufacturing Technology were the journals with the highest number of publications. Figure 3.3b shows the ten most cited journals, in this case, there is great emphasis on the International Journal of Production Research as well as several other journals that link studies of operations management, logistics, and technology.

Fig. 3.2 Publication temporal evolution

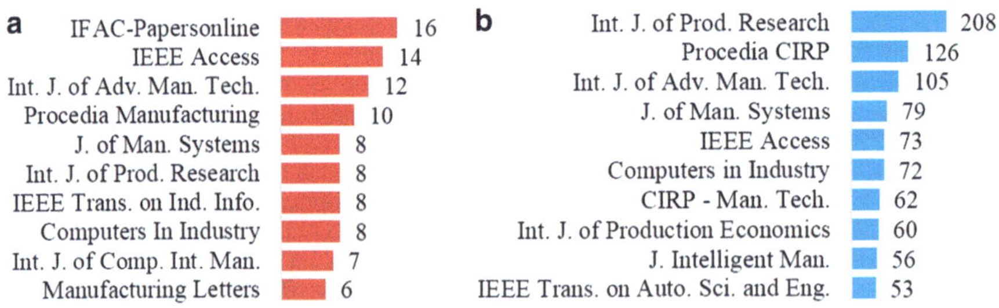

Fig. 3.3 (**a**) Journals of analyzed group. (**b**) Most cited journals

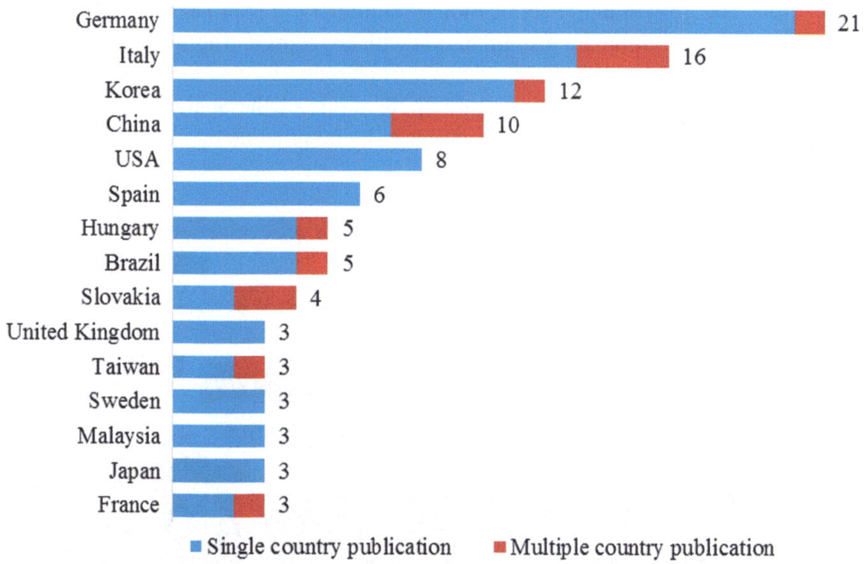

Fig. 3.4 Publications by country

Figure 3.4 shows the main countries which have associated studies in the research field, categorized by publications authored by only one country (in blue) and multiple countries (in red). Germany presents great prominence with 21 publications, followed by Italy with 16, Korea with 12, and the other countries with 10 or fewer publications. The high concentration of studies produced by institutions in Germany occurs mainly through the creation of the "Industrie 4.0" (I4.0) program by the government to promote the computerization of industrial processes (Thoben et al. 2017).

Figure 3.5 shows a keywords co-occurrence network by multidimensional scaling (Huang et al. 2005) using edge betweenness centrality clustering algorithm (Prell 2012). This analysis allows the identification of a main cluster of terms (in red) that deals with the intercession between the themes investigated in this research. As central terms "Industry 4.0", "simulation", "smart manufacturing" and "cyber-physical system" appear in the co-occurrence network. Other related terms such as Internet of Things, cloud computing, virtual reality, and big data demonstrate the connection between the research fields involving management, engineering, and computing disciplines.

Figure 3.6 illustrates the dynamics of evolution of the main themes over time. The frequency of keywords of the articles was analyzed, and the last 5 years were considered. It is important to highlight that some of the terms in the literature are used in similar contexts, the intention of this analysis is only to understand the dynamics of evolution of the concepts. It is possible to identify the growing interest for the development of research involving Industry 4.0. Another point evidenced is the maturation of the concept of Digital Twin in relation to the classic concept of simulation. The growth in the application of Digital Twin approaches is mainly due to technological development in conjunction with concepts and approaches proposed by Industry 4.0.

In the next section, the concepts and applications of the analyzed papers were discussed aiming at the identification of research and practice opportunities.

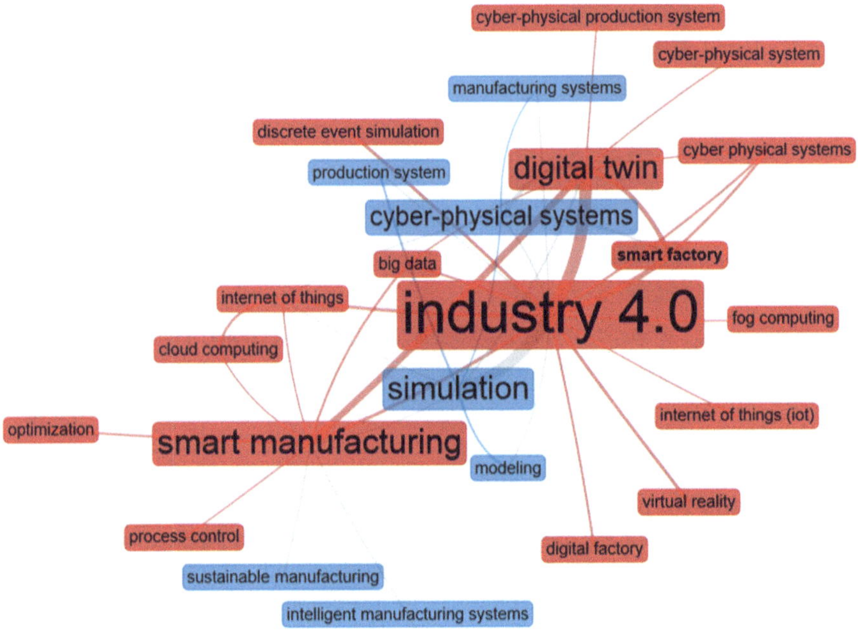

Fig. 3.5 Keywords co-occurrence network

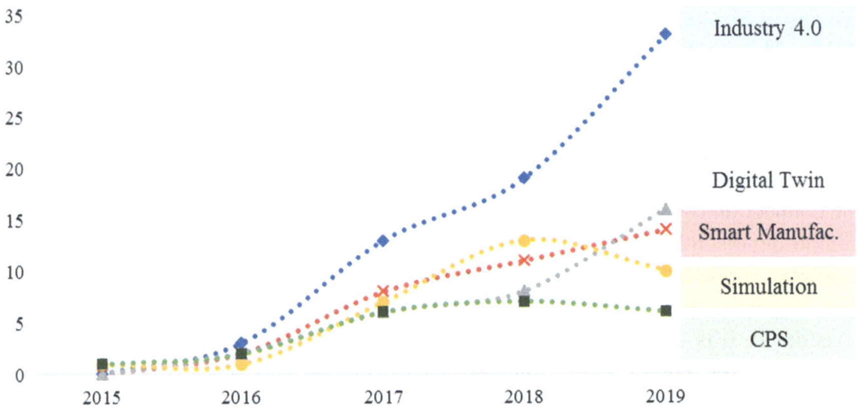

Fig. 3.6 Thematic evolution over time

3.3 Simulation and Digital Twin: Concepts and Applications

Several authors discuss the theoretical and conceptual aspects involving simulation models in production and logistic systems in the context of Industry 4.0. Turner et al. (2016) reviewed the literature involving discrete event simulation and virtual reality in the industry, the article addresses real-time integration, communication protocols, system design, and model application. The authors highlight the potential of application of these joint technologies in the industrial environment. Weyer et al. (2016) investigated the future of modeling and simulation applications by focusing on aspects of cyber-physical systems. The authors highlight the importance of the application of simulation models in the decision-making process and propose a framework for modeling cyber-physical systems based on literature. Polenghi et al. (2018) performed a review of surveys on the application of simulation models in manufacturing processes, the authors classified the articles and proposed an integration of models for simulation-based decision-making.

Simulation approaches have evolved in different stages: (1) simulation of a specific device based on special tools; (2) simulation of a generic device based on standard tools; (3) multilevel and multidisciplinary simulation. Currently, a new wave of transformation occurs with the possibility of developing Digital Twin models from real-time simulation (Tao et al. 2018; Qi and Tao 2018). Weyer et al. (2016) argue that there have been three waves of simulation models and recently a new paradigm called Digital Twin has been initiated with the possibility of incorporation data in real-time and digitalization of the industrial environment.

In the context of the application of Digital Twin models, some articles have reviewed the literature and the application requirements, Tao and Zhang (2017) investigated the application of Digital Twin models as a new paradigm in the direction of intelligent manufacturing. In this context, a Digital Twin can be defined as a multiphysics, multiscale, probabilistic, ultrafidelity simulation that reflects, in a timely manner, the state of a corresponding twin based on the historical data, real-time sensor data, and a physical model (Glaessgen and Stargel 2012). Figure 3.7 shows a Digital Twin conceptual model to real-time simulation.

Some practical applications involving the development of DT models are found in the literature, Wang et al. (2020) developed a DT application for material handling in a shop-floor environment in a manufacturing process in Southern China. The application of the model in the case study resulted in the reduction of energy consumption and better route optimization. Sujová et al. (2018) developed a DT model for 8 assembly lines, with 15 work points in each one. The main objective was to integrate the control of operations using simulation models with real data. The authors report the increased responsive and control capacity of the system with greater availability of data for synchronous decision-making.

The bibliometric analysis conducted allowed to understand the dynamics of evolution of the research area, as well as to analyze the main concepts related to the development of real-time simulation models in industry. In the next section the proposed model for the development of a real-time simulation approach is presented

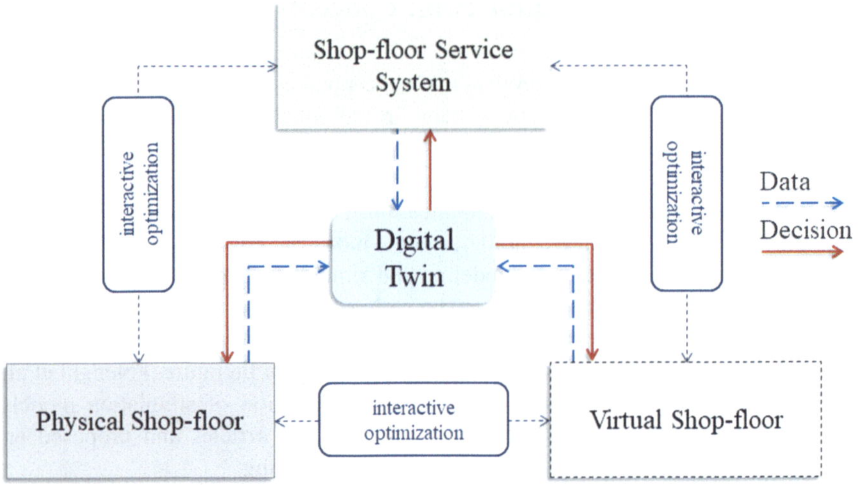

Fig. 3.7 Conceptual model for Digital Twin

and a case study with real industrial data is used to evaluate the model. The proposed framework objective is to serve as a generic model for implementation of Digital Twin solutions in dynamic production and logistics systems.

3.4 Building a Digital Twin for Planning and Control of Production

This section describes our approach of a Digital Twin for the planning and control of production. This Digital Twin is automatically synchronized to the state of the shop floor and optimizes the used dispatching rules for each machine in the system when a stochastic event occurs. This chapter first shows the definition of the Digital Twin applied to the problem of production planning and control and afterwards a description of the approach. This research is based on previous research of the working group as shown in Frazzon et al. (2018).

3.4.1 Definition of the Digital Twin

The basic idea of a Digital Twin is to build a digital representation of a physical object, i.e. a job shop production in a factory, which represents the real system and is updated automatically in case of changes. This scenario is shown in Fig. 3.8. On the left-hand side is the object in the real world, in this case the job shop production

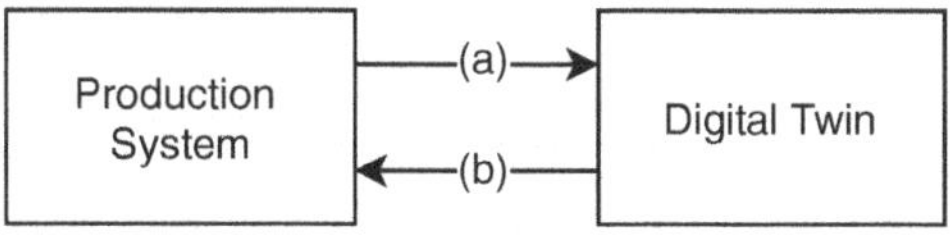

Fig. 3.8 Basic idea of a Digital Twin: information flows are bi-directional between the physical and digital object

system, and on the right-hand side is the Digital Twin, its digital representation. Between those two components is a bi-directional information flow from the real world object to the Digital Twin (a) and also from the Digital Twin to the real world object (b).

Following the definition of Kritzinger et al. (2018), there is a different intense of the integration of the information. On the lowest level no automatic data exchange between the real system and its digital representation is applied. All updates are done manually, there this type is called Digital Model. The first step towards a real Digital Twin is the automatic update of the digital representation (a), but not vice versa. In this case, changes in the real world object are automatically applied in the digital representation. This type is called Digital Shadow. To complete the Digital Twin it needs to also influence the real world object directly, so have an automatic information flow from the Digital Twin to the real world object (b).

So to implement a Digital Twin of a production system it needs to be updated regularly and in real-time with the current data from the production system. This data consists of the current machine status (idle, working, down), its current job and processing times. So it is possible to create a digital copy of the real system. The other way round the system also needs to apply the changes to the real production line, e.g. by applying a new schedule for the specific machine.

To ensure that the necessary data to build the Digital Twin is available, it needs to be gathered from different sources. For example, the current status of the machines from the machine execution system (MES) while the information about the jobs is inside an enterprise-resource-planning system (ERP).

3.4.2 Proposed Method for Production Planning and Control Using a Digital Twin

Figure 3.9 gives an overview of the proposed method. On the left side the real, physical production system is shown and on the right side the Digital Twin approach, which is the synchronized digital representation of the production system in a simulation model. This model is kept up-to-date with the current state of the real system and allows to optimize the dispatching rules in the real system. The arrows show the main communication between the components and is described in the next subsection. From the production system the current system state is regularly sent. On the one hand to a trigger function (a) and on the other hand to the simulation model

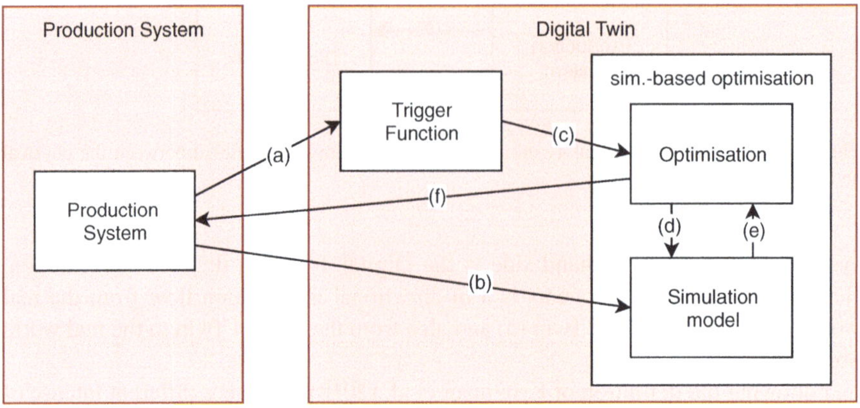

Fig. 3.9 Digital Twin approach for production planning and control

(b). The trigger function watches for changes in the production system. It reacts to events like new jobs, the absence of a worker, or the breakdown of a machine. In this case the simulation-based optimization is triggered (c), which then selects a new optimal set of dispatching rules for each individual machine. Additionally this method also triggers periodical optimizations, e.g. once per month. The simulation model is the Digital Twin of the real system. To keep this model synchronous it also gets the current system state (b) and updates its settings accordingly, e.g. broken machines of the real production system are disabled. If an optimization is triggered, the meta-heuristic, in this case a genetic algorithm, starts to reselect for each individual machine the dispatching rule. The meta-heuristic generates a population of possible solutions and uses the simulation model to evaluate these. In detail a possible configuration is determined by the algorithm and sent to the simulation model (d). The simulation model simulates this configuration for a given time and returns the key performance indicators (KPI) of this simulation run to the optimization (e). KPIs from the simulation are, e.g. the amount of tardy jobs or the flow time. This data is fed into the fitness function of the genetic algorithm and then a fitness is applied to the individuals. This fitness can be used to identify the current best solution for the given problem. The genetic algorithm will continue the optimization for some generations until it finishes due to its termination criterion. Then it returns the current best set of dispatching rules. This solution is sent to the production system and applied at the production (f). This process is also shown in Fig. 3.10.

For this research the production system was emulated in an emulation model which is a simulation model of the real production system. To delimit this model from the simulation model of the optimization it will be called emulation model. This emulation allows the authors to run experiments and optimize the method without inferring the activities in the real factory. During the experiments it was possible to let some machines break down and recover at previously defined times or

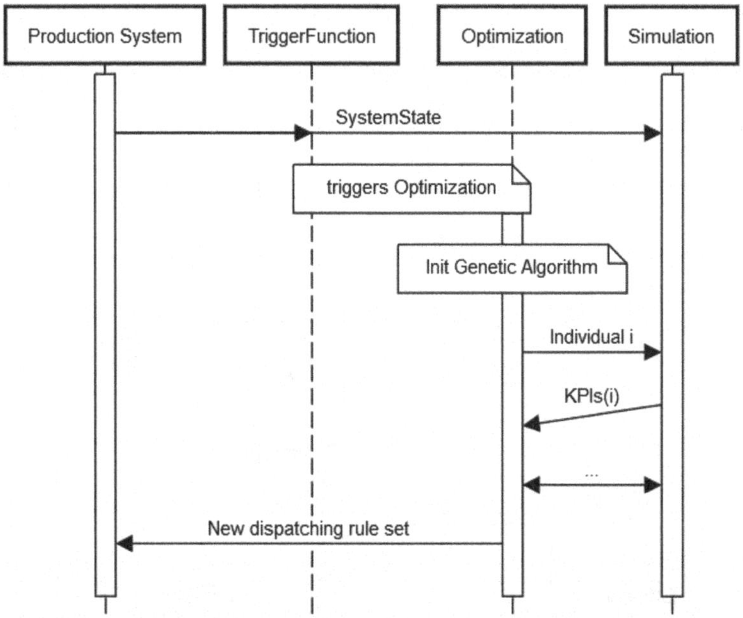

Fig. 3.10 Sequence diagram of the process

using stochastic processing and setup times. It also provides the possibility to rerun identical scenarios and test different strategies. The emulation also sends the data like it is described in the data exchange framework.

3.4.3 Data Exchange Framework

Figure 3.11 shows the Data Exchange Framework which shows the participants and different systems that interact to provide the data for this Digital Twin approach. Three participants are shown. On the left-hand side the suppliers which provide raw material and on the right-hand side the customers that demand products from the manufacturer, which is located in the middle. For this participant the figure shows the information systems that participate. Next to ERP and MES, which have already been mentioned earlier, a Production Data Acquisition System (PDA) and Machine Data Acquisition System (MDA) are shown. Both systems are for the automatic collection of current activities on the shop floor.

Central element of this framework is the MES which combines all the necessary data, stores it, and provides it as input for the Digital Twin approach. It is also responsible for applying the resulting, newly selected dispatching rules on the shop floor level.

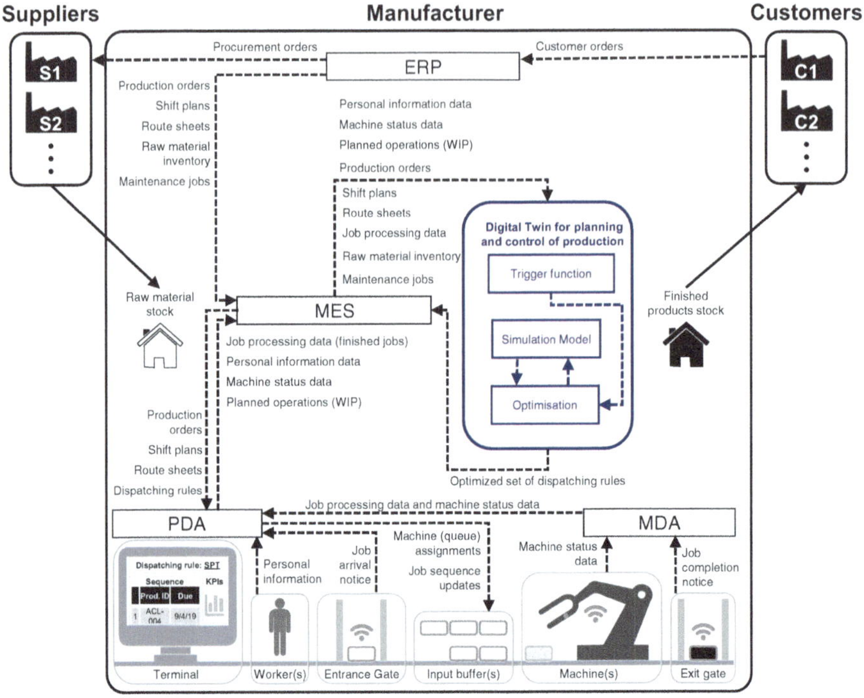

Fig. 3.11 Data Exchange Framework, continuation of Frazzon et al. (2018)

3.5 Use Case

The developed approach was applied in a job shop of a Brazilian supplier for the automotive industry. The performance of the individual dispatching rules, selected by the approach, will be compared to the currently used scheduling approach and the use of a static, system-wide dispatching rule. This chapter is structured as follows: first the scenario is described, afterwards the used approaches, the considered KPIs, and the experiment configuration are presented. The chapter closes with the results of the different approaches and a comparison of the methods.

The scenario is shown in Fig. 3.12. It contains four production lines. The horizontal main line (black) has two workstations with parallel machines (yellow) which are shared by jobs from two other lines (green, dashed and blue, dotted). The fourth line (red) produces parts that are assembled with parts of the main line. The whole scenario contains 20 workstations, which group 28 individual machines. Each workstation has between one and four machines and is grouped by a box in the graphic. A buffer is located in front of each workstation. On these lines 24 products are produced. Each product has a defined route that is basically one of the production lines but some jobs skip machines in the line. Therefore the amount of operations differs per product and takes between two and nine operations. The processing times

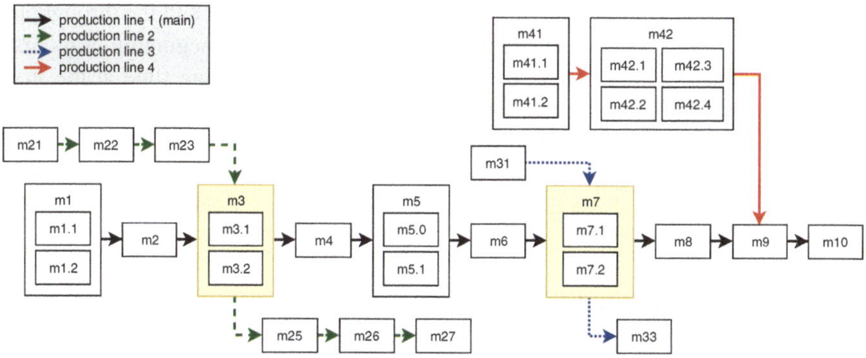

Fig. 3.12 Layout of the production line

are heterogenous for each product. Setup times are defined for the workstations m2 and m3 and applied on every change of the product.

Given monthly demand is splitted in weekly demands, as there are weekly deliveries to the customers. This demand is converted to jobs, using a given economic lot size which is a cost-optimal amount to be produced. There are 2570 jobs which are released monthly.

3.5.1 Experiment Details

This study compares the quality of the developed Digital Twin approach with two benchmarks. On the one hand the use of static dispatching rules at machine level and on the other hand a static, monthly calculated schedule. A dispatching rule is used to decide which of the arriving jobs at a machine should be produced next. As these rules just decide on the basis of the current queue state they are highly flexible, but they do not consider the overall system state and therefore only optimize the performance of an individual machine.

The other benchmark is a schedule which gives for each machine an ordered list of jobs to be processed. Following this list the production will be executed. This schedule optimizes the overall system but is not flexible if a job is delayed due to longer processing times at a previous operation. In this case a tardy job might lead to additional tardy jobs as capacities at machines are kept free to fulfill the job.

3.5.1.1 KPIs

These two benchmarks will be compared to the proposed Digital Twin approach by three KPIs. The first KPI is the number of tardy jobs. As each job has a due date, a job becomes tardy if it is finished after this due date. The number only counts the total number of tardy jobs, but not the total tardiness. As second KPI the throughput time is considered which is the duration the job takes from the beginning of its first operation until the end of its last operation. It omits the waiting time at the first workstation before the processing. The third KPI is the monthly working time usage. This is the mean value of the monthly maximum flow time of all jobs that have a due date in this month. It is like the mean of a monthly cMax calculation; as the overall cMax has the problem that it is highly influenced by the last month.

3.5.1.2 Experiment Configuration

The experiment is executed with stochastic influences. The processing and setup times are considered as stochastic values as the basic values are multiplied by a stochastic factor, drawn from a triangular distribution (with the parameters min 0.98, mode 1.01, and max 1.10). These processing times are recorded and used as mean values of the data from the last 6 months in the optimization method. Typically the times of individual machine breakdowns and reactivations are defined by distributions for the mean time between failure (MTBF) and mean time to repair (MTTR), but for this experiment they were defined static in advance to correspond to real breakdowns. Six failures occur on simple machines and five failures on a single machine in a machine group during the observation period. These breakdowns are between 1.4 and 6.5 working days, at a mean duration of 4.1 working days. The optimization uses a genetic algorithm that was run with the following settings: 100 individuals, 10 elitism individuals, mutation rate 10%, crossover rate 90%. The selection of parents is done by roulette wheel selection. However, each optimization is only running for five generations. In comparison to, e.g. 100 generations, our experiments have shown that the method in this case with fewer iterations provides better results. The population of the genetic algorithm, used for the optimization, is created using a strategy that reuses knowledge of a previously optimized population if available. Therefore a new population is initialized randomly, then 20% of the individuals are replaced by the best 20% individuals of the old population and additionally the benchmark rules are added, which have for all machines the same dispatching rule. For each dispatching rule we consider in the optimization, one benchmark individual is added. Orders for the next 3 months are taken into account during optimization. During the optimization 30 replications of the simulation model are used to determine the fitness of the individuals. The simulation model is adapted to the changed simulation state each time a new optimization is triggered. Replanning takes place both monthly and when events, like machine breakdowns or new job arrivals, occur. The total simulation time is 20 months, from which the first eight months are a transient phase and the last 12 months are the observation period.

3.5.1.3 Experiments

First the use of dispatching rules is evaluated. Therefore 28 dispatching rules were evaluated that were available in the used simulation software jasima. For each rule a separate experiment was executed where the dispatching rule was set at all machines. For each simulation 30 replications were executed and the mean values of the selected KPIs calculated. During this experiment stochastic influences are applied, but no optimization takes place. Afterwards these benchmark data were used to select the dispatching rules which will be listed as benchmark values in the comparison and also the rule set from that the optimization takes its individual rules.

Next step was to calculate the second benchmark, applying a static schedule. This schedule is generated monthly. For the benchmark the schedule was calculated using a planning method which utilizes a dispatching rule for the decision about the order in the schedule. This method runs all jobs in the current month with the rule and records the order of jobs. This generated schedule is then executed by a special dispatching rule, the ScheduleExecutor. This rule simply sequentially processes the schedule of the individual machines and only releases according to the schedule job operations for processing that are currently in the queue. This means that machines may not be able to process despite the fact that they are idle and orders are waiting in front of the machine because the machine is waiting for a particular job to arrive. For every previously selected dispatching rules a monthly schedule using this method has been calculated and then subsequently evaluated in 30 replications.

Afterwards the optimization of the individual machine dispatching rules has been executed. Therefore the selected dispatching rules have been considered as candidates. For each KPI an individual optimization is run as the method only considers one criterion in its fitness function. Each optimization is done in 10 replications and the results are mean values of these replications.

3.5.2 Results

The results for the three experiments that have been conducted are summarized below.

3.5.2.1 Benchmark Dispatching Rules

Table 3.1 shows for each KPI the benchmark results of the five best dispatching rules. The throughput time was converted from seconds to hours and the average monthly working time usage to working days. Taking into account the assumption that work is carried out in two shifts 5 days a week, the working time for a working day is 15 h. These best dispatching rules are Modified Due Date (MDD), Earliest Due Date (EDD), Modified Operation Due Date (MOD), Operation Due Date (ODD), Least (global) Slack (SLK), Shortest Processing Time (SPT), Shortest Remaining

Table 3.1 Average quality of the best five dispatching rules in each category from 30 replications

Number of tardy jobs		Throughput time [h]		Avg. monthly working time [d]	
Rule	KPI	Rule	KPI	Rule	KPI
MDD	99.9	**SPT**	53.9	**SLK**	19.5
EDD	119.6	**SRPT**	55.1	**EDD**	19.9
MOD	121.3	SI	60.0	**MDD**	19.9
ODD	124.0	**MDD**	73.7	**ODD**	20.0
SLK	187.9	FCFS	74.0	CR	20.5

Selected dispatching rules are bold

Processing Time (SRPT), SI × (SI), First Come First Served (FCSFS or FIFO), and Critical Ratio (CR). It shows that for each KPI a different rule is appropriate.

From these rules, those have to be selected that are used as benchmark for the optimization and also the rules that are used in the optimization to be selected for individual machines.

The results prove that the MDD rule is the best for the number of tardy orders. The EDD and ODD rules perform well and also have a good performance for the third criterion. For the throughput time the table shows that only the MDD rule is found in the best five for this KPI. This can be explained by the fact that the focus of these rules is different. While in the previous case the delays of orders were reduced, in this case the completion of the individual orders should take place as quickly as possible and therefore the throughput time of the orders should be minimized. The SPT rule is a typical representative here. The SRPT rule shows an even better quality and is therefore also selected. The other rules are significantly worse in the benchmark run and are therefore not considered further. The third criterion is the average monthly working time usage. The SLK rule provides the best result here. With EDD, MDD, and ODD, the remaining best rules for this criterion are also good for the first criterion. The rules MDD, EDD, ODD, SRPT, SPT, and SLK are therefore selected as rule set.

3.5.2.2 Benchmark Schedule

The second benchmark is the schedule. Table 3.2 shows an excerpt of the best scheduling results. These results are significantly worse than the application of dispatching rules because of the dynamic, stochastic influences in the production system that are not considered in the schedule as it is fixed. Again for each KPI a different schedule, based on a different dispatching rule, provides the best result. So schedules generated by EDD, SPT, and SLK are used as scheduling benchmarks.

3.5.2.3 Optimization

Table 3.3 shows the mean results for each optimization and the three KPIs. As expected for each KPI the corresponding optimization delivers the best results. A bit special is the optimization for the reduction of the tardy jobs as this one also reduces the monthly working time usage quite successful and to the nearly same level than the individual optimization for this KPI.

In the following these results will be considered and evaluated for each KPI.

The first criterion is the number of delayed orders. Table 3.4 lists the average number of tardy orders for 30 replications each for the selected dispatching rules, the three schedules, and the three optimizations according to the three criteria. The schedule selected for this target criterion, the schedule calculated by EDD, is regarded as the comparative value for the evaluation. It turns out that this schedule, with an average of 235.6 late orders, is of only mediocre quality. Some dispatching rules as

Table 3.2 Excerpt from the best schedule benchmark runs

Rule	Number of tardy jobs	Throughput time [h]	Avg. monthly working time usage [d]
EDD	235.6	90.6	20.7
MDD	237.7	89.8	20.7
SPT	252.4	67.9	22.6
SLK	299.6	90.0	20.5

Table 3.3 Results of the optimization for each of the three KPIs: [1] number of tardy jobs, [2] throughput time, and [3] average monthly working time usage

KPI	[1]	[2]	[3]
[1] Number of tardy jobs	92.3	75.66	19.89
[2] Throughput time	165.7	63.29	20.65
[3] Avg. monthly working time usage	167.1	75.75	19.87

Table 3.4 Results for the criterion "number of tardy jobs"

Criterion	Quality	Comp.
Number of tardy jobs	92.3	60.8%
MDD	99.9	57.6%
EDD	119.6	49.2%
ODD	124.0	47.4%
Throughput time	165.7	29.7%
Monthly working time usage	167.1	29.1%
SLK	187.9	29.2%
SPT	208.1	11.7%
SRPT	217.5	7.7%
Schedule EDD	235.6	–
Schedule MDD	237.7	−0.9%
Schedule SPT	252.4	−7.1%
Schedule SLK	299.6	−27.2%

well as the optimization according to two criteria provide better results. On average, the optimization gives the best result with 92.3 late orders, which are overall just 3.6% tardy jobs instead of 9.2% at a monthly schedule.

For the second criterion, the results mentioned in Table 3.5 are structured analogously. The schedule that was calculated using the SRPT rule is used here as the comparison value. The optimization for the corresponding target criterion is also better than the schedule here. Altogether, however, two dispatching rules are better than the optimization method. This is somewhat surprising, since the benchmark rules SRPT and SPT were also added to the population as benchmark individuals. Accordingly, the optimized solution should actually be much closer to the quality of the dispatching rules. Apparently, however, these rules are not of such high quality during the evaluation that they are selected as a corresponding set of rules. Further research is necessary to get an insight why this is happening.

For the last criterion, when looking at the results in Table 3.6, it is noticeable that the differences between the results seem to be much smaller here than for the other criteria. The maximum deviation here is only 5.0%, but in absolute terms

Table 3.5 Results for criterion "throughput time"

Criterion	Quality	Comp.
SPT	53.87	20.7%
SRPT	55.08	18.9%
Throughput time	63.29	6.8%
Schedule SPT	67.91	–
MDD	73.70	−8.5%
Number of tardy jobs	75.66	−11.4%
Monthly working time usage	75.75	−11.5%
EDD	76.00	−11.9%
ODD	76.49	−12.6%
SLK	76.56	−12.7%
Schedule SLK	89.98	−32.2%
Schedule EDD	90.58	−33.4%

Table 3.6 Results for criterion "monthly working time usage"

Criterion	Quality	Comp.
SLK	19.5	5.0%
Monthly working time usage	19.87	3.3%
Number of tardy jobs	19.89	3.2%
EDD	19.90	3.1%
MDD	19.93	3.0%
ODD	19.99	2.7%
Schedule SLK	20.54	–
Schedule EDD	20.67	−0.6%
SRPT	22.35	−8.5%
SPT	21.76	−5.9%
Schedule SRPT	22.64	−10.2%

this is 1.04 working days, which corresponds to 1.04 days $\cdot$15 h/day = 15.6 h. This observation makes it clear that the deviation is definitely more extensive. Effectively, the optimizations according to two criteria, both the corresponding and the number of delayed orders, do quite well here, but unfortunately cannot keep up with the SLK benchmark rule. As with the previous criterion, further adjustments to the optimization method are therefore required. The similar quality of the optimization according to the number of delayed orders indicates that a small number of delayed orders also lead to a good use of the available working time.

Taking all three optimization criteria into account, it can be concluded that in all cases the results of the method are better than those of a previously calculated schedule that is not adjusted with current system data during its execution. For the criterion number of delayed orders, the optimization finds a better rule selection that has a significantly better quality than a simple, identical dispatching rule on all machines. In this case, the additional effort for the selection of the dispatching rules is especially appropriate.

3.6 Conclusion and Outlook

The recent development of real-time approaches with the advent of Industry 4.0 technologies has been driving applications of simulation models with direct links to operational data. This new scenario enables the development of Digital Twin models, which represent a robust approach for the optimization and data analysis in industrial contexts. In the consulted literature, a clear research direction embraces the development of Digital Twin concepts and approaches, along with their application in real scenarios. The maturation of the Digital Twinning concept seems to go hand in hand with the development of Industry 4.0, both in the literature and in practice. However, there is no consolidated reference model in the literature that guides the application of simulation and optimization models for real-time data treatment for synchronous and data-oriented decision-making.

The data exchange framework of the proposed Digital Twin approach for the production planning and control aims to contribute to this gap in literature and practice to enable the development of real-time simulations models. The presented use case allows for the evaluation of the proposed approach, demonstrating promising results for future adoption. The approach is able to improve all KPIs in comparison to a static schedule that is not considering the current system date. For the amount of late jobs the selection of individual dispatching rules even outperforms a static dispatching rule on all machines. Also the reselection of a new set of rules is quite fast. For the given scenario it takes around 20 s on a normal computer which is much faster than the approach the industrial partner is currently using for its scheduling.

The current approach lacks some aspects of the real world. Currently no maintenance strategy is considered. To be more realistic the method should consider preventive maintenance and planned maintenance jobs. Additionally, raw material is not considered. It is assumed that there is always enough material in the stocks to fulfil the demand of the customers. But in reality some production stops might be originated from missing raw material. Both aspects need to be taken into account in further research.

The described method provides a way to build a real-time simulation model of a production system with high-resolution data from the real process. This model could also be used for different purposes, e.g. for the simulation of alternative system designs in the case of the renewal or enlargement of the production or to test, e.g. different strategies for maintenance in the system without interfering the real production.

Overall this data-driven approach offers possibilities to use the newly available data from the shop floor to improve the operational performance and to foster value creation.

Acknowledgments This work is funded by the German Research Foundation (DFG) under reference number FR 3658/1-2 and also by CAPES (Coordenação de Aperfeiçoamento de Pessoal de Nível Superior—Brasil) under reference number 99999.006033/2015-06, in the scope of the BRAGECRIM program.

References

Agostino, I., Sousa, S., Frota, P., Daher, R., & Souza, A. (2019). Modeling and simulation of operations: A case study in a port terminal of vale S/A. In *New global perspectives on industrial engineering and management* (pp. 91–99). New York: Springer. https://doi.org/10.1007/978-3-319-93488-4_11.

Bergeron, D., Jamali, M. A., & Ait-Kadi, D. (2009). Coupling simulation and optimization for an integrated production and maintenance planning in manufacturing systems. In *IFAC Proceedings Volumes, 42*(4) (pp. 1091–1095).

Borshchev, A. (2013). *The big book of simulation modeling: Multimethod modeling with AnyLogic 6*. Chicago: AnyLogic North America

Cardin, O., & Castagna, P. (2009). Using online simulation in holonic manufacturing systems. *Engineering Applications of Artificial Intelligence, 22*(7), 1025–1033.

Cardin, O., & Castagna, P. (2011). Proactive production activity control by online simulation. *International Journal of Simulation and Process Modelling, 6*(3), 177–186.

Dolgui, A., Ivanov, D., Potryasaev, S., Sokolov, B., Ivanova, M., & Werner, F. (2019). Blockchain-oriented dynamic modelling of smart contract design and execution in the supply chain. *International Journal of Production Research 58*(7), 2184–2199. https://doi.org/10.1080/00207543.2019.1627439.

Frazzon, E. M., Hartmann, J., Makuschewitz, T., & Scholz-Reiter, B. (2013). Towards socio-cyber-physical systems in production networks. *Procedia CIRP, 7*, 49–54.

Frazzon, E. M., Kück, M., & Freitag, M. (2018). Data-driven production control for complex and dynamic manufacturing systems. *CIRP Annals - Manufacturing Technology, 67*(1), 515–518. https://doi.org/10.1016/j.cirp.2018.04.033.

Glaessgen, E., & Stargel, D. (2012). The digital twin paradigm for future NASA and US air force vehicles. In: *53rd AIAA/ASME/ASCE/AHS/ASC Structures, Structural Dynamics and Materials Conference 20th AIAA/ASME/AHS Adaptive Structures Conference 14th AIAA* (p. 1818).

Heger, J., Grundstein, S., & Freitag, M. (2017). Online-scheduling using past and real-time data. An assessment by discrete event simulation using exponential smoothing. *CIRP Journal of Manufacturing Science and Technology, 19*, 158–163.

Huang, J. J., Tzeng, G. H., & Ong, C. S. (2005). Multidimensional data in multidimensional scaling using the analytic network process. *Pattern Recognition Letters, 26*(6), 755–767.

Ivanov, D., Dolgui, A., Das, A., & Sokolov, B. (2019). Digital supply chain twins: Managing the ripple effect, resilience, and disruption risks by data-driven optimization, simulation, and visibility. In *Handbook of ripple effects in the supply chain* (pp. 309–332). New York: Springer.

Ivanov, D., Dolgui, A., & Sokolov, B. (2019). The impact of digital technology and industry 4.0 on the ripple effect and supply chain risk analytics. *International Journal of Production Research, 57*(3), 829–846.

Kück, M., Ehm, J., Freitag, M., Frazzon, E. M., & Pimentel, R. (2016). A Data-driven Simulation-based Optimisation Approach for Adaptive Scheduling and Control of Dynamic Manufacturing Systems. In J. P. Wulfsberg, M. Fette, & T. Montag (Eds.), *Advanced Materials Research* ((Eds) ed., pp. 449–456). Pfaffikon, Switzerland: Trans Tech Publications. https://doi.org/10.4028/www.scientific.net/AMR.1140.449

Kück, M., Broda, E., Freitag, M., Hildebrandt, T. & Frazzon, E. M. (2017). Towards Adaptive Simulation-Based Optimization to Select Individual Dispatching Rules for Production Control. Chan, W. K. V., DÁmbrogio, A., Zacharewicz, G., Mustafee, N., Wainer, G. and Page, E. (Eds). In Proceedings of the 2017 Winter Simulation Conference, IEEE, (pp. 3852–3863), https://doi.org/10.1109/WSC.2017.8248096

Kusiak, A. (2018). Smart manufacturing. *International Journal of Production Research, 56*(1–2), 508–517.

Kritzinger, W., Karner, M., Traar, G., Henjes, J., & Sihn, W. (2018). Digital twin in manufacturing: A categorical literature review and classification. *IFAC-PapersOnLine, 51*(11), 1016–1022. https://doi.org/10.1016/j.ifacol.2018.08.474.

Lee, J., Bagheri, B., & Kao, H. A. (2015). A cyber-physical systems architecture for industry 4.0-based manufacturing systems. *Manufacturing Letters, 3*, 18–23.

Megahed, F. M., & Jones-Farmer, L. A. (2015). Statistical perspectives on "big data". In *Frontiers in statistical quality control* (Vol. 11, pp. 29–47). New York: Springer.

Monostori, L., Kádár, B., Bauernhansl, T., Kondoh, S., Kumara, S., Reinhart, G., et al. (2016). Cyber-physical systems in manufacturing. *CIRP Annals, 65*(2), 621–641.

Panetto, H., Iung, B., Ivanov, D., Weichhart, G., & Wang, X. (2019). Challenges for the cyber-physical manufacturing enterprises of the future. *Annual Reviews in Control.* (47) 200–213. https://doi.org/10.1016/j.arcontrol.2019.02.002

Peres, F. A. P., & Fogliatto, F. S. (2018). Variable selection methods in multivariate statistical process control: A systematic literature review. *Computers & Industrial Engineering, 115*, 603–619.

Polenghi, A., Fumagalli, L., & Roda, I. (2018). Role of simulation in industrial engineering: Focus on manufacturing systems. *IFAC-PapersOnLine, 51*(11), 496–501.

Prell, C. (2012). *Social network analysis: History, theory and methodology*. Thousand Oaks: Sage.

Qi, Q., & Tao, F. (2018). Digital twin and big data towards smart manufacturing and industry 4.0: 360 degree comparison. *IEEE Access, 6*, 3585–3593.

Qiu, T., Chen, S., Wu, L., & Wang, Y. (2001). Petri net based modeling and analysis for welding flexible manufacturing cell. *China Welding, 10*(1), 1–6.

Sujová, E., Střihavková, E., & Čierna, H. (2018). An analysis of the assembly line modernization by using simulation software. *Manufacturing Technology, 18*(5), 839–845. https://doi.org/10.21062/ujep/180.2018/a/1213-2489/MT/18/5/839

Tao, F., Qi, Q., Liu, A., & Kusiak, A. (2018). Data-driven smart manufacturing. *Journal of Manufacturing Systems, 48*, 157–169.

Tao, F., Zhang, H., Liu, A., & Nee, A. Y. (2018). Digital twin in industry: State-of-the-art. *IEEE Transactions on Industrial Informatics, 15*(4), 2405–2415.

Tao, F., & Zhang, M. (2017). Digital twin shop-floor: A new shop-floor paradigm towards smart manufacturing. *IEEE Access, 5*, 20418–20427.

Thoben, K. D., Wiesner, S., & Wuest, T. (2017). "Industrie 4.0" and smart manufacturing-A review of research issues and application examples. *International Journal of Automation Technology, 11*(1), 4–16.

Tu, M., Lim, M. K., & Yang, M. F. (2018). IoT-based production logistics and supply chain system–part 1: Modeling IoT-based manufacturing supply chain. *Industrial Management & Data Systems, 118*(1), 65–95.

Turner, C. J., Hutabarat, W., Oyekan, J., & Tiwari, A. (2016). Discrete event simulation and virtual reality use in industry: New opportunities and future trends. *IEEE Transactions on Human-Machine Systems, 46*(6), 882–894.

Wang, L., Törngren, M., & Onori, M. (2015). Current status and advancement of cyber-physical systems in manufacturing. *Journal of Manufacturing Systems, 37*, 517–527.

Wang, W., Zhang, Y., & Zhong, R. Y. (2020). A proactive material handling method for CPS enabled shop-floor. *Robotics and Computer-Integrated Manufacturing, 61*, 101849.

Weyer, S., Meyer, T., Ohmer, M., Gorecky, D., & Zühlke, D. (2016). Future modeling and simulation of CPS-based factories: an example from the automotive industry. *IFAC-PapersOnLine, 49*(31), 97–102.

Zheng, Y., Yang, S., & Cheng, H. (2019). An application framework of digital twin and its case study. *Journal of Ambient Intelligence and Humanized Computing, 10*(3), 1141–1153.

Zhong, R. Y., Xu, X., Klotz, E., & Newman, S. T. (2017). Intelligent manufacturing in the context of industry 4.0: A review. *Engineering, 3*(5), 616–630.

Zülch, G., Jonsson, U., & Fischer, J. (2002). Hierarchical simulation of complex production systems by coupling of models. *International Journal of Production Economics, 77*(1), 39–51.

Chapter 4
Adaptive Scheduling in the Era of Cloud Manufacturing

D. Mourtzis

Abstract Industry 4.0 enables the transition of traditional manufacturing models to the digitalized paradigm, creating significant economic opportunities through market reshaping. Scheduling is a key field of manufacturing systems. Academia and industry are closely collaborating for producing enhanced solutions, taking advantage of multiple criteria. Initially, the scheduling problem was dealt with more simplistic methods resulting in static solutions; however, with the evolution of digital technologies, scheduling became more dynamic to the company's environmental changes. As Information and Communication Technologies (ICT) became mainstream and systems were integrated, rescheduling and adaptive scheduling became the cornerstones of Smart Manufacturing. These technologies have been further advanced to yield more reliable results in a shorter period of time. The efficient design, planning, and operation of manufacturing systems and networks can be achieved with the adoption of cyber physical systems (CPS) in conjunction with the Internet of Things (IoT) and cloud computing. The transition to Smart Manufacturing is achieved with the adoption of cutting-edge digital technologies and the integration state-of-the-art manufacturing assets. Consequently, this chapter presents an opportunity for tracking the evolution of scheduling techniques during the last decade, as well as for extracting insightful and meaningful inferences from the application of innovative solutions in industrial use cases.

4.1 Introduction

In the realm of the fourth industrial revolution, global manufacturing is reshaping. The current market era is by far the most competitive landscape in the history of humanity. The modern market is driven by ever-increasing demand for high-quality,

D. Mourtzis (✉)
Laboratory for Manufacturing Systems and Automation, Department of Mechanical Engineering and Aeronautics, University of Patras, Rio Patras, Greece
e-mail: mourtzis@lms.mech.upatras.gr

B. Sokolov et al. (eds.), *Scheduling in Industry 4.0 and Cloud Manufacturing*, International Series in Operations Research & Management Science 289,
https://doi.org/10.1007/978-3-030-43177-8_4

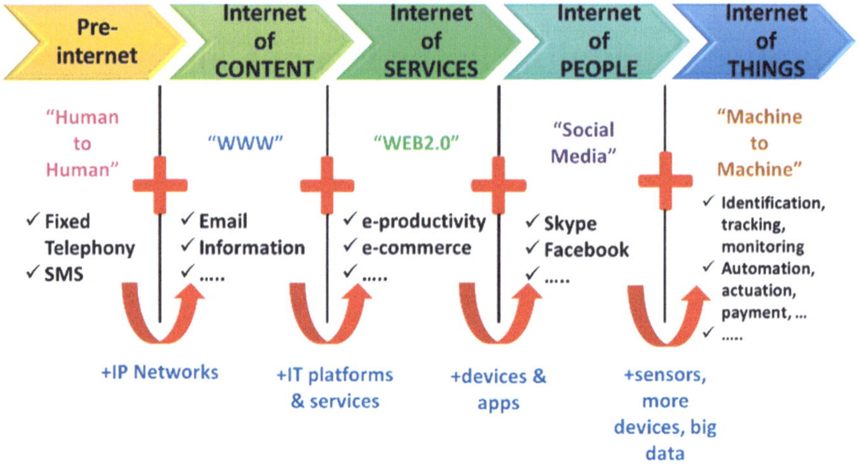

Fig. 4.1 Internet of Things (IoT) evolution. (Adapted by Mathew 2014)

high-quantity, and variety of products. As such, engineers are motivated to constantly improve existing technologies, as well as discover new techniques with the aim of improving business models, keep productivity and revenue rates constantly high, and most importantly, ensure high competitiveness. In the last few decades, with the emergence of the Internet, the market and the industry have become global (Fig. 4.1).

The number of research studies in the field of Information and Communication Technologies (ICT) has greatly increased during the last decade. With the advent of Industry 4.0, also known as the fourth industrial revolution, very interesting and innovative technologies have been brought to the foreground, including cloud manufacturing (CMfg) and Digital Twin (DT), and by extension, existing technologies, for instance, Scheduling, have flourished as well. However, due to numerous inhibitors reported in scientific publications, the full potential of CMfg and DT has not yet been achieved. On the positive side, the adoption of the Industry 4.0 paradigm is laying solid foundations, on top of which the aforementioned technologies will be successfully bridged. This chapter will present the current advances in the theory of scheduling methodology to CMfg under the umbrella of Industry 4.0 in order to build an adequate theoretical background. In continuation, insightful feedback will be parsed from the practical implementation in real-life modern industrial use cases. Finally, based on the knowledge acquired, recommendations on bridging CMfg to Scheduling and DT will be made.

4.2 State of the Art

4.2.1 Smart and Cloud Manufacturing

With the advent of the fourth industrial revolution, followed by the introduction of digitalization and automation in the industrial domain, cloud technologies have become a necessity for modern manufacturing systems. Li et al. (2010) define Cloud Manufacturing (CMfg) as an advanced manufacturing paradigm for improving resource utilization and MFG system efficiency in order to constantly meet the largely diverse customer needs. In more detail, CMfg is described as a computer- and service-oriented manufacturing model that has been developed from enterprise information technologies and existing advanced manufacturing models under the support of Internet of Things, cloud computing, advanced computing technologies and virtualization, and service-oriented technologies. The same definition is validated by Kumar et al. (2016).

In the existing literature, one can find innovations and techniques based on cloud technologies (Dimitris and Vlachou 2018; Doukas et al. 2014; Mourtzis and Vlachou 2016). The increased demand for customized products has driven original equipment manufacturers (OEMs) to shift to decentralization as presented by Mourtzis et al. (2012a). The authors in that publication suggest a methodology for identifying efficient supply chains in decentralized manufacturing environments.

Cloud manufacturing is a term that was first coined by Li et al. (2010). They define CMfg as an advanced manufacturing paradigm that is capable of improving resource utilization while efficiently responding to diverse customer needs. The emergence of cloud computing could be considered the major enabler for the replacement of traditional manufacturing methods from CMfg (Xu 2012). The increasing demand for customized products leads companies to reshape their business models, in order to maintain the high quality of the products as well as to integrate personalization into their production models (Zhang et al. 2019). Consequently, the research for new advanced manufacturing techniques and approaches occurs. From the previously published works presented by Rauschecker et al. (2011), Wang and Xu (2013) and Lu and Rob (2019), it can be derived that, with the adoption of CMfg, customers become effectively part of the design and production phases, with their specialized requirement being taken into consideration. However, the adoption of CMfg is not a path of roses. On the contrary, it poses great challenges for production engineers. In the study presented by Ahn et al. (2019), the composition of tasks in a CMfg environment has been investigated. To that end, the authors have proposed graph-based algorithms for the extraction of important task data, which are then modeled using the Kano model, in order to conclude that the variable neighborhood search algorithm performs better in contrast to the genetic algorithm and the simulated annealing algorithms.

Another crucial aspect of CMfg is the Local Pickup and Deliver Problem (LPDP). In the published literature, a long list of contributions can be found suggesting optimization algorithms for addressing the aforementioned issue. Chen

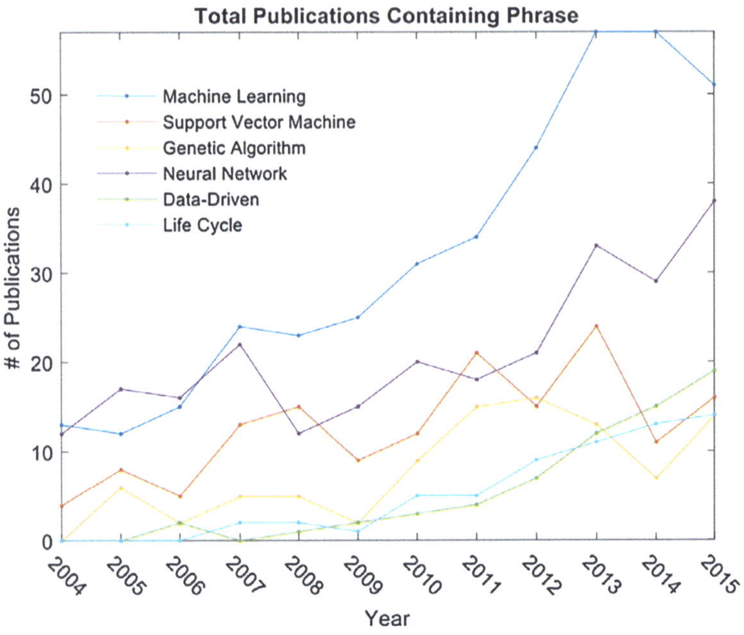

Fig. 4.2 Visualization of interest in Machine Learning by year. (Adapted from Sharp et al. 2018)

et al. (2013) contribute to the body of knowledge by proposing a data mining-based Dispatching System. Among the benefits observed, the proposed system offers a simulation platform for strategic decision-making and analysis; the results yielded from historical data can be further utilized with an optimization algorithm for improving the outcome of the LPDP.

Equally important to the above-mentioned technologies are Machine Learning (ML) techniques. Machine Learning is a subset of the Artificial Intelligence (AI) domain, which has gained a certain advantage over the last decades as a result of technological improvement. Indeed, ML can form a strategic solution to the integration of advanced manufacturing processes. In the extensive review presented by Sharp et al. (2018), they have investigated in depth the adoption of ML in the field of Smart Manufacturing. Additionally, Kusiak (2018) presents a comprehensive vision of smart manufacturing. The interest in different AI techniques in the timespan of a decade is being reflected, namely Machine Learning, the Support Vector Machine, the Genetic Algorithm, the Artificial Neural Networks, the Data-Driven Techniques, and the Life Cycle, are depicted in Fig. 4.2 in Sharp et al. (2018).

Task composition in cloud manufacturing is related to the selection of appropriate services from the CM platform and the combination of them to process the task. The main objective is to achieve the expected performances (Wu et al. 2013). The composition problem is one of the most important functional issues in CM, with different performance measures, and several researchers have discussed it. Tao et al.

(2013) considered time, cost, energy, reliability, maintainability, trust, and function similarity for the problem, and developed a parallel intelligent algorithm. Zhou and Yao (2017) considered price, time, availability, and reputation as performance measures, and an aggregated level of them by using weight for each measure and finally combined the modified artificial bee colony and cuckoo search algorithms for the problem under investigation. Moreover, Ahn et al. (2018) consider a task-guided acyclic graph and proposed graph-based algorithms to obtain the cost, time, quality, and reliability of a task with multiple patterns of composition. Furthermore, they model the problem of task composition by introducing the cost and time of execution as performance attributes, and quality and reliability as basic attributes in the Kano model.

4.2.2 Advances in Scheduling Methodologies

MFG systems have been well defined by Chryssolouris in his book (2006) as "the combination of humans, machinery and equipment that are bound by a common material and information flow." What is more, in this book, the importance of scheduling for design, control and operation of MFG systems is highlighted among with the main issues troubling engineers. These issues can be summarized as *(1) the requirement generation, (2) the processing complexity, (3) the scheduling criteria, and (4) the scheduling environment*. Consequently, with recent technological advances, and the changes in the global market schema, research efforts have focused on delivering innovative solutions for elevating the existing techniques in the topic of scheduling—to that end, scheduling a multi-faceted problem providing fertile ground for extensive research. A common practice in scheduling modeling is the provision of an energy-efficient production plan by incorporating machine reliability. As such, Chen et al. (2019), based on the ant-colony optimization method, have developed a mathematical programming model aiming at minimizing tardiness cost as well as energy cost. Moreover, a comprehensive review of the state-of-the-art and new trends, as well as the major historical milestones in the development of simulation engineering for manufacturing systems and recent approaches to business and a study in key industrial areas, is presented by Mourtzis (2019). Additionally, the state-of-the-art and the research challenges regarding scheduling in cloud manufacturing are discussed by Yongkui et al. (2019). This study summarizes the fundamentals of scheduling in cloud manufacturing, the typical work on service composition in cloud manufacturing, and the typical work on scheduling in cloud manufacturing, with related scheduling issues and with research challenges, techniques, and methodologies. Following the same topic, a new decision-making schema that uses an efficient screening procedure, named Smart Scheduling, with the aim to yield flexible and efficient production schedules taking advantage of the features of the aspects of centralized and decentralized manufacturing systems, is presented by Daniel et al. (2019).

Continuously, maintenance activities are a cornerstone of manufacturing systems. However, maintenance can induce a great amount of uncertainty, as the occurrence of malfunction in many cases cannot be predicted or even avoided efficiently. What is more, with the increased complexity of modern manufacturing equipment, maintenance activities could be characterized as multi-mode. Moradi and Shadrokh (2019) present a heuristic algorithm that has been designed and developed by them, which takes into consideration the effect of entities on system reliability, as well as their importance value, for choosing maintenance activities in the production schedule. Mourtzis et al. (2017b) proposed a Product Service System (PSS) based on a cloud platform for condition-based predictive maintenance. The PSS is supported by a shop-floor monitoring service as well as an Augmented Reality (AR) application.

Machines are a vital part, especially in custom IoT manufacturing environments. A study to bridge the literature gap for new algorithms that reap the benefits of such technologies is done by Xu et al. (2019). The authors by taking advantage of the benefits of IoT technologies, have developed a framework, where workers with stochastic workloads, arrive arbitrarily in a machine. Therefore with the provision of an IoT enabled online tool, engineers can calculate the workers' workload in real-time and by extension can effectively balance future uncertainties.

Mathematical optimization methods for scheduling problems have been studied from different perspectives over the past decades, whereby significant progress can be made in the development of robust theoretical models and efficient techniques for solutions. Different scheduling problems with complex logical and terminal hybrid constraints, non-stationary system execution as well as complex interrelationships between process design dynamics, capacity utilization, and device configurations need further study and implementation of a wide range of methodological approaches. Optimal regulation is one of these methods. To that end, Ivanov et al. (2018b) present a two-objective study. More specifically, a survey of the applications of optimum control for production planning, supply chain, and Industry 4.0 systems with an emphasis on the total deterministic rule is described. As mathematical programming methods that characterize schedules as trajectories, optimal control techniques take a different perspective. For optimal control, optimal control models, qualitative methods for performance analysis, and computational methods were considered. Finally, the survey enables the classification of models with only terminal constraints with application to master output scheduling, models with hybrid terminal-logical constraints with applications for the short-term job and flow shop scheduling, and hybrid structural-terminal-logical constraints with applications for specialized assembly systems such as Industry 4.0. Zhou et al. (2018) present a solution to the dynamic CMfg environment with randomly arriving tasks. To that end, an event-triggered adaptive task scheduling method is presented.

Supervisory Control and Data Acquisition (SCADA) is a significant technology that enables data collection from one or more distant facilities with simultaneous limited control instructions to send to the facilities (Boyer 2004). Although this technology is not new to the market, a methodology for deriving scheduling decisions is introduced by Mourtzis et al. (2014) for the exploitation of machine information in near-real-time machine condition monitoring. Even though SCADA systems have already been implemented in manufacturing to a large extent, it is important to support increased data exchange rates with external ICT tools and, as a consequence, enhanced data security and integrity practices (Nicholson et al. 2012). As such, Mousavi and Siervo (2017) present a real-time data-acquisition technology, coupled with discrete event simulation, to accurately measure the performance metrics of key manufacturing systems to allow performance management and the prediction of destabilizing factors.

As far as supply chain (SC) simulation is concerned, an extended presentation of the supply chain control logic is done by Makris et al. (2011). Moreover, Mourtzis (2011) states that modern manufacturing companies must collaborate with their business partners through their business processes such as development, production, distribution, and after-sales service. In addition, a model has been introduced in the form of an Internet-enabled technology system, providing a range of features including digital organization, scheduling, and monitoring to promote collaboration and flexible planning and monitoring through extended manufacturing companies. Furthermore, one simulation method applied to the example of the supplier choice is highlighted by Mourtzis (2016).

Regarding the design and planning of manufacturing networks, existing systems also fail to address the numerous issues of manufacturing network management in a systematic comprehensive way as they apply to data and context not directly related to manufacturing (e.g., long-term strategic planning vs. short-term operational scheduling), as presented by Si et al. (2011).

4.3 Applied Innovations of Scheduling in Cloud Manufacturing

Indeed, scheduling has been heavily researched in the last few decades. Concretely with the introduction of digitalization, engineers have grabbed the opportunity to improve the existing techniques of scheduling and invent new ones. However, research has expanded from the theoretical talk to the practical implementation of the introduced solutions to the adoption from the early adopters. Furthermore, the influence of scheduling advances in modern MFG systems across many real-life applications will be presented in this chapter.

One of the most important results of IoT's emergence is data generation in increasing volume, variety, and speed, which is also called big data. Analyzing this data is the basis for the modern mass customization range, which implies meeting the needs of individualized consumer markets. The analysis of this data is necessary for the enhancement of knowledge repositories and the improvement of decision-making (Renu et al. 2013). Additionally, next-generation smart manufacturing and service systems using big data analytics are described by Shukla et al. (2019). Moreover, a conceptual framework and typical application scenarios addressing the role of big data in supporting smart manufacturing is presented by Tao et al. (2018). Beyond the positive consequences of the IoT paradigms and Industry 4.0 examples, a crucial challenge that occurs is the complexity. Many approaches have been developed for the quantification of complexity, based on heuristics, statistics, and probabilities

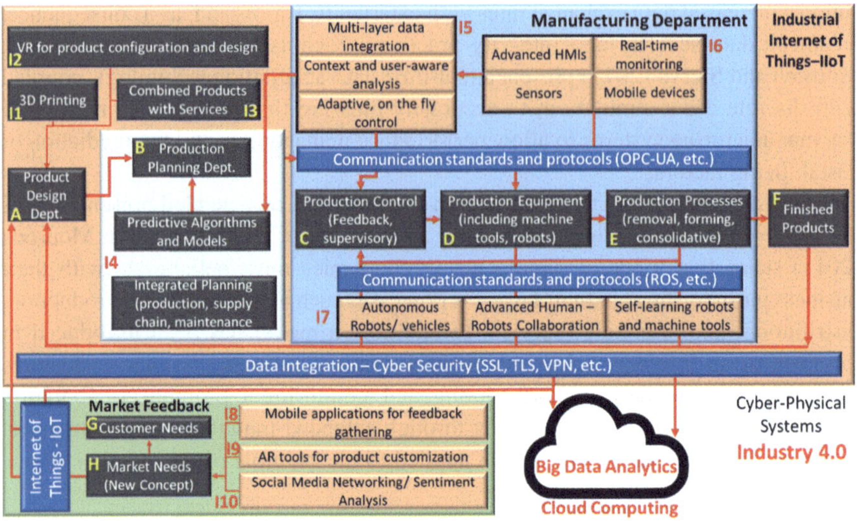

Fig. 4.3 Industry 4.0 model. (Adapted from Mourtzis et al. 2019)

(ElMaraghy et al. 2012; Mourtzis et al. 2013b). Mourtzis et al. (2019) aims to bridge the literature gaps regarding the complexity quantification approaches. The Industry 4.0 model, which is based on the Information Theory and quantitatively evaluates the complexity and capacity of Industry 4.0 systems, is presented in Fig. 4.3.

4.3.1 Decision-Making Applications for Smart Scheduling

Decentralized decision-making and real-time reaction to unexpected developments, frequently happening in both manufacturing and consumer environments, are two important factors that affect the flexibility required to meet demand by a production chain. Papakostas et al. (2012) present an agent-based approach to tackle real-time and hierarchical decision-making problems in manufacturing. Over the past few decades, the global market landscape has changed, and hierarchical mass production seems unable to fulfill the changing demand demands imposed by globalization. Toward this shift, the performance and viability of centralized and decentralized production networks under heavy production are analyzed by Mourtzis et al. (2012b). Discrete-event simulation models of automotive production networks have been developed to evaluate their quality under highly diversified consumer demand. Moreover, for the assessment, multiple conflicting user-defined parameters were used, including lead time, final product price, versatility, the annual volume of production, and environmental impact due to product transport. Additionally, the mass customization paradigm, in combination with efficient manufacturing configurations, is presented by Mourtzis et al. (2013a). This work presents the design and operation of manufacturing networks based on a multi-objective decision-

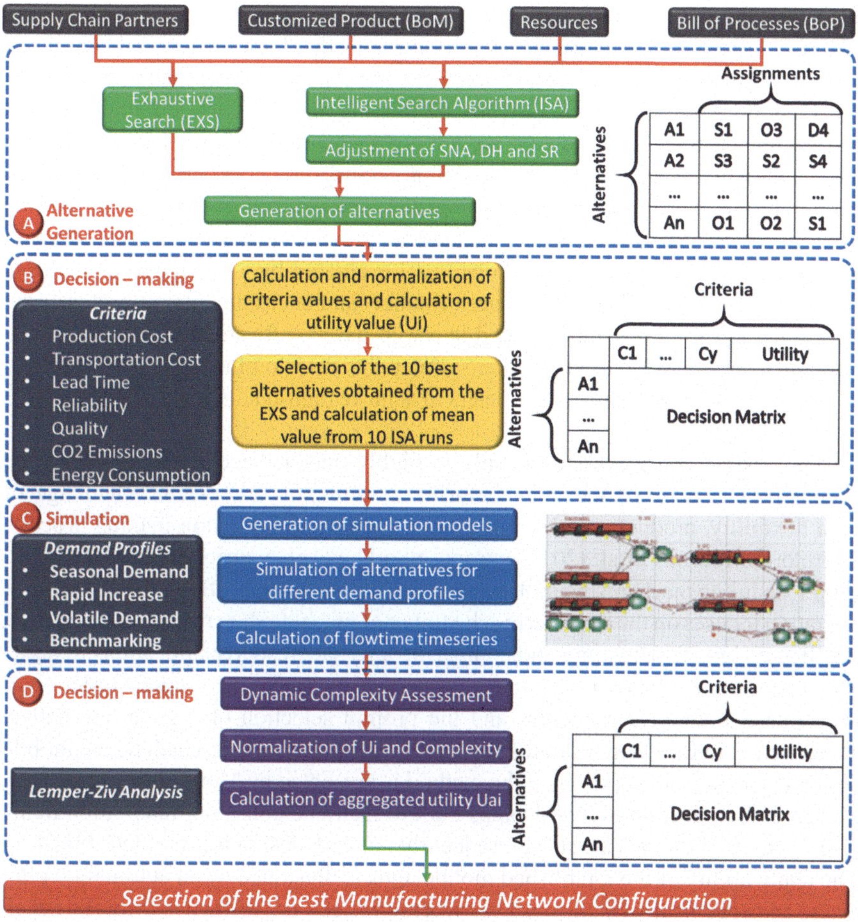

Fig. 4.4 Workflow of the proposed design and operation method. (Adapted from Mourtzis et al. 2013a)

making and simulation approach. The architecture tool for the design and planning of centralized and decentralized production networks is illustrated in Fig. 4.4.

The rising need for higher brand customization, in combination with demand uncertainty, needs efficient ways to model network configurations for manufacturing. A tool to help decision-making in practical network design issues of manufacturing which explores the quality and feasibility of hierarchical and distributed production networks under intense product customization is done by Mourtzis et al. (2015c, d). In the decision-making process, two approaches are used: an exhaustive search and a smart search algorithm. Multiple conflicting user-defined parameters are used to assess alternative production and transport schemes, including lead time, cost of production, versatility, annual volume of production, and environmental impact. The performance of the smart search method is investigated using statistical design of experiments (SDoE). In addition, an intelligent algorithm calibration technique is provided. Toward the globalization paradigm, an investigation of the performance of decentralized manufacturing networks through the Tabu Search and Simulated metaheuristic methods, in conjunction with the Artificial Intelligence method, is analyzed by Mourtzis et al. (2015b). Following the shift toward high-demand and highly customized products, an intelligent method that uses three variable control parameters and can be used to define secure, globalized production network configurations capable of manufacturing mass-customized goods is described by Mourtzis et al. (2014). The decision support system enables alternative production network configurations to be developed and evaluated by a collection of multiple conflicting user-defined cost, time, performance, and environmental impact criteria. The suggested solution, implemented in a web-based software method, is explored through a probabilistic framework to direct the decision-maker in selecting the values of the variable control parameters to achieve high-quality network designs for manufacturing.

The product complexity, especially in highly personalized markets, affects the overall performance of the manufacturing systems. To address the challenge of high flexibility, production scheduling is a vital part of a decision-making process. Therefore, Mourtzis et al. (2015a) proposed a knowledge-enriched short-term job-shop scheduling mechanism, implemented into a mobile application. The principle of operation focuses on the short-term scheduling of machine shop resources through an intelligent algorithm creating and comparing alternate resource allocations to tasks. The adaptive calculation procedure leads to the requirements of a new order so as to evaluate alternative schedules and the prompt selection of a good alternative. Moreover, it offers the possibility to present the derivative schedule on mobile devices. The overall workflow of the method is presented by Mourtzis et al. (2016a). Moreover, a collection of mobile apps built to facilitate consumer integration in the design phase of the service and subsequently network design is presented (Fig. 4.5). The applicability of the established mobile apps is illustrated by an automobile pilot event, and specifically by customizing accessories and car aesthetics (Mourtzis et al. 2016d).

4.3.2 Cloud-Based Scheduling Approaches

Another approach to address the issue of disturbances on manufacturing shop-floors and the increasing number of product variants uses adaptive and flexible process planning methods. Mourtzis et al. (2016b) describe a system-oriented, two-service cloud-based software platform. The first system generates non-linear process plans and a genetic algorithm. The second service collects data from machine tools through sensors, operator inputs, and device schedules. The process planning system requires as inputs the status, some selected machine requirements and the availability time periods of machine tools. These monitored data are provided through an information fusion technology (Fig. 4.6).

Mourtzis et al. (2016c) present a framework consisting of a mobile application supported by Augmented Reality technology and a manufacturing network design

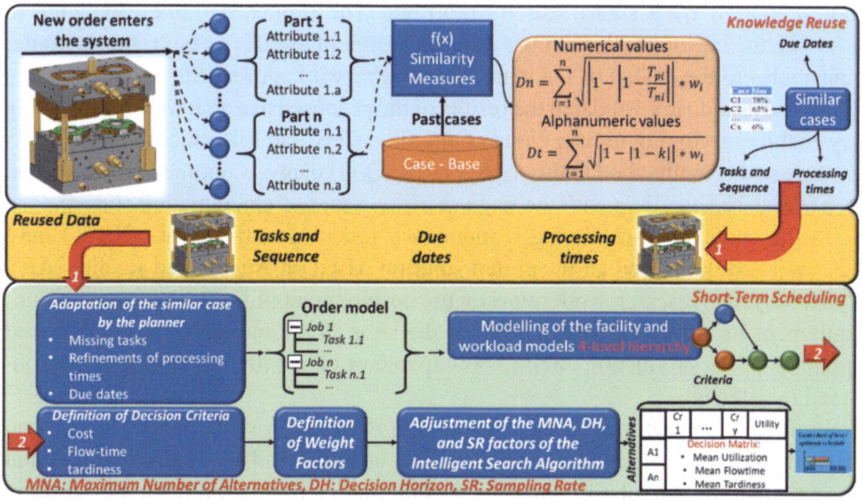

Fig. 4.5 Workflow of the knowledge-enriched short-term scheduling (KES) method. The method consists of two mechanisms, namely: (1) the knowledge extraction and reuse mechanism and (2) the short-term scheduling mechanism. (Adapted from Mourtzis et al. 2016a)

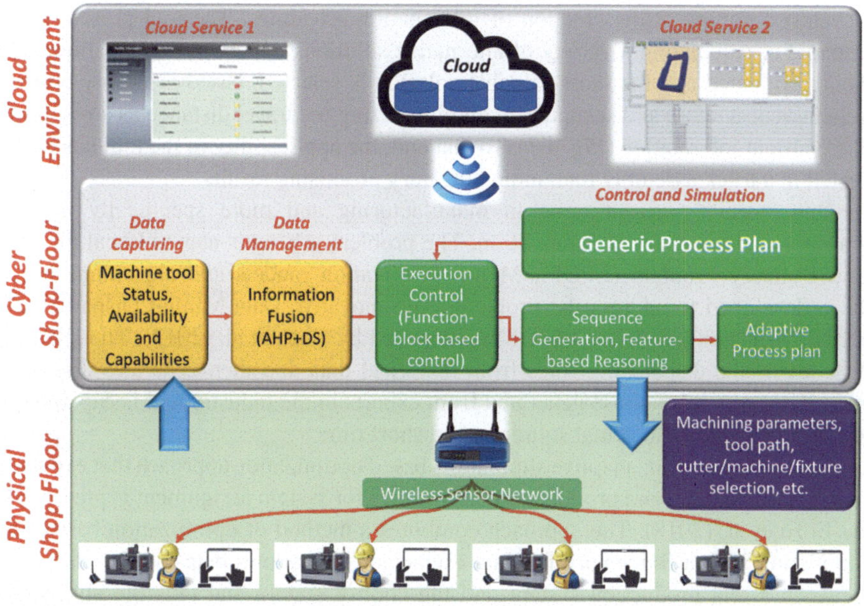

Fig. 4.6 Overall architecture of the proposed framework. (Adapted from Mourtzis et al. 2016b)

tool supported by a smart search algorithm. The proposed framework aims to integrate the customer ideas in the product design phase. As a result, the customer contributes to the design of the manufacturing network. A real-life application in a white-goods industry validated the results of the proposed research work. Combining key enabling technologies like cloud computing, CPS, and IoT, a theoretical CBCPS is suggested. Finally, the Quality of Services (QoS) analysis outlined in the second step is undertaken along with the main security issues of cloud manufacturing.

In the context of scheduling maintenance tasks, Mourtzis et al. (2017c) have designed and developed an integrated solution based on Augmented Reality (AR). Concretely, this research work relies on the development of DaQ (Data Acquisition) equipment for retrieving real data from the shop floor, and through an AR-based application, the production engineer is capable of efficiently scheduling condition-based maintenance.

In many manufacturing plants, production is based on routine tasks, which can cause disorders related to workload. A widely accepted and adopted solution to that issue is given by job rotation. Besides the elimination of the disorders, the variation in tasks granted from job rotation allows operators to develop multiple skills, as a result of the variation in demand in the different tasks. However, due to the mainly physical limitation of the human operators, careful scheduling is a requirement for the avoidance of bottlenecks in the production. As such, the engineer is responsible for even distribution of workload among the available workforce. In the research paper presented by Michalos et al. (2011), a web-based tool has been designed and developed aiming at the creation of job rotation schedules. The tool presented is capable of calculating a variety of alternative solutions which, in continuation, are being evaluated and the most valuable solution is selected, based primarily on the repetitiveness score and secondarily on minimizing the travel distance as a result of the constant job rotation (Fig. 4.7). To that end, the applicability of the tool has been tested in a truck assembly line, indicating very promising results.

One more significant issue in manufacturing and more specifically in the transportation industry is competition. The problem arising is named Local Pickup and Delivery Problem (LPDP). Many optimization models and algorithms have been developed to address the problem. An innovative Data-Mining Dispatching-based Dispatching System (DMDS) is proposed by Chen et al. (2013). The DMDS first learns the dispatching rules from historical data and then generates dispatch solutions, as good as those generated from experts in the industry and is capable of generating better than actual solutions in a short time.

A new data-driven, adaptive simulation-based optimization approach that extracts automated dispatch and production control rules for system assignment is presented by Enzo et al. (2018). The approach combines a method of optimization based on simulation with a framework for data exchange. This allows data to be exchanged between a real production system and the control system and thus the ability to react in real time to system changes. In a real-world application and specifically in a production line of a Brazilian manufacturer of mechanical parts for the automotive industry, the approach achieved significantly higher performance than a considered company's current scheduling method.

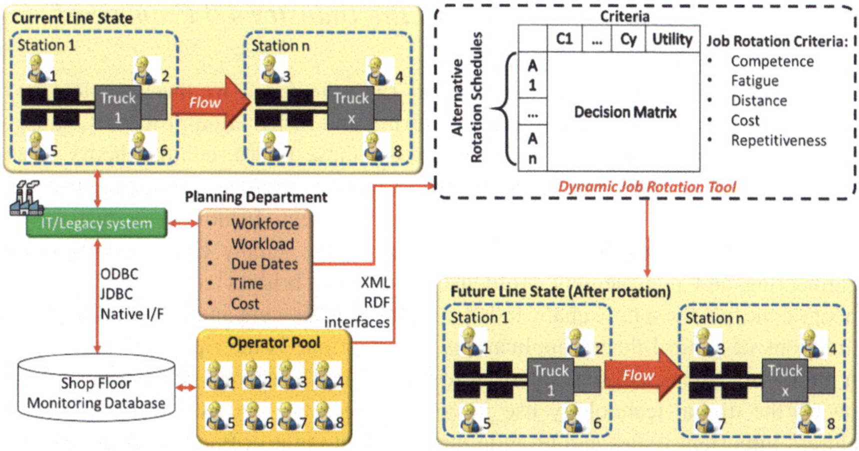

Fig. 4.7 Dynamic job rotation tool architecture. (Adapted from Michalos et al. 2011)

Mourtzis et al. (2017a) describe how the traditional manufacturing system is converted into the Industry 4.0. Furthermore, they suggest a modeling and a quantification methodology that includes metrics from the Information Theory for estimating the complexity and the capacity of Industry 4.0 systems. Additionally, the application of the proposed Industry 4.0 methodology and the analysis of these measures before and after transitioning to Industry 4.0 are also evaluated and validated in a case study from Robotics industry (Mourtzis et al. 2019).

At the flow-shop scheduling where scheduling is interconnected with the control function, non-deterministic issues were considered by Ivanov et al. (2016). Important factors such as temporary machine unavailability, pre-emptions, and processing time fluctuations and technological constraints in an integrated manner were included. The novelty of this study yields in the approach which is a dynamic decomposition of an NP-hard combinatorial scheduling problem. The decomposition is based on the model developed and an algorithm for optimal control of the output of MP-mixed operations. With the assistance of sophisticated scientific methodology, the theoretical contribution of this study aims at increasing the scheduling quality. The proposed novelty of this study consists of a detailed theoretical analysis of time-based decomposition and computational complexity with a continuous flow and discrete assignments application to flow-shop scheduling. Moreover, for the simultaneous solution of assignment and flow control tasks, a dynamic model and an algorithm were developed. This approach is not designed to outperform heuristics or MP algorithms but extends them to the following aspects. The results obtained can obviously be included for optimal programs (e.g., schedules) in general iterative search procedures. To that end, the main contribution of the study can be summarized to the extension of the results to existing scheduling models: dynamics of the execution of the operations, machine non-stationarity, and the use of continuous variables for the continuous flow scheduling.

4.3.3 Manufacturing Networks in the Industry 4.0 Framework

Continuously increasing uncertainties and risks, multiple feedback cycles, and complexities threaten traditional manufacturing and distribution processes, supply chains, and Industry 4.0 networks. To address these issues, control theory is an important field of study that leads to further insights into the management of the problems in operations and supply chain management. Ivanov et al. (2018a, b) investigate the applicability of control theory in the supply chain operations to engineering and management problems. This study bridges control and process theory concepts for supply chain and operations management and identifies two new directions of control theory applications. The first one is the ripple effect analysis in the supply chains and scheduling in Industry 4.0. The second path is the analysis toward the digital technology use in control theoretic models. In an attempt to address important issues and provide insight about the processes in supply chains, logistics and Industry 4.0 networks, the second direction defines and systematizes various channels for the implementation of control theory to operations and supply management during the last six decades. Extended collaboration between control engineers and supply chain experts would add more complexity to complex planning and models, and improve performance in manufacturing and logistics processes, supply chains, and Industry 4.0 networks. Last but not least, the trends toward control intellectualization and its progress toward the analytics of supply chain management are examined. Finally, a Blockchain-oriented adaptive model for smart contract design and execution in the supply chain is presented by Dolgui et al. (2019).

Manufacturing enterprises under the Industry 4.0 concept face the challenge of manufacturing highly customized goods in small lot sizes. The digitization of production processes is one way to respond to the ever-changing demands that improve resource flexibility. The solution to these challenges is the integration of the Internet of Things (IoT) technologies in automated customized shopping so as to create a sustainable environment of connected distribution and selling points. An analysis of IoT technologies and systems, which are the drivers and pillars of smart manufacturing data-driven developments, is done in the review presented by Yang et al. (2019). The evolution of the internet from computer networks to human networks into the new age of smart and linked production networks (e.g., materials, sensors, devices, people, goods, and supply chain), as well as the challenges and opportunities arising from the Internet of Manufacturing Things (IoMT), is presented by the authors of the study.

4.3.3.1 Augmented Reality and Cloud Technology

This can be achieved through the development of an Augmented Reality mobile application for effective remote maintenance supported by Cloud technology and smart assembly/disassembly algorithms that enable the automated generation of the AR scenes and increase the level of automation. The increased level of automation

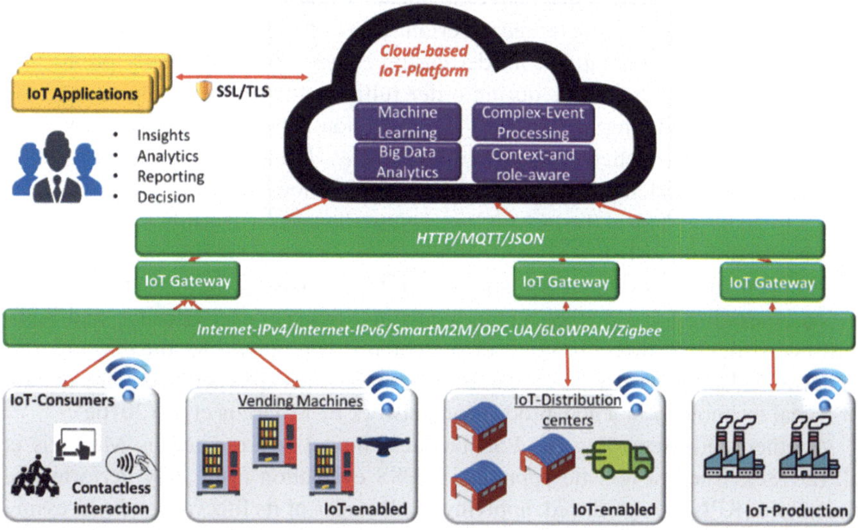

Fig. 4.8 The proposed Internet of Things platform. (Adapted from Mourtzis et al. 2018b)

and the usability of the approach enable the provision of added-value solutions, as presented by Mourtzis et al. (2018a). The implementation of the IoT vending platform is depicted in Fig. 4.8. This innovative IoT platform aims to support automated customized shopping. The IoT platform consists of various components capable of detecting the requirements of customers, integrating data from different levels, analyzing it, and producing customized products adaptively and efficiently. The main benefits lie in the reduction of waste and the increased product customization level. Modern production systems will meet an increasing need for highly customized goods and turbulent situations on their shop floors. Complexity and uncertainty are increasing as a result of these new requirements for producers, leading to difficulties in decision-making. In most manufacturing systems, short-term planning is one of the main everyday decision-making functions (Chryssolouris et al. 1992).

4.3.3.2 Internet of Things (IoT) and Product Service Systems (PSSs) Platforms

Product Service System (PSS) customization is required in Industry 4.0's new production paradigm. Nevertheless, minimal literature work is found in the configuration of PSS and the Industry 4.0 PSS inquiry. To that end, Mourtzis et al. (2018a) suggest a framework for quantifying the difficulty of PSS configuration, taking into account aspects of Industry 4.0. The proposed metrics are applied in a real industrial case study from a large laser machining industry, with the aim of evaluating the various PSS alternatives in terms of complexity. Given today's customer requirements in rising market segments, when using their global production networks, companies face increased uncertainty. In order to meet long-term corporate goals, it becomes essential to consider customer requirements in early production planning stages and processes during order fulfillment. A holistic approach to the development, scheduling, and control of production networks is sought to address identified political, tactical, and operational issues. Integrating product allocation and production, and supply network (re-)design, followed by assigning customer orders to plants and local (re-)scheduling, is introduced as a decision-support model (Hochdörffer et al. 2018).

The new frugal design paradigm is evolving toward a new business model by incorporating local market requirements and offering cost-effective and high-customer-value solutions. To this end, a technique is proposed for the development of networks by means of a smart search algorithm, which aims at the implementation of frugal technology in a new production network by Mourtzis et al. (2016e).

Another study regarding the PSS business model is presented by Mourtzis et al. (2018c), where a holistic approach to PSS evaluation using Key Performance Indicators (KPIs) is introduced, applying to all stages of its lifecycle and addressing issues from both provider and consumer perspectives. A software tool has been developed to assist decision-makers in selecting suitable KPIs to track for PSS analysis and to provide data collection, storage, processing, and visualization capabilities for KPIs.

The literature review highlights the need for planning adaptability and agility. In order to identify flexible and feasible solutions, scheduling systems that leverage last-time changes occurring in a dynamic environment and consider shop floor situations are required. As such, Mourtzis et al. (2015a) provided an adaptive scheduling strategy, based on the availability of machine tools, capable of generating flexible and feasible schedules across the cloud to make the best use of machine tools' availability. Among the main advantages of the methodology proposed by Mourtzis et al. (2015a) is the adaptation of the short-term scheduling as well as the decentralized shop-floor control. In the manufacturing environment mentioned in the case study, the smart search algorithm used by the scheduling application revealed its output among other commonly used dispatching rules. The viability of the method suggested is confirmed by the case study findings. Through the suggested system, adaptability to changes and near real-time shop-floor awareness were achieved. The developed cloud platform is more appropriate for fast response systems and highly customized products as it enables interaction between the different IT tools and multi-user data access.

4.3.3.3 Cloud-Based Maintenance Frameworks and Applications

Over the years, maintenance and its costs have continued to attract the attention of production management as unplanned failures decrease process performance as well as investment returns. Advanced maintenance technologies collecting and processing information from the shop floor can minimize costs and increase an enterprise's profitability. To that end, a condition-based approach to preventive maintenance

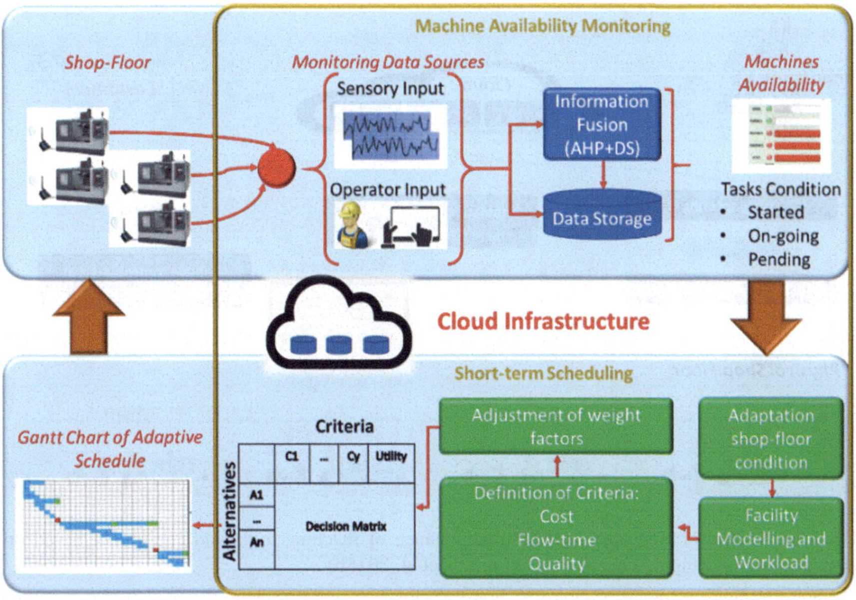

Fig. 4.9 Architecture of the proposed framework. (Adapted from Mourtzis et al. 2015a)

incorporated into a system for device monitoring is presented by Mourtzis et al. (2015a) (Fig. 4.9).

The following approach acquires data from machine tools on the shop floor and analyzes them to support condition-based preventive maintenance operations through an information fusion technique. The suggested solution is built into a cloud-based software system. The service collects and stores data relating to the use of machine tools and equipment, such as their total processing time and machining time per unit, and measures the estimated remaining useful life of components. It also provides alerts to machine tool operators and service departments, and allows mobile technology to connect with them. The framework refers to a case study including information from a machining SME (Mourtzis et al. 2015b) (Fig. 4.10).

4.3.4 Communication Platforms Toward Machine Shop 4.0

As it has been stated before, thephilosophy of Industry 4.0 promotes the digitalization of production systems. This leads to a transformation of existing production systems and creates new skills in monitoring and control subjects. New design methods and standards of interaction should be used to help this transition. For this reason, Fig. 4.11 depicts a framework through a general computer model, a structure for the design of CNC machine tools and lathes. More specifically, the architecture

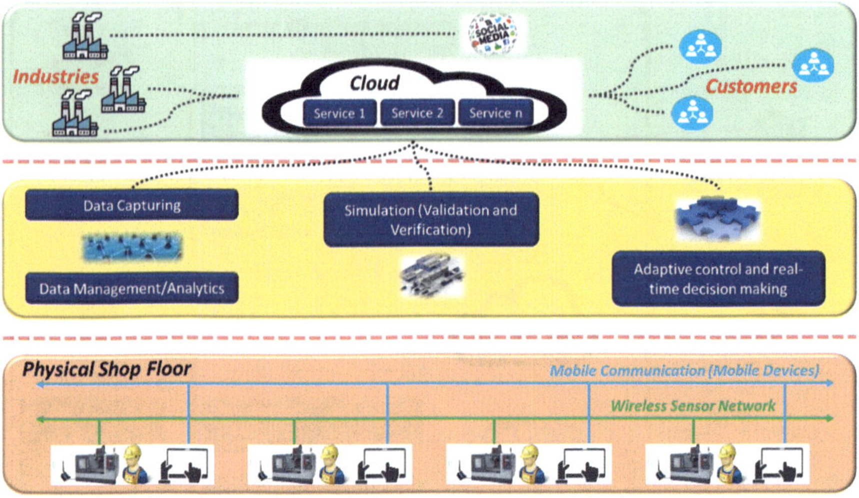

Fig. 4.10 A cloud-based approach for maintenance of machine tools and equipment based on shop-floor monitoring. (Adapted from Mourtzis et al. 2015b)

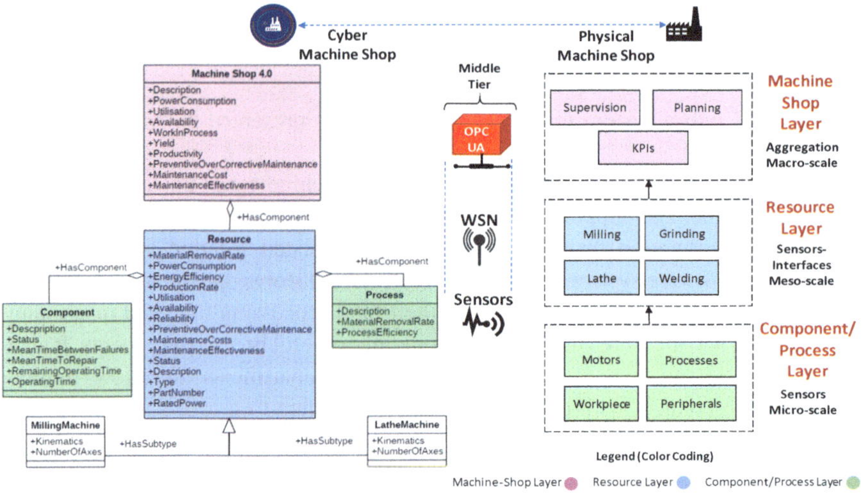

Fig. 4.11 The overview of the Machine Shop 4.0. (Adapted from Mourtzis et al. 2018e)

is based on the communications model of Open Platform Communications–Unified Architecture (OPC-UA) to provide machine shops with macroscopic and microscopic views of the principle of Machine Shop 4.0. Using OPC-UA, different systems can be combined after semantic modeling. In addition, a data acquisition system is being created to advance obsolete machine tools into the digitized period. Therefore, in this comprehensive model, machine tools without networking capabilities can be incorporated (Mourtzis et al. 2018e).

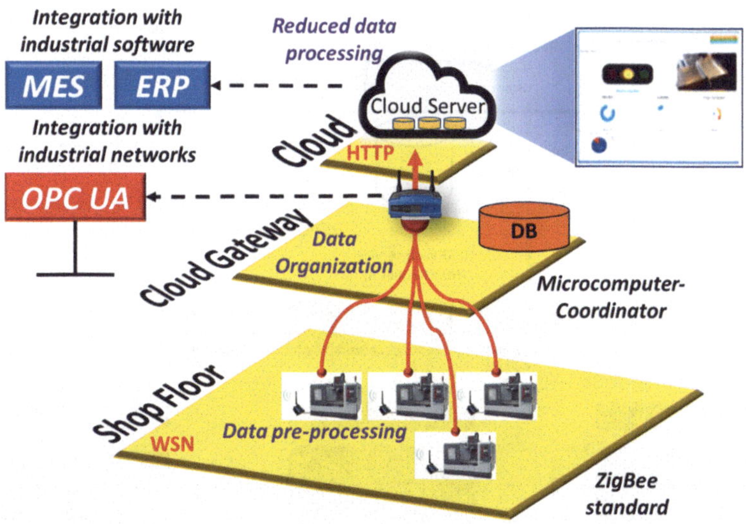

Fig. 4.12 The Internet of Things (IoT) based monitoring system. (Adapted from Mourtzis et al. 2018b)

Manufacturing processes were turned into virtual economies with the start of the fourth industrial revolution (Industry 4.0). The Internet of Things (IoT) and other emerging technologies play a major role in this transition. Smart sensor systems are expected to link their assets to the virtual world in order to move manufacturing companies toward IoT. To address this issue, a monitoring system for shop floor control following the IoT model is presented in Fig. 4.12.

The abovementioned proposed monitoring system consists of a data acquisition device (DAQ) capable of capturing data from machine tools quickly and efficiently and transmitting data via a wireless sensor topology to a cloud gateway. For further storage and analysis, the tracked data is moved to a cloud server. The developed framework follows the IoT model in terms of linking the physical to the cyber world and providing capabilities for integration with established industrial systems. Therefore, the standard of the open platform communication-unified architecture (OPC-UA) is used to support the integration of the proposed monitoring system with other IT resources in a company. The proposed system improves the interoperability, performance, and connectivity of the system, providing useful information from the monitoring system that can be further evaluated and transformed into flexible independent planning and control systems (Mourtzis et al. 2018b).

As it was previously mentioned, manufacturing is moving into the next step through the Industry 4.0 concept, that of digitalization. Industry 4.0 enables traditional manufacturing processes to be converted into new digitalized systems, creating significant economic opportunities through market reshaping. The implementation of the Internet of Things (IoT) in manufacturing would make it possible to schedule and manage production systems effectively and adaptively. To address this issue, an

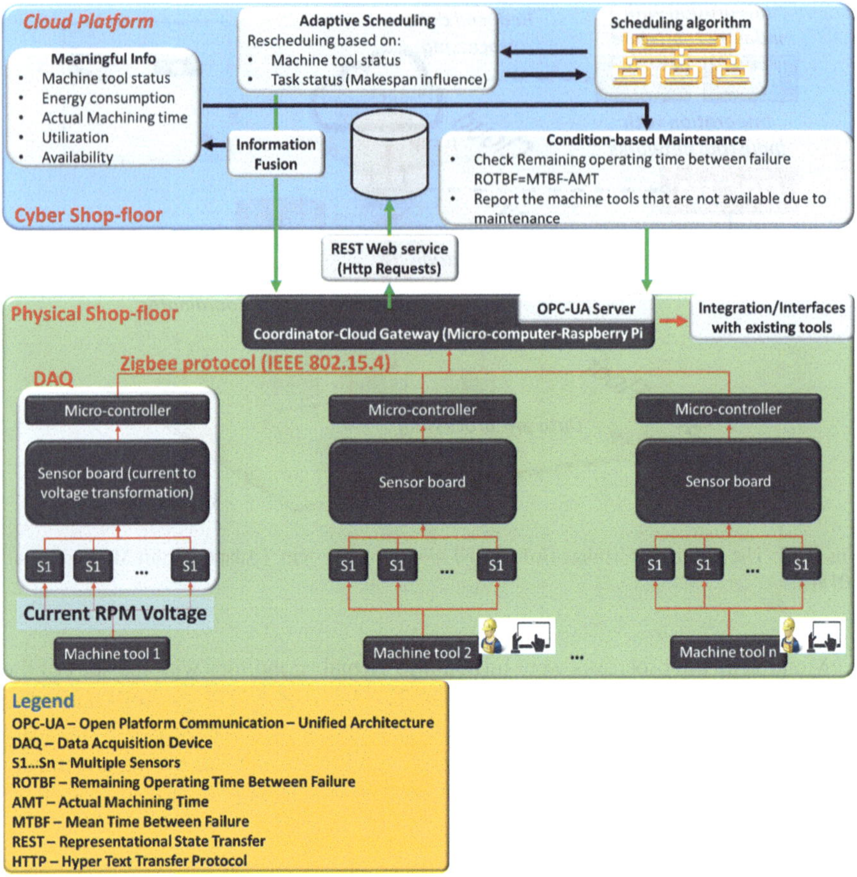

Fig. 4.13 The cloud-based cyber-physical system. (Adapted from Mourtzis et al. 2018d)

automated shop-floor planning and condition-based maintenance cloud-based cyber-physics program is presented by Mourtzis et al. (2018d). The proposed system (Fig. 4.13) has shown that a cost-effective and reliable collection, storage, and evaluation of real-time data can be implemented from the shop floor. The main contributions of the system can be summarized as follows: (1) a cost-effective monitoring system capable of collecting and transmitting data from different sources through an established wireless sensor network and communication protocols, (2) a multi-criteria decision-making algorithm for adaptive scheduling that can give us insight into real-time data from different shop floor sources and perform precise and effective production scheduling, (3) a real cyber-physical system consisting of various modules capable of communicating all together enabling interoperability and developed in a cloud environment supported by various technologies aimed at moving toward Industry 4.0, digitalization, and IoT, and (4) an overall process that can be readily applied to various types of companies and implemented in this case in a mold manufacturing industry, delivering highly satisfactory results compared to traditional methods.

4.4 Discussion

This chapter has investigated and presented the latest advances in scheduling in the cloud manufacturing scheme. Through the review of the available publications, useful conclusions can be drawn, regarding the advances of the abovementioned topic of investigation. It can be stressed that Industry 4.0 is a big enabler, and the industrial domain has to take advantage of the given opportunities in order to survive and take a greater piece of the global market. However, the transition to the full digitalization of the production is an issue that still has to be addressed. In order to do so, research focus has been set on the topics of Smart Manufacturing, Cloud Manufacturing, and Scheduling. The latter, i.e. Scheduling, as examined in the previous paragraphs, is a multi-faceted problem that requires special attention from the research community.

By and large, industry is constantly getting modernized with the integration of cutting-edge technologies. During the last decade, digital technologies have advanced considerably, thus unveiling new opportunities for the optimization of the modeling, design, and operation of manufacturing systems and networks. In the current era of digitalization, researchers are putting effort on developing solutions based on Artificial Intelligence algorithms. Indeed, AI can be considered a mainstream topic that will elevate the Smart Manufacturing paradigm.

References

Ahn, G., Park, Y.-J., & Hun, S. (2018). Performance computation methods for composition of tasks with multiple patterns in cloud manufacturing. *International Journal of Production Research, 57*(2), 1–14. https://doi.org/10.1080/00207543.2018.1451664.

Ahn, G., Park, Y. J., & Hur, S. (2019). Performance computation methods for composition of tasks with multiple patterns in cloud manufacturing. *International Journal of Production Research, 57*(2), 517–530. https://doi.org/10.1080/00207543.2018.1451664.

Boyer, S. (2004). *SCADA: Supervisory control and data acquisition* (3rd ed.). Research Triangle Park, NC: ISA—The Instrumentation, Systems and Automation Society.

Chen, W., Song, J., Shi, L., Pi, L., & Sun, P. (2013). Data mining-based dispatching system for solving the local pickup and delivery problem. *Annals of Operations Research, 203*, 351–370. https://doi.org/10.1007/s10479-012-1118-1.

Chen, L., Wang, J., & Xu, X. (2019). An energy-efficient single machine scheduling problem with machine reliability constraints. *Computers & Industrial Engineering, 137*, 106072. https://doi.org/10.1016/j.cie.2019.106072.

Chryssolouris, G. (2006). *Manufacturing systems: Theory and practice* (2nd ed.). New York: Springer.

Chryssolouris, G., Pierce, J., & Dicke, K. (1992). A decision-making approach to the operation of flexible manufacturing systems. *International Journal of Flexible Manufacturing Systems, 4*(3–4), 309–330.

Daniel, R., Fernando, T., & Mariano, F. (2019). Industry 4.0: Smart scheduling. *International Journal of Production Research, 57*, 3802–3813. https://doi.org/10.1080/00207543.2018.1504248.

Dolgui, A., Ivanov, D., Potryasaev, S., Sokolov, B., Ivanova, M., & Werner, F. (2019). Blockchain-oriented dynamic modelling of smart contract design and execution in the supply chain. *International Journal of Production Research*. https://doi.org/10.1080/00207543.2019.1627439.

Doukas, M., Psarommatis, F., & Mourtzis, D. (2014). Planning of manufacturing networks using an intelligent probabilistic approach for mass customised products. *International Journal of Advanced Manufacturing Technologies, 74*, 1747. https://doi.org/10.1007/s00170-014-6121-z.

ElMaraghy, W., ElMaraghy, H., Tomiyama, T., & Monostori, L. (2012). Complexity in engineering design and manufacturing. *CIRP Annals, 61*(2), 793–814. https://doi.org/10.1016/j.cirp.2012.05.001.

Enzo, F., Mirko, K., & Michael, F. (2018). Data-driven production control for complex and dynamic manufacturing systems. *CIRP Annals—Manufacturing Technology, 67*, 515–518. https://doi.org/10.1016/j.cirp.2018.04.033.

Hochdörffer, J., Buergin, J., Vlachou, E., Zogopoulos, V., Lanza, G., & Mourtzis, D. (2018). Holistic approach for integrating customers in the design, planning, and control of global production networks. *CIRP Journal of Manufacturing Science and Technology, 23*, 98–107. https://doi.org/10.1016/j.cirpj.2018.07.004.

Ivanov, D., Dolgui, A., Sokolov, B., Frank, W., & Marina, I. (2016). A dynamic model and an algorithm for short-term supply chain scheduling in the smart factory industry 4.0. *International Journal of Production Research, 54*, 386–402. https://doi.org/10.1080/00207543.2014.999958.

Ivanov, D., Dolgui, A., Suresh, S., & Sokolov, B. (2018a). Scheduling in production, supply chain and Industry 4.0 systems by optimal control. *International Journal of Production Research, 57*(2), 411–432.

Ivanov, D., Sethi, S., Dolgui, A., & Sokolov, B. (2018b). A survey on control theory applications to operational systems, supply chain management, and Industry 4.0. *Annual Reviews in Control, 46*, 134–147. https://doi.org/10.1016/j.arcontrol.2018.10.014.

Kumar, A., Shankar, R., Choudhary, A., & Thakur, L. S. (2016). A big data MapReduce framework for fault diagnosis in cloud based manufacturing. *International Journal of Production Research, 54*(23), 7060–7073.

Kusiak, A. (2018). Smart manufacturing. *International Journal of Production Research, 56*(1–2), 508–517. https://doi.org/10.1080/00207543.2017.1351644.

Li, B. H., Zhang, L., Wang, S. L., Tao, F., Cao, J. W., Jiang, X. D., Song, X., & Chai, X. D. (2010). Cloud manufacturing: A new service- oriented networked manufacturing model. *Computer Integrated Manufacturing Systems, 16*(1), 1–7.

Lu, Y., & Rob, X. X. (2019). Cloud-based manufacturing equipment and big data analytics to enable on-demand manufacturing services. *Computer Integrated Manufacturing, 57*, 92–102. https://doi.org/10.1016/j.rcim.2018.11.006.

Makris, S., Zoupas, P., & Chryssolouris, G. (2011). Supply chain control logic for enabling adaptability under uncertainty. *International Journal of Production Research, 60*(1), 453–456. https://doi.org/10.1080/00207543.2010.508940.

Mathew, S. (2014). Architecting networks, Supporting IoT, Alcatel-Lucent. Retrieved November 4, 2019, from https://www.slideshare.net/usmanusb/mimos-iot-twg-day1-session-ii-2nd-speaker-mathew-al.

Michalos, G., Makris, S., & Mourtzis, D. (2011). A web based tool for dynamic job rotation scheduling using multiple criteria. *CIRP Annals, 60*(1), 453–456. https://doi.org/10.1016/j.cirp.2011.03.037.

Moradi, H., & Shadrokh, S. (2019). A robust reliability-based scheduling for the maintenance activities during planned shutdown under uncertainty of activity duration. *Computers & Chemical Engineering, 130*, 106562. https://doi.org/10.1016/j.compchemeng.2019.106562.

Mourtzis, D. (2011). Internet-based collaboration in the manufacturing supply chain. *CIRP Journal of Manufacturing Science and Technology, 4*(3), 296–304. https://doi.org/10.1016/j.cirpj.2011.06.005.

Mourtzis, D. (2016). Challenges and future perspectives for the life cycle of manufacturing networks in the mass customisation era. *Logistics Research 9.1*, 1–20. https://doi.org/10.1007/s12159-015-0129-0.

Mourtzis, D. (2019). Simulation in the design and operation of manufacturing systems: State of the art and new trends. *International Journal of Production Research*. https://doi.org/10.1080/00207543.2019.1636321.

Mourtzis, D., & Vlachou, E. (2016). Cloud-based cyber-physical systems and quality of services. *The TQM Journal, 28*(5), 704–733. https://doi.org/10.1108/TQM-10-2015-0133.

Mourtzis, D., & Vlachou, E. (2018). A cloud-based cyber-physical system for adaptive shop-floor scheduling and condition-based maintenance. *Journal of Manufacturing Systems, 47*(589), 179–198. https://doi.org/10.1016/j.jmsy.2018.05.008.

Mourtzis, D., Doukas, M., & Psarommatis, F. (2012a). A multi-criteria evaluation of centralized and decentralized production networks in a highly customer-driven environment. *CIRP Annals, 61*(1), 427–430. https://doi.org/10.1016/j.cirp.2012.03.035.

Mourtzis, D., Doukas, M., & Psarommatis, F. (2012b). Design and planning of decentralised production networks under high product variety demand. *Procedia CIRP, 3*, 293–298. https://doi.org/10.1016/j.procir.2012.07.051.

Mourtzis, D., Doukas, M., & Psarommatis, F. (2013a). Design and operation of manufacturing networks for mass customization. *CIRP Annals, 62*(1), 467–470. https://doi.org/10.1016/j.cirp.2013.03.126.

Mourtzis, D., Doukas, M., Psarommatis, F., & Panopoulos, N. (2013b). Design of multi-stage manufacturing networks for personalized products using metaheuristics. In A. Azevedo (Ed.), *Advances in sustainable and competitive manufacturing systems* (Lecture notes in mechanical engineering) (pp. 1263–1276). Cham: Springer. https://doi.org/10.1007/978-3-319-00557-7_103.

Mourtzis, D., Doukas, M., Vlachou, A., & Xanthopoulos, N. (2014). Machine availability monitoring for adaptive holistic scheduling: A conceptual framework for mass customization. *Procedia CIRP, 25*(Part C), 406–413. https://doi.org/10.1016/j.procir.2014.10.056.

Mourtzis D, Vlachou E., Doukas M., Kanakis N., Xanthopoulos N., Koutoupes A. (2015a). Cloud-based adaptive shop-floor scheduling considering machine tool availability. In *Proceedings of the ASME 2015 International Mechanical Engineering Congress & Exposition, IMECE 2015, November 13–19, Houston, Texas*.

Mourtzis, D., Vlachou, E., Milas, N., & Xanthopoulos N. (2015b). A cloud-based approach for maintenance of machine tools and equipment based on shop-floor monitoring. In *48th CIRP Conference on Manufacturing Systems—CIRP CMS 2015, June 24–26, Ischia, Italy*.

Mourtzis, D., Doukas, M., & Psarommatis, F. (2015c). Design of manufacturing networks for mass customisation using an intelligent search method. *International Journal of Computer Integrated Manufacturing, 28*(7), 679–700. https://doi.org/10.1080/0951192X.2014.900867.

Mourtzis, D., Doukas, M., & Psarommatis, F. (2015d). A toolbox for the design, planning and operation of manufacturing networks in a mass customisation environment. *Journal of Manufacturing Systems, 36*, 274–286. https://doi.org/10.1016/j.jmsy.2014.06.004.

Mourtzis, D., Doukas, M., & Vlachou, E. (2016a). A mobile application for knowledge-enriched short-term scheduling of complex products. *Logistics Research, 9*, 3. https://doi.org/10.1007/s12159-015-0130-7.

Mourtzis, D., Vlachou, E., Xanthopoulos, N., Givehchi, M., & Wang, L. (2016b). Cloud-based adaptive process planning considering availability and capabilities of machine tools. *Journal of Manufacturing Systems, 39*, 1–8. https://doi.org/10.1016/j.jmsy.2016.01.003.

Mourtzis, D., Vlachou, E., Giannoulis, C., Siganakis, E., & Zogopoulos, V. (2016c). Applications for frugal product customization and design of manufacturing networks. *Procedia CIRP, 52*, 228–233. https://doi.org/10.1016/j.procir.2016.07.055.

Mourtzis, D., Doukas, M., & Vandera, C. (2016d). Smart mobile apps for supporting product design and decision-making in the era of mass customization. *International Journal of Computer Integrated Manufacturing, 30*, 1–18. https://doi.org/10.1080/0951192X.2016.1187295.

Mourtzis, D., Vlachou, E., Boli, N., Gravias, L., & Giannoulis, C. (2016e). Manufacturing networks design through smart decision making towards frugal innovation. *Procedia CIRP, 50*, 354–359. https://doi.org/10.1016/j.procir.2016.04.166.

Mourtzis, D., Fotia, S., & Vlachou, E. (2017a). Quantification of Industry 4.0 manufacturing complexity based on information theory. In *Seventh IESM Conference, October 11–13, Saarbrucken, Germany*.

Mourtzis, D., Vlachou, A., & Zogopoulos, V. (2017b). Cloud-based augmented reality remote maintenance through shop-floor monitoring: A product-service system approach. *Journal of Manufacturing Science and Engineering, 139*, 061011. https://doi.org/10.1115/1.4035721.

Mourtzis, D., Vlachou, E., Zogopoulos, V., & Fotini, X. (2017c). Integrated production and maintenance scheduling through machine monitoring and augmented reality: An Industry 4.0 approach. In H. Lödding, R. Riedel, K. D. Thoben, G. von Cieminski, & D. Kiritsis (Eds.), *The path to intelligent, collaborative and sustainable manufacturing. APMS 2017. IFIP Advances in Information and Communication Technology* (Vol. 513). Cham: Springer. https://doi.org/10.1007/978-3-319-66923-6_42.

Mourtzis, D., Fotia, S., Boli, N., et al. (2018a). Product-service system (PSS) complexity metrics within mass customization and Industry 4.0 environment. *The International Journal of Advanced Manufacturing Technology, 97*, 91. https://doi.org/10.1007/s00170-018-1903-3.

Mourtzis, D., Milas, N., & Katerina, E. (2018b). An internet of things-based monitoring system for shop-floor control. *Journal of Computing and Information Science in Engineering, 18*, 021005. https://doi.org/10.1115/1.4039429.

Mourtzis, D., Papatheodorou, A. M., & Sophia, F. (2018c). Development of a key performance indicator assessment methodology and software tool for product-service system evaluation and decision-making support. *Journal of Computing and Information Science in Engineering, 18*, 041005. https://doi.org/10.1115/1.4040340.

Mourtzis, D., Vlachou, K., & Zogopoulos, V. (2018d). 51st CIRP Conference on Manufacturing Systems. An IoT-based platform for automated customized shopping in distributed environments. *Procedia CIRP, 72*, 892–897. https://doi.org/10.1016/j.procir.2018.03.199.

Mourtzis, D., Milas, N., & Athinaios, N. (2018e). Towards machine shop 4.0: A general machine model for CNC machine-tools through OPC-UA. *Procedia CIRP, 78*, 301–306. https://doi.org/10.1016/j.procir.2018.09.045.

Mourtzis, D., Milas, N., & Athinaios, N. (2018f). Towards machine shop 4.0: A general machine model for CNC machine-tools through OPC-UA. *Procedia CIRP, 78*, 301–306. https://doi.org/10.1016/j.procir.2018.09.045.

Mourtzis, M., Fotia, S., Boli, N., & Vlachou, E. (2019). Modelling and quantification of industry 4.0 manufacturing complexity based on information theory: A robotics case study. *International Journal of Production Research, 57*, 6908–6921. https://doi.org/10.1080/00207543.2019.1571686.

Mousavi, A., & Siervo, H. R. A. (2017). Automatic translation of plant data into management performance metrics: A case for real-time and predictive production control. *International Journal of Production Research, 55*(17), 4862–4877. https://doi.org/10.1080/00207543.2016.1265682.

Nicholson, A., Webber, S., Dyer, S., Patel, T., & Janicke, H. (2012). SCADA security in the light of cyber-warfare. *Computers & Security, 31*(4), 418–436. https://doi.org/10.1016/j.cose.2012.02.009.

Papakostas, N., Mourtzis, D., Michalos, G., Makris, S., & Chryssolouris, G. (2012). An agent-based methodology for manufacturing decision making: A textile case study. *International Journal of Computer Integrated Manufacturing, 25*(6), 509–526. https://doi.org/10.1080/0951192X.2011.637963.

Rauschecker, U., Meier, M., Muckenhirn, R., Yip, A., Jagadee-san, A., & Corney, J. (2011). *Cloud-based manufacturing-as-a-service environment for customized products*. Dublin: IIMC, International Information Management Corporation.

Renu, R. S., Mocko, G., & Koneru, A. (2013). Use of big data and knowledge discovery to create data backbones for decision support systems. *Procedia Computer Science, 20*, 446–453.

Sharp, M., Ak, R., & Hedberg, T. (2018). A survey of the advancing use and development of machine learning in smart manufacturing. *Journal of Manufacturing Systems, 48*(Part C), 170–179. https://doi.org/10.1016/j.jmsy.2018.02.004.

Shukla, N., Tiwari, M. K., & Beydoun, G. (2019). Next generation smart manufacturing and service systems using big data analytics. *Computers & Industrial Engineering, 128*, 905–910. https://doi.org/10.1016/j.cie.2018.12.026.

Si, X. S., Wang, W., Hu, C. H., & Zhou, D. H. (2011). Remaining useful life estimation—A review on the statistical data-driven approaches. *European Journal of Operational Research, 213*(1), 1–14.

Tao, F., LaiLi, Y., Xu, L., & Zhang, L. (2013). FC-PACO-RM: A parallel method for service composition optimal-selection in cloud manufacturing system. *IEEE Transactions on Industrial Informatics, 9*(4), 2023–2033.

Tao, F., Qi, Q., Liu, A., & Kusiak, A. (2018). Data-driven smart manufacturing. *Journal of Manufacturing Systems, 48*(Part C), 157–169. https://doi.org/10.1016/j.jmsy.2018.01.006.

Wang, X. V., & Xu, X. W. (2013). An interoperable solution for cloud manufacturing. *Robotics and Computer Integrated Manufacturing, 29*(4), 232–247. https://doi.org/10.1016/j.rcim.2013.01.005.

Wu, D., Greer, M. J., Rosen, D. W., & Schaefer, D. (2013). Cloud manufacturing: Drivers, current status, and future trends. In *International Manufacturing Science and Engineering Conference Collocated with the 41st North American Manufacturing Research Conference, Madison, WI.*

Xu, X. (2012). From cloud computing to cloud manufacturing. *Robotics and Computer-Integrated Manufacturing, 28*(1), 75–86. https://doi.org/10.1016/j.rcim.2011.07.002.

Xu, J., Tran, H. M., Gautam, N., & Bukkapatnam, S. (2019). Joint production and maintenance operations in smart custom-manufacturing systems. *IISE Transactions, 51*(4), 406–421. https://doi.org/10.1080/24725854.2018.1511938.

Yang, H., Kumara, S., Bukkapatnam, S., & Tsung, F. (2019). The internet of things for smart manufacturing: A review. *IISE Transactions, 51*(11), 1190–1216. https://doi.org/10.1080/24725854.2018.1555383.

Yongkui, L., Lihui, W., Xi, W., Xun, X., & Lin, Z. (2019). Scheduling in cloud manufacturing: State-of-the-art and research challenges. *International Journal of Production Research, 57*, 4854–4879. https://doi.org/10.1080/00207543.2018.1449978.

Zhang, Z., Wang, X., Zhu, X., Cao, Q., & Tao, F. (2019). Cloud manufacturing paradigm with ubiquitous robotic system for product customization. *Robotics and Computer-Integrated Manufacturing, 60*(2019), 12–22. https://doi.org/10.1016/j.rcim.2019.05.015.

Zhou, J., & Yao, X. (2017). A hybrid approach combining modified artificial bee colony and cuckoo search algorithms for multi-objective cloud manufacturing task composition. *International Journal of Production Research, 55*(16), 1–20.

Zhou, L., Zhang, L., Sarker, B. R., Laili, Y., & Ren, L. (2018). An event-triggered dynamic scheduling method for randomly arriving tasks in cloud manufacturing. *International Journal of Computer Integrated Manufacturing, 31*(3), 318–333. https://doi.org/10.1080/0951192X.2017.1413252.

Chapter 5
Cloud Material Handling Systems: Conceptual Model and Cloud-Based Scheduling of Handling Activities

Fabio Sgarbossa, Mirco Peron, and Giuseppe Fragapane

Abstract Nowadays, the implementation of cloud manufacturing technologies epitomizes the avant-garde in production systems. This affects several aspects of the management of these production systems, in particular scheduling activities, due to the possibility provided by cloud manufacturing of having real-time information about the stages of a product life cycle and about the status of all services. However, so far, cloud manufacturing has mainly focused on machines, with limited interest in material handling systems. This shortfall has been addressed in this study, where a new material-handling paradigm, called Cloud Material Handling System (CMHS) and developed in the Logistics 4.0 Lab at NTNU (Norway), has been introduced. With CMHS, the scheduling of the Material Handling Modules (MHMs) can be optimized, increasing the flexibility and productivity of the overall manufacturing system. To achieve this, the integration of advanced industry 4.0 technologies such as Internet of Things (IoT), and in particular Indoor Positioning Technologies (IPT), Cloud Computing, Machine Learning (ML), and Artificial Intelligence (AI), is required. In fact, based on the relevant information provided on the cloud platform by IPT and IoT for each product, called Smart Object (SO) (position, physical characteristics and so on), an Intelligent Cognitive Engine (ICE) can use ML and AI to decide, in real time, which MHM is most suitable for carrying out the tasks required by these products based on a compatibility matrix, on their attributes, and on the defined scheduling procedure.

5.1 Introduction

In recent years, connectivity and exchange of information have played an important role in the world of manufacturing (Alcácer and Cruz-Machado 2019). Researchers

F. Sgarbossa (✉) · M. Peron · G. Fragapane
Department of Mechanical and Industrial Engineering Valgrinda, NTNU—Norwegian University of Science and Technology, Trondheim, Norway
e-mail: fabio.sgarbossa@ntnu.no; mirco.peron@ntnu.no; giuseppe.fragapane@ntnu.no

B. Sokolov et al. (eds.), *Scheduling in Industry 4.0 and Cloud Manufacturing*,
International Series in Operations Research & Management Science 289,
https://doi.org/10.1007/978-3-030-43177-8_5

and practitioners have focused their attention on new technologies (such as cyber physical systems [CPS], IoT, and cloud computing) that enable the integration of production with network connectivity (Xu 2012; Lee et al. 2018; Ivanov et al. 2016, 2019; Panetto et al. 2019; Fragapane et al. 2020; Battini et al. 2018). In this way, the traditional rigid and automated operational processes can evolve in fully connected and flexible systems, highly valuable in a market characterized by unpredictable changes. In fact, connecting different machines in production, material-handling systems in warehouses, and equipment in laboratories can form networks capable of dynamic reconfiguration and high flexibility, and can provide global feedback in order to achieve high efficiency.

Lately, the impact of COVID-19 outbreak on global industry has demonstrated the role of fully connected and flexible manufacturing system in contributing to the resilience of the supply chains. Companies with integrated manufacturing network seem to be better positioned at the crisis time and for the future recovery (Ivanov 2020; Ivanov and Das 2020; Ivanov and Dolgui 2020).

Most effort has been put toward the connectivity of different machines in production systems, with the establishment of cloud manufacturing. By connecting different machines with cloud services and managing them in a centralized way, the production systems can adapt, in real time, to new demands with increased flexibility (Liu et al. 2019). This new manufacturing paradigm influences several aspects, one of which is scheduling.

Scheduling is the process of arranging, controlling, and optimizing work or workloads. Therefore, in terms of allocating resources/services to tasks, monitoring, controlling, and optimizing task execution, the possibility provided by cloud manufacturing of having real-time information about all the stages of a product life cycle and about the status of all the services is of great impact. Considering the allocation of resources to different tasks, Liu et al. described the process as a sum of different sub-steps, at the end of which an optimized schedule is generated based on the real-time status information (availability, position, etc.) of the machines required to process an order (Liu et al. 2019; Dolgui et al. 2019).

As mentioned earlier, connectivity of different machines has been already developed, improving scheduling activities. However, a step forward toward the optimization of these new manufacturing scheduling procedures would be the implementation of cloud services for material-handling equipment. However, due to the complexity and variability of possible performed tasks, available resources, and equipment, the scheduling and planning of material-handling equipment has been mainly treated as separate from the manufacturing systems with which they operate.

Material handling is defined as the movement, storage, protection, and control of materials throughout the manufacturing and distribution process (including their consumption and disposal). It involves providing the right amount of the right material, in the right condition, at the right place, in the right position, in the right sequence, and for the right cost, by the right method(s). The main objective of material handling is to perform it safely, efficiently, at low cost, in a timely manner, accurately, and without damage.

The typical structure of today's material-handling systems is a mix of different equipment with various levels of automation (Furmans and Gue 2018). It is still largely dominated by manual and mechanized systems, such as manual carts and industrial vehicles (i.e., pallet trucks, forklifts), in which humans still play an important decisional role. In other cases, automated solutions, such as Automated Guided Vehicles (AGVs) or Automated Storage and Retrieval Systems (AS/RS), are implemented with their own decentralized control systems. Since they are not connected to each other, a multilevel hierarchical control system is necessary to coordinate the different sub-systems and allow the products to be moved from one point to the next within the manufacturing system. Typically, decision-making processes such as scheduling are distributed over the levels of systems interacting with each other (such as the local PLCs, the material flow controller, the Warehouse Management System [WMS], Manufacturing Execution System [MES], and the Enterprise Resource Planning [ERP] system).

Decisions, processes, and activities in material-handling systems show great dependencies and should not be seen as isolated, independent procedures. Materials handling should be seen within a system context. The systems concept is particularly helpful because it identifies and analyzes the interrelation within a system. Blanchard and Fabrycky define a system as "a set of interrelated components working together with the common objective of fulfilling some designated need" (Blanchard and Fabrycky 1990). The efficient scheduling of the material handling equipment (MHE) has a strong effect on the productivity, profitability, and flexibility of the manufacturing systems, especially in a low-volume high-variety context (Zangaro et al. 2019; Zennaro et al. 2019a, b). Some examples are a machine waiting for the product to process since the forklift driver is not available, or a machine being blocked because the unit loads in the unloading station are still waiting to be transported to the next production phase. The availability in recent years of industry 4.0 technologies, such as IPT as part of IoT, motion tracking and control, and cloud computing is making MHE one of the most feasible solutions for increasing the flexibility of manufacturing systems.

By extending the definition of cloud manufacturing to handling activities, a new kind of paradigm, called a Cloud Material Handling System (CMHS), has been introduced and developed by the authors in the Logistics 4.0 Laboratory at the Norwegian University of Science and Technology. This is Norway's first logistics laboratory that merges digital technologies with traditional production and logistics systems, enabling researchers, practitioners, engineers, pioneers, students, and other enthusiasts to come together and collaborate on common ground.

As described previously, the typical aim of cloud manufacturing is to deliver on-demand manufacturing services to customers based on orders received via the Internet. With the CMHS, this aim can be adapted to the efficient management and scheduling of MHE. Based on the same concept as cloud manufacturing, the CMHS has the scope to satisfy consumers' requests (handling of unit loads) through the available resources (MHE) in a cloud environment, reducing the complexity of a multilevel hierarchical control system and increasing the overall flexibility and productivity of the manufacturing system. The CMHS has been developed mostly for applications within a single factory, but it can be also extended to a multi-factory environment where the logistics activities are, for example, external transportation.

For this purpose, the CMHS needs information that is not typically collected and used in cloud manufacturing. According to the definition of material handling (movement, storage, protection, and control), the most important information required from the system is the real-time locations of products and MHE. This lack can be filled by implementing IPT, allowing real-time localization of the products/unit loads and Material Handling Equipment in a cloud platform. IPT is in fact a technology that continuously determines in real time the position of something or someone in a physical space (Hightower and Borriello 2001b). As part of IoT, according to Gu et al. (2009), an IPT can provide different kinds of data, including position, travel path, time, speed, and required activities. Depending on the applications, there are several different kinds of IPT. They will be described in the next section, focusing on their characteristics and on their pros and cons. In Sect. 5.3, the CMHS will be introduced and described, while in Sect. 5.4, the characteristics of scheduling in the CMHS will be described and compared to the traditional method. Finally, conclusions, future research, and perspectives will be provided in Sect. 5.5.

5.2 Indoor Positioning Technologies

IPT, as part of the IoT, is defined as the sub-system that permits a mobile device to determine its position and that renders this position available for position-based services (Gu et al. 2009). These position-based services can bring benefits in several environments, such as hospitals, where the position of the equipment needs to be known to efficiently use the medical resources, supermarkets, where the customers want to know the fastest path to reach the desired products, and large museums, where tourists are interested in knowing the location of the artworks they are interested in Tesoriero et al. (2008). In the last few years, IPT has gained interest also in the industrial field (Zuin et al. 2018). In particular, material-handling processes are those that can benefit the most from this system. In fact, given the need for handling huge volumes of products with very short lead times, a material-handling system has to eliminate all the inefficiencies, such as delays in the search for the required product in the warehouse, errors in the storage or in the picking of an item, and waste of time during the travelling of carts and operators.

As stated by Curran et al. (2011), there are several available technologies to identify, in real time, the location and flow of material depending on the required performances. There is not a single "best solution" that is suitable for each scenario. Therefore, in recent years, different types of IPTs have been introduced depending on the desired level of accuracy, coverage area, robustness, scalability, cost, and complexity (Gu et al. 2009). They can use one or several positioning technologies (i.e., triangulation, fingerprinting, proximity, and vision analysis (Kaemarungsi and Krishnamurthy 2004; Hightower and Borriello 2001)), but the majority leverage the triangulation method, where once the coordinates X and Y of the three reference elements *A*, *B*, and *C* are known, the position can be calculated by using either the length or the directions of at least three vectors from the respective reference points (Gu et al. 2009).

However, the classification of these IPTs is not usually based on the positioning technology used. Instead, it is usually based on either network features (Deak et al. 2012) or on their hardware requirements (Liang et al. 2013). Another classification can be based on the main technology used to determine the location, which may include infrared (IR) signals, ultrasound waves, vision-based analysis, and radio frequency.

Infrared positioning systems. This is one of the most common positioning systems since IR emitters are small and lightweight. The system architecture is simple, and it performs positioning estimation in a very accurate way (with an accuracy of several mm). It was used by Pinto et al. for the localization of small mobile robots, and they reported high accuracies (0.06 m and 7°) and fast responsiveness in the localization (less than 40 ms) (Pinto et al. 2015). Similar results were reported in Zhang et al. (2010), where an accuracy of 3 mm and a responsiveness of 3 ms were reported. However, they also reported a limited coverage area (7 m). In fact, one of the main limitations is related to the application environment of this system, as it requires the absence of interferences and obstacles. Nevertheless, this can also be an advantage; the inability of the IR beams to penetrate the walls ensures that it is possible to limit the signal inside a specific room (Mainetti et al. 2014). Other limitations are the high cost of the hardware, the short-range signal transmission between devices, and the interference from florescent light and sunlight (Fernando et al. 2003).

Ultrasound positioning systems: This system uses ultrasound signals to measure the distance of a mobile target from a fixed-point receiver. Despite its low cost, the diffusion of this IPT is hampered by the low accuracy (several cm) and the short range (from 2 to 10 m). These drawbacks have been shown to be overcome by using a large number of transmitters on the ceiling (Woodman and Harle 2010). However, in doing so, one of the main advantages of this IPT (low cost) would be offset. Both studies reported the system to be highly susceptible to noise sources, which affect its reliability and accuracy.

Vision-based positioning systems: This system is based on the use of fixed or mobile cameras that can cover a large area at a low installation cost. The tracked person and/or device does not need to wear or carry any device. It has been successfully used as a tracking system for rolled milled plates with an accuracy of 0.5 m (Tratnig et al. 2007). However, some inaccuracies were reported, especially in a dynamic environment and in environments characterized by many interference sources. These systems are, in fact, preferable for quality control and inventory-level assessments (Wu et al. 2009). In addition, the difficulty in tracking several persons and/or devices has also limited its use (Zuin et al. 2018).

Radio frequency position systems: There are several systems that use radio waves, such as Wi-Fi, Radio Frequency Identification (RFID), and Ultra-Wide Band (UWB). The Wi-Fi system is often used to monitor the movements of mobile devices due to its low cost. In fact, Wi-Fi is widely spread inside the facilities and thus extra software are not required. Accuracy is quite low (between 20 and 40 m) due to problems of refraction, scattering, and multi-path fading in the indoor propagation of Wi-Fi signals. This can be improved by increasing the number of access points and adding wireless routers (Chen et al. 2014), reaching an accuracy of 2 m (Han and He 2018), or by using particular filters, such as Unscented Kalman Filter (Khan et al. 2017). RFID is a means of storing and retrieving data through radio waves to an RF-compatible integrated circuit (Ni et al. 2003). These systems can cover large distances since they do not require line-of-sight (i.e., they can easily travel through walls and human bodies). Moreover, RFID leverages small and light tags and this allows for a unique identification of equipment and persons. However, this system needs several infrastructure components in the working area and they suffer from multi-path distortion of radio signals reflected by walls and obstacles (Gu et al. 2009). UWB technology uses pulses with a very short duration (less than 1 ns) to overcome the limitation of the multi-path distortion of radio signals that affects RFID systems. In addition, the use of multiple bands of frequencies simultaneously (each transmitting its own signal, as opposed to RFID, which uses just a single portion of the frequency spectrum) overcomes RFID's line-of-sight requirement and thus increases accuracy. That was shown by Regattieri and Santarelli, who tested a UWB system in different industrial environments and reported an accuracy of 1 m (Regattieri and Santarelli 2013). These results agreed with those reported in Zuin et al. (2018), who found that 70% of measurements of a moving target showed the gap between the tracked position and the real one as lower than 0.40 m.

By using IPT, it is thus possible to have real-time information about products/operators/MHE location in an easy and fast way, representing a prerogative for the development of the CMHS, since it requires the real-time location of all the elements of the system.

5.3 Cloud Material Handling System

Uber is revolutionizing the world of urban logistics (i.e., people mobility and product delivery) by providing peer-to-peer ridesharing, ride service hailing, food delivery, and a bicycle sharing system. Its success is based on the cloud storage and smart use of huge amounts of data regarding the real-time locations of consumers and cars/bicycles/scooters.

Using Uber is very simple both for consumers and drivers. The phone app helps the users in setting the pickup location (using GPS), setting the destination, and requesting a car. It also supports the drivers in managing requests and receiving payments. However, there is a huge amount of data processed in a cloud platform, even when the driver has no passengers or products out for delivery. All this data is stored and leveraged to predict supply and demand, as well as for setting fares and gathering information about issues such as bottlenecks, traffic jams, and shortcuts.

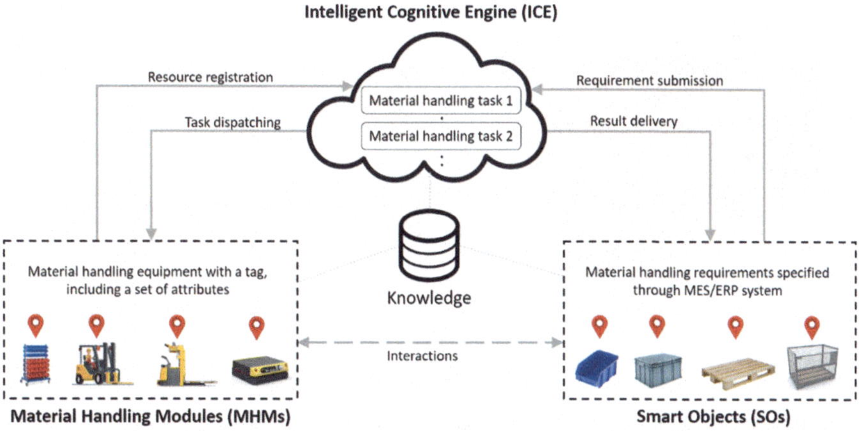

Fig. 5.1 Operation model of Cloud Material Handling System

The concept behind the CMHS developed by the authors at the Logistics 4.0 Laboratory at NTNU is similar to the taxi-hailing web app. In this case, the "consumers" are the unit loads that require a specific service from the system (typically to be transported from one point to another), while the "cars" and "drivers" are the MHE (forklifts, manual trolleys, conveyors, etc.) with different capabilities (capacity, cost, speed, time, service level, etc.). The cloud platform has a cognitive engine able to dynamically assign the requests to the available resources based on knowledge gained over time.

Similarly to cloud manufacturing (Liu et al. 2019; Zhang et al. 2014), the operation model of the Cloud Material Handling System has a direct effect on the characteristics of the scheduling. Figure 5.1 depicts the operation model developed by the authors at the base of the CMHS. Adapting the concept introduced by Zhang et al. (2014), it consists of three categories of stakeholders: SOs, Material Handling Modules (MHMs), and ICE, sharing a common knowledge of the system.

SOs: These are the unit loads that require handling by the MHE (Furmans and Gue 2018). They are uniquely identified through a tag containing a dynamic set of attributes, such as material flow or process of the object, the current and next step of the process, or its physical characteristics (size, weight, fragility, etc.). The tag is localized through one of the IPTs described in the previous section (Fig. 5.2).

Material handling modules (MHMs): These have the capabilities to perform one or more physical functions related to handling and storing, as required by the SOs. MHMs are typical MHE, such as forklifts, conveyors, cranes, trolleys and carts, AGVs, AMRs, shelving units, storage racks, or ASR/RS (Furmans and Gue 2018). They can be purely manual or mechanized equipment, where the operators still play an important role, like driving, loading, or unloading, but also fully automated equipment with automatized functions. Like the SOs, they are classified in the CMHS with a tag including a set of attributes, such as

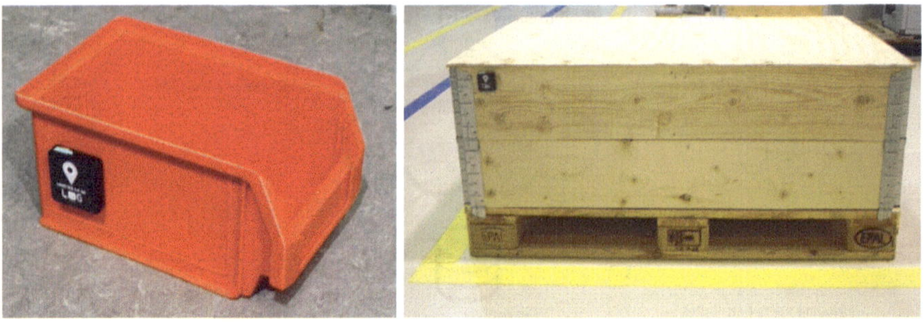

Fig. 5.2 Examples of SOs

dimensions, capacity, autonomy, automation level, and cost. They also have a set of functions they are able to perform. According to Furmans and Gue (2018), four classes can be identified: holder (storing an SO, such as shelves and racks); mover (moving an SO, such as forklifts); picker-placer (picking and placing an SO, such as human pickers or robot arms); and unitized-separator (putting together more SOs, such as a palletizer, or creating more SOs from an initial single one, such as a depalletizer). They are connected to the ICE, communicating their locations in real time thanks to IPT. Their availability and lists of attributes/functions are shared with the other stakeholders in the cloud platform, through which they receive jobs. In case they require the human to be active, such as manual trolleys or traditional pallet jacks or forklifts, the interaction with the ICE is based on the use of industrial smartphones or devices, through which the operators receive the scheduled assigned tasks after processing in the ICE (Fig. 5.3).

ICE: This is responsible for the management of the cloud platform where all the information is collected and elaborated (Greis et al. 2019). It receives and stores all the information about the available MHMs and their attributes and functions. These characteristics are matched with the handling requirements sent by the active SOs through an MHM-SO compatibility matrix. The ICE is an AI-based computing system, and thanks to several ML algorithms, it can predict how the global CMHS performs in different situations. In the ICE, the models, rules, and algorithms for scheduling are implemented. Their performance, such as the percentage of loaded travel time, the percentage of added value time, or service level, is assessed in real-time in order to be improved over time through what-if scenarios simulation (Fig. 5.4).

5.4 Scheduling in Cloud Material Handling System

The real-time localization of the SOs and MHMs due to the IPT implementation, and the sharing of their attributes/functions along with positions are enabling a

Fig. 5.3 Examples of MHMs

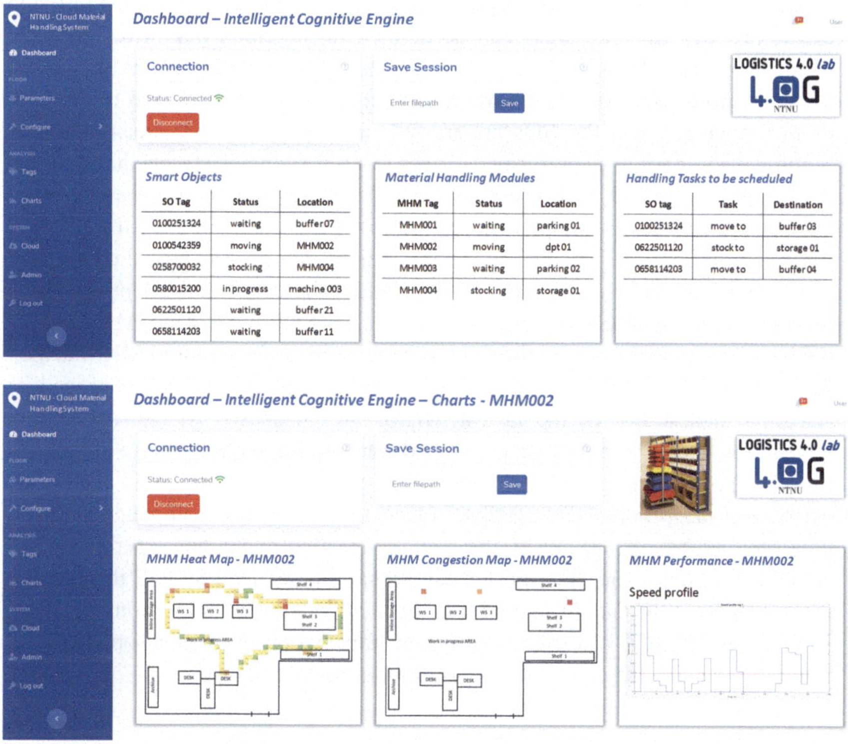

Fig. 5.4 Screenshots of the ICE dashboard

new way of scheduling and control of all the components in the system. In this section, the procedure for assignment of the required jobs by SOs to the different available MHMs is described, and the characteristics of the scheduling in CMHS are presented in comparison to the scheduling in a traditional environment. The industrial implications and advantages in its application will be illustrated and finally some research opportunities are discussed.

5.4.1 Procedure of Scheduling

There are three different phases for the scheduling in CHMS: order/task release, scheduling and delivery, and performance assessment.

Order/task release: The scheduling process begins when SOs send their requirement of a specific service to ICE. The specific service may be the need for the SOs to be stored, moved, transferred, picked, placed, unitized, separated, and/or a combination of these needs. It can be triggered manually by the operator of the machine/MHM where the SO is produced/moved. IPTs can automatize this step of the process. Specific areas can be defined in the cloud platform; when the location of the SOs (using IPT) is within one of these areas, it could be linked to the claiming of service from MHMs. Other areas could be used as destination of the service; when the SOs are moved into those areas, it means that the required tasks have been performed. These tasks are defined a priori thanks to the connection to the ERP system and material flows/production processes.

Scheduling: Every SO is uniquely identified by a tag containing all the relevant information (characteristics such as size, weight, the next step of the process). Based on these characteristics, the ICE can decide the most suitable type of MHM, leveraging the MHM-SO compatibility matrix. Once the type of MHM that fulfils the handling requirements of the SO has been defined, the required MHM is then found among those available according to the pre-defined scheduling procedure (i.e., the closest MHM, the fastest MHM, the cheapest). AI and scheduling algorithms derived from those developed for cloud manufacturing will support this phase of the process in the case of more complex environments.

Delivery and performance assessment: Once the MHM has completed its task, data such as the average speed, the path, congestion phenomena, percentage of added value, time, and other performances can be obtained by ICE through the extrapolation of the data from the tag associated with the SO under consideration. In this way, the ICE can use the acquired information in order to optimize the future tasks using the acquired data as input in its ML algorithms.

5.4.2 Characteristics of Scheduling in CMHS

The scheduling of material-handling tasks is a decision-making process at an operational level. In today's manufacturing and services industries, the coordination of these tasks is mainly hierarchical in structure, with centralized or decentralized control (Scholz-Reiter and Freitag 2007). Several decision stages are thereby, to a different degree, interconnected and automated. Low degree of such decisions have led to fixed assignment using either simple scheduling rules as FIFO (First in first out) or EDD (Earliest Due Date), weighted priority rules, or heuristics scheduling rules. The complexity of the material flow and the push toward increased automation have led to standardized multilevel control architectures and information flow for material-handling systems (Furmans and Gue 2018). The standards contain the tasks to be performed on each level and the communication protocols between the levels. The decision-making process is normally distributed over several system levels: PLC, ERP, MES, process, and machine control. Within a company using material-handling systems, there can typically be found between five and eight levels of systems interacting with each other. Multilevel design and scheduling is often motivated by a company's low capability for real-time information sharing. The introduction of CMHS changes the characteristics of scheduling as follows:

Knowledge-sharing-based scheduling: The introduction of the cloud platform and its ICE modified the way of doing scheduling. In the CMHS, all the equipment are at the same level as the handled products. This allows for real-time adaptability of the MHMs to the current requirements of the systems with which they operate. Each module and object are individual stakeholders, autonomous decision-makers, and interest-independent entities (Liu et al. 2019).

Many MHMs to many jobs scheduling: The sharing of information about locations and attributes of MHMs and SOs enables the distribution of multiple jobs to multiple integrated resources. In traditional MHS, each job is managed individually from one single resource, or it is treated as a sequence of elementary jobs to be executed by single resources.

Dynamic and complex scheduling: Scheduling in CMHS is more complex due to the two characteristics described previously. In this case, the application of AI and ML can support the scheduling, learning from previous experience, and assessing the performance of each single module of the system. Rules, models, and algorithms are developed based on the huge amount of data collected over time.

The CMHS allows for more efficient scheduling since all MHMs are shared among all the SOs requiring a handling activity (compared to traditional MHS where, for example, equipment is limited to a specific area or specific material flow and unit loads). Simple examples of its benefits are MHMs idle while no SOs require them in the area can move to another area to serve other SOs and share their availability with the rest of the CMHS; delay in handling an SO due to over-utilizing of MHMs can be reduced using the other similar equipment available in other areas of the systems. The basic concept is based on a shared equipment, which allows reducing the total number of required modules as compared to a case with fixed assignments. Therefore, the average usage of the MHMs is higher, as is the service level and responsiveness of the system. Moreover, the introduction of AI and ML in the ICE allows for increased efficiency in every scheduled task thanks to the learning effect based on the huge amount of data collected over time. The big data available (about the travelled path, delivery time, travel time, service level, etc.) can provide useful feedback on the efficiency of the scheduling, enabling the AI and ML to improve future scheduling.

Scheduling in CMHS still faces many challenges. For example, it is still very important to consider how to manage the different interests and objectives of the different MHMs and SOs in the system. The scheduling could be based on a market mechanism for coordinating the activities of the MHMs that pursue their own interest. The market model could be used for solving dynamic job processing and scheduling problems. Conflicts between individual requirements could be resolved by negotiating and bargaining on simple terms such tasks, due dates, and prices (Márkus et al. 1996). New game theory, negotiation methodology, cooperation, and coordination among the MHMs are just some examples of new research areas for more efficient scheduling. Another example is how to define the optimal level of control between a fully centralized system with ICE as the main actor and a fully decentralized/holonic system where the MHMs and SOs act as individual intelligent agents. Several control models from other fields, such as cloud manufacturing, can be applied to support the scheduling.

5.5 Conclusions

Sharing real-time information on MHE and unit loads has enabled and changed the decision-making process for scheduling of material-handling activities from a hierarchically structured and centralized control system to a cloud-based one. With the introduction of the CMHS, the scheduling of material-handling activities can be optimized, leading to increasing flexibility and productivity of the overall manufacturing system. Current literature does not specify how production systems should be adapted from a material-handling perspective to enable cloud manufacturing. We contribute to the existing literature by introducing a conceptual model of CMHS and application at NTNU's Logistics 4.0 laboratory, reflecting a manufacturing environment. The current state of the IPTs and CMHS has been described. A special focus has been given to scheduling, highlighting the differences and benefits for future material-handling systems. Future research should investigate how AI and ML can further improve scheduling and consecutive activities of routing and dispatching, which have not been investigated until now.

References

Alcácer, V., & Cruz-Machado, V. (2019). Scanning the Industry 4.0: A literature review on technologies for manufacturing systems. *Engineering Science and Technology, an International Journal, 22*(3), 899–919. https://doi.org/10.1016/J.JESTCH.2019.01.006.

Battini, D., Calzavara, M., Persona, A., Sgarbossa, F., Visentin, V., & Zennaro, I. (2018). Integrating mocap system and immersive reality for efficient human-centred workstation design. *IFAC-PapersOnLine, 51*(11), 188–193.

Blanchard, B. S., & Fabrycky, W. J. (1990). *Systems engineering and analysis* (Pearson International Edition) (4th ed.). Englewood Cliffs, NJ: Prentice Hall. Retrieved from http://sutlib2.sut.ac.th/sut_contents/H104135.pdf.

Chen, L.-H., Wu, E. H.-K., Jin, M.-H., & Chen, G.-H. (2014). Intelligent fusion of Wi-Fi and inertial sensor-based positioning systems for indoor pedestrian navigation. *IEEE Sensors Journal, 14*(11), 4034–4042. https://doi.org/10.1109/JSEN.2014.2330573.

Curran, K., Furey, E., Lunney, T., Santos, J., Woods, D., & McCaughey, A. (2011). An evaluation of indoor location determination technologies. *Journal of Location Based Services, 5*(2), 61–78. https://doi.org/10.1080/17489725.2011.562927.

Deak, G., Curran, K., & Condell, J. (2012). A survey of active and passive indoor localisation systems. *Computer Communications, 35*(16), 1939–1954. https://doi.org/10.1016/J.COMCOM.2012.06.004.

Dolgui, A., Ivanov, D., Sethi, S. P., & Sokolov, B. (2019). Scheduling in production, supply chain and Industry 4.0 systems by optimal control. *International Journal of Production Research, 57*(2), 411–432.

Fernando, X. N., Krishnan, S., Sun, H., & Kazemi-Moud, K. (2003). Adaptive denoising at infrared wireless receivers. In B. F. Andresen & G. F. Fulop (Vol. 5074, p. 199). International Society for Optics and Photonics. https://www.spiedigitallibrary.org/conference-proceedings-of-spie/5074/0000/Adaptive-denoising-at-infrared-wireless-receivers/10.1117/12.486354.short

Fragapane, G., Ivanov, D., Peron, M., Sgarbossa, F., & Strandhagen, J. O. (2020). Increasing flexibility and productivity in industry 4.0 production networks with autonomous mobile robots and smart intralogistics. Annals of operations research, 1–19.

Furmans, K., & Gue, K. (2018). A framework for modeling material handling with decentralized control. In *Progress in Material Handling Research, January*. Retrieved from https://digitalcommons.georgiasouthern.edu/pmhr_2018/23.

Greis, N. P., Nogueira, M. L., Schmitz, T., & Dillon, M. (2019). Manufacturing-Uber: Intelligent operator assignment in a connected factory. In *Ninth IFAC Conference on Manufacturing Modelling, Management and Control, Berlin*. Retrieved from https://ifac.papercept.net/conferences/conferences/MIM19/program/MIM19_ContentListWeb_3.html.

Gu, Y., Lo, A., & Niemegeers, I. (2009). A survey of indoor positioning systems for wireless personal networks. *IEEE Communications Surveys & Tutorials, 11*(1), 13–32. https://doi.org/10.1109/SURV.2009.090103.

Han, X., & He, Z. 2018. A wireless fingerprint location method based on target tracking. In *2018 12th International Symposium on Antennas, Propagation and EM Theory (ISAPE)* (pp. 1–4). IEEE. https://doi.org/10.1109/ISAPE.2018.8634177.

Hightower, J., & Borriello, G. (2001). Location sensing techniques. Retrieved from http://kom.aau.dk/group/05gr999/reference_material/locationtechniques/hightower2001techniques.pdf.

Ivanov, D. (2020). Predicting the impacts of epidemic outbreaks on global supply chains: A simulation-based analysis on the coronavirus outbreak (COVID-19/SARS-CoV-2) case. *Transportation Research Part E: Logistics and Transportation Review, 136*, 101922.

Ivanov D., Dolgui A. (2020). Viability of intertwined supply networks: Extending the supply chain resilience an-gles towards survivability. A position paper motivated by COVID-19 outbreak. Accepted for Publication in International Journal of Production Research.

Ivanov D., Das A. (2020). Coronavirus (COVID-19 / SARS-CoV-2) and supply chain resilience: A research note. Accepted for Publication in International Journal of Integrated Supply Management.

Ivanov, D., Sokolov, B., Dolgui, A., Werner, F., & Ivanova, M. (2016). A dynamic model and an algorithm for short-term supply chain scheduling in the smart factory Industry 4.0. *International Journal of Production Research, 54*(2), 386–402.

Ivanov, D., Dolgui, A., & Sokolov, B. (2019). The impact of digital technology and Industry 4.0 on the ripple effect and supply chain risk analytics. *International Journal of Production Research, 57*(3), 829–846.

Kaemarungsi, K., & Krishnamurthy, P. (2004). Properties of indoor received signal strength for WLAN location fingerprinting. In *The First Annual International Conference on Mobile and Ubiquitous Systems: Networking and Services, 2004. MOBIQUITOUS 2004* (pp. 14–23). IEEE. https://doi.org/10.1109/MOBIQ.2004.1331706.

Khan, M., Kai, Y. D., & Gul, H. U. (2017). Indoor Wi-Fi positioning algorithm based on combination of location fingerprint and unscented Kalman filter. In *2017 14th International Bhurban Conference on Applied Sciences and Technology (IBCAST)* (pp. 693–698). IEEE. https://doi.org/10.1109/IBCAST.2017.7868127.

Lee, C. K. M., Yaqiong, L., Ng, K. K. H., Ho, W., & Choy, K. L. (2018). Design and application of internet of things-based warehouse management system for smart logistics. *International Journal of Production Research, 56*(8), 2753–2768. https://doi.org/10.1080/00207543.2017.1394592.

Liang, J. Z., Corso, N., Turner, E., & Zakhor, A. (2013). Image based localization in indoor environments. In *2013 Fourth International Conference on Computing for Geospatial Research and Application* (pp. 70–75). IEEE. https://doi.org/10.1109/COMGEO.2013.11.

Liu, Y., Wang, L., Wang, X. V., Xu, X., & Lin, Z. (2019). Scheduling in cloud manufacturing: State-of-the-art and research challenges. *International Journal of Production Research, 57*(15–16), 4854–4879. https://doi.org/10.1080/00207543.2018.1449978.

Mainetti, L., Patrono, L., & Sergi, I. (2014). A survey on indoor positioning systems. In *2014 22nd International Conference on Software, Telecommunications and Computer Networks (SoftCOM)* (pp. 111–120). IEEE. https://doi.org/10.1109/SOFTCOM.2014.7039067.

Márkus, A., Kis Váncza, T., & Monostori, L. (1996). A market approach to Holonic manufacturing. *CIRP Annals, 45*(1), 433–436. https://doi.org/10.1016/S0007-8506(07)63096-0.

Ni, L. M., Liu, Y., Lau, Y. C., & Patil, A. P. (2003). LANDMARC: Indoor location sensing using active RFID. In *Proceedings of the First IEEE International Conference on Pervasive Computing and Communications, 2003 (PerCom 2003)* (pp. 407–415). IEEE Computer Society. https://doi.org/10.1109/PERCOM.2003.1192765.

Panetto, H., Iung, B., Ivanov, D., Weichhart, G., & Wang, X. (2019). Challenges for the cyber-physical manufacturing enterprises of the future. *Annual Reviews in Control, 47*, 200–213.

Pinto, A. M., Moreira, A. P., & Costa, P. G. (2015). A localization method based on map-matching and particle swarm optimization. *Journal of Intelligent & Robotic Systems, 77*(2), 313–326. https://doi.org/10.1007/s10846-013-0009-2.

Regattieri, A., & Santarelli, G. (2013). Manufacturing logistics and packaging management using RFID. In *Radio frequency identification from system to applications*. Rijeka: InTech. https://doi.org/10.5772/53890.

Scholz-Reiter, B., & Freitag, M. (2007). Autonomous processes in assembly systems. *CIRP Annals, 56*(2), 712–729. https://doi.org/10.1016/J.CIRP.2007.10.002.

Tesoriero, R., Gallud, J., Lozano, M., & Penichet, V. (2008). Using active and passive RFID technology to support indoor location-aware systems. *IEEE Transactions on Consumer Electronics, 54*(2), 578–583. https://doi.org/10.1109/TCE.2008.4560133.

Tratnig, M., Reisinger, J., & Hlobil, H. (2007). A vision-based material tracking system for heavy plate rolling mills. In D. Fofi & F. Meriaudeau (Eds.) (Vol. 6356, p. 63560B). International Society for Optics and Photonics. https://www.spiedigitallibrary.org/conference-proceedings-of-spie/6356/63560B/A-vision-based-material-tracking-system-for-heavy-plate-rolling/10.1117/12.736728.short?SSO=1&tab=ArticleLink.

Woodman, O. J., & Harle, R. K. 2010. Concurrent scheduling in the active bat location system. In *2010 Eighth IEEE International Conference on Pervasive Computing and Communications Workshops (PERCOM Workshops)* (pp. 431–437). IEEE. https://doi.org/10.1109/PERCOMW.2010.5470631.

Wu, C. H., Ip, W. H., & Chan, C. Y. (2009). Real-time distributed vision-based network system for logistics applications. *International Journal of Intelligent Systems Technologies and Applications, 6*(3/4), 309. https://doi.org/10.1504/IJISTA.2009.024260.

Xu, X. (2012). From cloud computing to cloud manufacturing. *Robotics and Computer-Integrated Manufacturing, 28*(1), 75–86. https://doi.org/10.1016/J.RCIM.2011.07.002.

Zangaro, F., Finco, S., Battini, D., & Zennaro, I. (2019). An optimization model for the storage assignment of the reference under ergonomics constraints. In *Proceedings of the Summer School Francesco Turco, September 11–13, 2019, Brescia, IT.*

Zennaro, I., Finco, S., Battini, D., & Persona, A. (2019a). Big size highly customised product manufacturing systems: A literature review and future research agenda. *International Journal of Production Research, 57*(15–16), 5362–5385.

Zennaro, I., Calzavara, M., Finco, S., Battini, D., & Persona, A. (2019b). A new part feeding model for big size and highly customized product assembly systems. In *25th International Conference on Production Research Manufacturing Innovation: Cyber Physical Manufacturing, August 9–15, 2019, Chicago, US.*

Zhang, D., Xia, F., Yang, Z., Yao, L., & Zhao, W. 2010. Localization technologies for indoor human tracking. In *2010 Fifth International Conference on Future Information Technology* (pp. 1–6). IEEE. https://doi.org/10.1109/FUTURETECH.2010.5482731.

Zhang, L., Luo, Y., Tao, F., Li, B. H., Ren, L., Zhang, X., Guo, H., Cheng, Y., Hu, A., & Liu, Y. (2014). Cloud manufacturing: A new manufacturing paradigm. *Enterprise Information Systems, 8*(2), 167–187. https://doi.org/10.1080/17517575.2012.683812.

Zuin, S., Calzavara, M., Sgarbossa, F., & Persona, A. (2018). Ultra-wide band indoor positioning system: Analysis and testing of an IPS technology. *IFAC-PapersOnLine, 51*(11), 1488–1492. https://doi.org/10.1016/j.ifacol.2018.08.292.

Chapter 6
Coupling Robust Optimization and Model-Checking Techniques for Robust Scheduling in the Context of *Industry 4.0*

Pascale Marangé, David Lemoine, Alexis Aubry, Sara Himmiche, Sylvie Norre, Christelle Bloch, and Jean-François Pétin

Abstract This chapter presents a generic methodology when considering robustness in production systems of *Industry 4.0*. It is the first milestone for coupling Operations Research models for robust optimization and Discrete Event Systems models and tools for property checking. The idea is to iteratively call Operations Research and Discrete Event Systems Models for converging towards a solution with the required robustness level defined by the decision-maker.

6.1 Introduction

In recent years, a new type of industry is emerging that aims to be more adaptable, agile, and flexible. This industry called "Industry 4.0" promises to adapt to the personalized needs of customers, thanks to the integration and generalization of new Information and Communication Technologies (IoT, Big Data, RFID, Digital Twin, etc.) into the production system such that new features can emerge:

- dynamical adaptation to the high market volatility and the need for tailor-made product solutions.

P. Marangé (✉) · A. Aubry · S. Himmiche · J.-F. Pétin
Université de Lorraine, CRAN, CNRS 7039, Nancy, France
e-mail: Pascale.Marangé@univ-lorraine.fr

D. Lemoine
IMT Atlantique, LS2N UMR CNRS 6004, Nantes, France
e-mail: david.lemoine@imt-atlantique.fr

S. Norre
Université Clermont Auvergne, LIMOS UMR CNRS 6158, Aubière, France
e-mail: sylvie.norre@uca.fr

C. Bloch
Université Bourgogne Franche-Comt., FEMTO-ST Institute, CNRS 6174, Montbéliard, France
e-mail: christelle.bloch@univ-fcomte.fr

B. Sokolov et al. (eds.), *Scheduling in Industry 4.0 and Cloud Manufacturing*, International Series in Operations Research & Management Science 289,
https://doi.org/10.1007/978-3-030-43177-8_6

- Communication with other systems and their environment.
- Distributed intelligence: each component is able to sense and to decide.

However some enablers are needed to support the realization of this new paradigm (Panetto et al. 2019). In particular, mass customization in shorter and shorter delay leads to a difficulty in knowing the quantity and type of demand, the flow of products and their fluctuations and thus increases perturbations into the production model due to the great diversity of the manufactured products and their shortened life cycle. Perturbations can be categorized as follows: (1) Uncertainties: as the difference between predicted and actual information (uncertainties about the volume of demand, the duration of operations, etc.); (2) Hazards are defined by the occurrence of uncontrollable Event in production or in the environment (machine failure, urgent order, etc.).

To incorporate perturbations into the problem, different types of models exist in the literature. (Ierapetritou and Jia 2007) have listed the three common models for integrating perturbations into production models: delimited form or scenario description, probability description or stochastic models, and fuzzy modeling.

The main disadvantage of stochastic models was the need to have knowledge about historical data for identifying the right probability distribution and its parameters. However, *Industry 4.0* and integration of big data technologies promise to have access to data coming from the shop-floor such that this historical data and their analysis should help to build the right stochastic model of perturbations.

Such models can be integrated into classical Operations Research Models for Robust Optimization (Bertsimas and Sim 2004). The Operations Research models and associated solution tools are particularly efficient for tending towards the optimal solution despite the complexity of the problem. But the counterpart of this efficiency is often a dedicated static model whose price of adaptation when considering a new characteristic can be very high. By essence, production systems in Industry 4.0 will be highly dynamical and reconfigurable. Discrete Event Systems (DES) models and tools are particularly efficient to capture and model the dynamics of a system through the modeling of states and Event (Cassandras and Lafortune 2009).

The objective of this chapter is to present a generic methodology to assess the impact of perturbations into production systems in order to define a solution with a good balance between performance and robustness. This methodology is the first milestone for combining the advantages of robust optimization and Discrete Event Systems models and tools. The idea beside is to iteratively call robust optimization and Discrete Event Systems models for reaching the robustness level required by the decision-maker. This methodology is shown to be relevantly applied in the context of robust production scheduling when considering uncertainties on operation execution durations.

This chapter is built as follows: the first section presents the generic hybrid approach between Operations Research models for robust optimization and Discrete Event Systems models and tools for property verification. The second section presents the instantiation of this methodology to a scheduling problem under perturbations in a workshop with parallel machines. The third section illustrates and discusses the results on a use case. Finally the last section concludes the chapter by recalling the obtained results and by opening the discussion considering general considerations about perturbations and *Industry 4.0*.

6.2 A Hybrid Approach for Optimization Under Perturbations

In this section, we begin by introducing Linear Programming and robustness, then Discrete Event Systems concepts are also presented. We finish by describing the proposed methodology to deal with the robustness level wanted by the decision-maker for a solution, thanks to a combination of a robust linear programming approach and Discrete Event Systems Models.

6.2.1 Linear Programming and Robustness

Linear programming is one of the most powerful tools in Operations Research. It allows to model a wide variety of practical problems (particularly in logistics) and is often able to solve them to optimality. Among these logistic problems, we can quote scheduling, production planning, vehicle routing, time tabling, etc.

According to Papadimitriou and Steiglitz (1998), a Mixed Integer Linear Programming (MILP) can be expressed as:

$$\text{Maximize} \quad \sum_{j=1}^{n} c_j x_j \tag{6.1}$$

s.t.

$$\sum_{j=1}^{n} a_{ij} x_j \leq b_i \quad \forall i = 1, \cdots, m \tag{6.2}$$

$$x_j \in \mathbb{N} \quad \forall j = 1, \cdots, p \tag{6.3}$$

$$x_j \in \mathbb{R}_+, \quad \forall j = p+1, \cdots, n \tag{6.4}$$

where

- $(c_j)_{j=1\ldots,n}, (b_i)_{i=1\ldots,m}, (a_{ij})_{(i,j)=\{1\ldots,m\}\times\{1\ldots,n\}}$ are real variables which represent the problem's parameters (for instance, costs, distances, capacities, etc.).
- $X = (x_j)_{j=1\ldots,n}$ are the decision variables. They represent the solution we seek to determine.

- Function (6.1) is a linear form which represents the criterion we seek to optimize (in this case, maximize. For instance, it can be some logistics costs, customer's satisfaction, etc.).
- Equation (6.2) is a set of affine constraints that any solution of the modeled problem must satisfy (it describes the problem specificities).
- Equations (6.3) and (6.4) are integrity and positivity constraints.

For more information about linear programming, readers can also refer to Chvátal (1983), Wolsey (1998), and Nemhauser and Wolsey (1999).

Usually, when using such modeling, all parameters are assumed to be well known and deterministic. Nevertheless, this situation is very rarely encountered in real life. Therefore, solutions determined by this method may be unrealistic in practice. To avoid this, one possibility is to introduce uncertainties on parameters in order to better model reality and to try to find solutions able to absorb these perturbations without unreasonably degrading their quality. This kind of approach is usually referred to as robust (Billaut et al. 2013).

In linear programming, several robust approaches have been designed depending on the type of parameters on which the uncertainties fall on. Here, we focus on issues where uncertainties are related to (a_{ij}) parameters. More precisely, we assume that each parameter a_{ij} takes its values in a bounded interval $\left[\bar{a}_{ij} - \hat{a}_{ij}, \bar{a}_{ij} + \hat{a}_{ij}\right]$. That is to say that there is a random real variable ζ_{ij} which takes its values in $[-1, 1]$ such that

$$a_{ij} = \bar{a}_{ij} + \zeta_{ij}\hat{a}_{ij}$$

Thus, according to these assumptions, a MILP that takes into account these uncertainties can be formalized as follows:

$$\text{Maximize} \sum_{j=1}^{n} c_j x_j \tag{6.5}$$

s.t.

$$\sum_{j=1}^{n} \bar{a}_{ij} x_j + \sum_{j=1}^{n} \zeta_{ij} \hat{a}_{ij} x_j \leq b_i \quad \forall i = 1, \cdots, m \tag{6.6}$$

$$x_j \in \mathbb{N} \quad \forall j = 1, \cdots, p \tag{6.7}$$

$$x_j \in \mathbb{R}_+, \quad \forall j = p+1, \cdots, n \tag{6.8}$$

where $\sum_{j=1}^{n} \zeta_{ij} \hat{a}_{ij} x_j$ models the uncertainty in constraint (6.6).

The main idea of robust approaches presented in this chapter is to try to reasonably protect oneself from this uncertainty by taking into account the risk, thanks to a set of deterministic functions $\left(\beta_i^{\Omega_i}(x)\right)_{i=1,\cdots,m}$, where $(\Omega_i)_{i=1,\cdots,m}$ are parameters tuned in order to meet the degrees $\left(\Gamma_i^{ref}\right)_{i=1,\cdots,m}$ of protection the decision-maker wants to implement, depending on the criticality of the constraint.

In other words, $(\Omega_i)_{i=1,\cdots,m}$ have to be set up to be sure that the probability that the uncertainty does not exceed $\beta_i^{\Omega_i}(x)$ is greater or equal to Γ_i^{ref}, for $i = 1, \cdots, m$:

$$\mathbb{P}\left[\sum_{j=1}^{n} \zeta_{ij}\hat{a}_{ij}x_j \leq \beta_i^{\Omega_i}(x)\right] \geq \Gamma_i^{ref}, \quad \forall i = 1, \cdots, m \tag{6.9}$$

Thus, if one solution $(\Omega_i)_{i=1,\cdots,m}$ can be set up such that Eq. (6.9) is satisfied, solving the following optimization problem will ensure to have a solution $X = (x_j)_{j=1,\ldots,n}$ which can resist to uncertainty with degrees wanted by the decision-maker:

$$\text{Maximize} \sum_{j=1}^{n} c_j x_j \tag{6.10}$$

s.t.

$$\sum_{j=1}^{n} \bar{a}_{ij}x_j + \beta_i^{\Omega_i}(x) \leq b_i \quad \forall i = 1, \cdots, m \tag{6.11}$$

$$x_j \in \mathbb{N} \quad \forall j = 1, \cdots, p \tag{6.12}$$

$$x_j \in \mathbb{R}_+, \quad \forall j = p+1, \cdots, n \tag{6.13}$$

In Bertsimas and Sim (2004), the authors propose to use the following set of functions:

$$\beta_i^{\Omega_i}(x) = \max_{\sum_{j=1}^{n}|\zeta_{ij}| \leq \Omega_i} \left(\sum_{j=1}^{n} \zeta_{ij}\hat{a}_{ij}x_j\right) \tag{6.14}$$

and they prove that this non-linear formulation can be linearized and is equivalent to a MILP. Thus traditional Linear Programming technics can be used for solving the initial problem. This kind of MILP is called a Robust Linear Programming Model.

Nevertheless, tuning $\Omega = (\Omega_i)_{i=1,\cdots,m}$ for satisfying (6.9) can be very difficult. In Bertsimas and Sim (2004), the authors show that if for all i, each ζ_{ij} is independent and symmetrically distributed in $[-1, 1]$, Ω_i can be analytically determined. But, such a hypothesis is not often verified in real industrial problems.

6.2.2 Discrete Event Systems Models for Evaluating Solution Robustness

Industrial systems can be modeled by Discrete Event Systems (DES) that allow a representation of the behavior of a system by considering the state and Event that allow it to evolve. The event is seen as an instantaneous occurrence of an action or phenomenon in the system environment. Changes due to the event can be deterministic when the behavior is known with certainty or stochastic when the occurrence of an event can lead to different states. These modeling tools can be Petri Nets, State Automata, Statecharts, Bayesian Networks (Cassandras and Lafortune 2009).

To model the behavior of industrial systems and perturbations, we should be able to represent many dynamic characteristics such as the communication between elements of the workshop (jobs, resources), the time, and the probabilistic behavior of perturbations. Many stochastic Discrete Event Systems languages allow the modeling of these characteristics. For instance, Stochastic Petri Nets (Chiola et al. 1993), Stochastic Automata (Alur and Dill 1994), Stochastic Automata Networks (Plateau and Atif 1991).

The language chosen here is the Stochastic Timed Automata (STA). In fact, it is an extension of the well-known Timed Automata (Alur and Dill 1994) which is enriched with shared variables, synchronizing Event and probabilistic characteristics (Larsen et al. 1997).

Definition 6.1 Formally a Stochastic Timed Automaton is presented as the following n-tuple:
$A = (L, V, E, C, Inv, Pr, T, L_m, l_0, v_0)$ where

- L is a finite set of locations.
- V is a finite set of variables.
- E is a finite set of synchronizing Event
- C is a finite set of clocks.
- Inv is a set of invariants (conditions in location).
- Pr is a set of probabilities: (i) discrete for the set of transitions (from a location, probabilistic transitions allow to attend different locations l_i with a given probability p_i, with $\sum p_i = 1$). (ii) Continuous for the variables (the crossing condition of a transition is defined randomly by a probability distribution).
- T is a finite set of transitions $(l, e, g, m, l') \in L \times E \times G \times M \times L$, where l and l' are, respectively, the starting and arriving locations. On a transition, three optional elements are defined: (i) a guard (condition on variables) g from the set of guards G, (ii) an update (on variables) m from the set of updates M, and (iii) a synchronizing event e from the set E.
- $L_m \subseteq L$ is the set of marked locations.
- $l_0 \in L$ is the initial location of the automaton.
- v_0 is the initialization vector of variables.

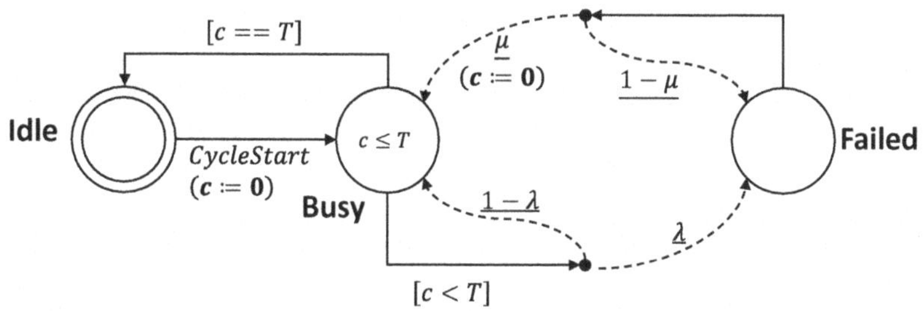

Fig. 6.1 A Stochastic Timed Automaton representing a machine

The elements of a STA can be graphically represented as follows (see example in Fig. 6.1). Locations are represented by vertices and transitions by arcs. An initial active location is represented by a double vertex. The invariants are represented inside the associated vertex (location). Guards are represented between brackets "[ ]". Synchronizing Event is represented in italics. The update of variables and clocks are represented between parenthesis "( )". Discrete probabilities are modeled by dotted arcs and associated probabilistic values are underlined. For continuous probabilities, they are directly linked with the definition of variables.

The automaton $MACHINE$ in Fig. 6.1 represents the behavior of a machine that can be subjected to failures. For this purpose, the machine can be represented by three states: Idle, Busy, Failed. Moreover, the failure rate is represented by λ and the repair rate by μ.

Initially, the MACHINE is in the location **Idle** waiting for the $CycleStart$ event. After the occurrence of this event, the local clock c is initialized ($c := 0$) and the MACHINE becomes **Busy**, i.e. is used for executing a cycle (that lasts normally T time units). Before T times units, a failure may occur with a probability λ (the MACHINE reaches the location **Failed**) or the machine continues its cycle with the probability $1 - \lambda$. In the location **Failed**, the MACHINE is repaired with the probability μ and, in this case, the cycle restarts to zero ($c := 0$). In this example, a failure has a big impact because the cycle restarts from zero after being repaired.

Discrete Event systems models can be used either to control, i.e. to inhibit certain state transitions to avoid unwanted behaviors, or to evaluate performance, i.e. to check properties such as the reachability of a state or the execution of an events sequence. This "Model-Checking" ability can be used for evaluating the impact of a perturbation on a system.

Properties that are traditionally desirable for an industrial system concern the reliability, maintainability, and safety. And DES models and tools are usually used for evaluating these properties. For instance, in Morel et al. (2009), Reliability is defined as *the ability of a device or system to perform a required function under stated conditions for a specified period of time*. This property is often measured by the probability $R(t)$ that a system will operate without failure before time t (depending on the failure rate λ), i.e. the probability that the Time To Failure TTF is greater than the time t:

$$R(t) = \mathbb{P}(TTF > t) \tag{6.15}$$

Now, we will show how DES models and tools can be used for evaluating reliability. In the example in Fig. 6.1, reliability can be the probability that the **Failed** state will be never reached before a cycle time T (meaning $TTF > T$, such that the failure does not happen during the cycle, if not the cycle has to restart from zero). For stochastic DES models, such a property can be expressed in PCTL (Probabilistic Computation Tree Logic). This language is a probabilistic extension of CTL (Computation Tree Logic) (Baier and Kwiatkowska 1998). This type of logic allows to express properties like "*What is the probability that the model is in the state* $Failed$*, in the precise interval [0,T]?*" This question can be transcribed in PCTL as in the expression (6.16).

$$P = ?[F \leq T \text{ ``}MACHINE.Failed''] \tag{6.16}$$

where P =? means that we want to assess the probability that the property that is inside the brackets [ ] is reached. This property can be translated as follows:

- $F \leq T$ means "There exists in the future in a time that is less or equal to T."
- "MACHINE.Failed" means "a state where the stochastic timed automata MACHINE is in the marked location Failed."

Finally the obtained result assesses $1 - R(T)$. So Model-Checking can be used for evaluating the reliability of a system. And we could do the same for the maintainability and safety.

The implementation of property verification is done by model-checking. The input of the model-checker is a system model and a property. At the output, the model-checker indicates whether the property is checked and, if not, a counter-example is returned (i.e., an example that shows that the property is not checked). In the case of stochastic system modeling, model-checking can be done numerically or statistically:

- Numerical model-checking uses accurate valuation methods to determine the probability value of a property. This type of model control ensures the accuracy of the given solution, but it is not suitable for large problems.
- Statistical model-checking generates different execution paths and verifies, after each execution, the satisfaction of a property. Statistical model-checking is similar to the Monte Carlo simulation. Monte Carlo simulation is a method for estimating a numerical quantity using random numbers. At each simulation step, the expectation of the variable is calculated and the simulation stops when the statistical parameters are satisfied. This avoids the combinatorial explosion and is therefore adapted to check real systems (Ballarini et al. 2011).

Actually, reliability can be seen as a robustness property. The notion of robustness has different definitions in literature that converge to the same idea: a robust system should maintain or guarantee some performances despite perturbations and variations generated by the system or its environment (Billaut et al. 2013).

When considering perturbations modeled as stochastic variables, the concept of "service level" can be used for assessing the robustness (Dauzères-Pérès et al. 2010). A service level $\mathcal{SL}$ is defined as *the probability that a criterion is smaller (resp. larger) or equal to a given value*. Thus, the assessed robustness level $\mathcal{SL}$ can be translated as the probability $\mathbb{P}$ that a system state z is lower than a value z_{max} (or larger than a value z_{min}) as in the following equation:

$$\mathcal{SL} = \mathbb{P}(z \leq z_{max}) \tag{6.17}$$

We can see that reliability falls within this definition. DES models and tools are thus good candidates for evaluating the robustness level of a system.

6.2.3 Proposed Methodology for Combining the Two Approaches

As said before, fine-tuning the $\Omega = (\Omega_i)_{i=1,\cdots,m}$ for satisfying (6.9) can be very difficult in general cases. Therefore, we propose a methodology for iteratively and numerically tuning Ω, thanks to Discrete Event Systems Models and associated Model-Checking tools. Figure 6.2 sketches the proposed approach.

Here, we suppose that:

- the problematic we want to solve and which relies on the system-of-interest can be formalized by a Mixed Integer Linear Programming model,
- the system-of-interest and its dynamic we want to study can be modeled, thanks to a Discrete Event Systems Model (DES),
- the decision-maker is able to define the different robustness indicators $\left(\Gamma_i^{ref}\right)_{i=1,\ldots,m}$ for each constraint i that must be satisfied by the solution X.

To determine the parameters $(\Omega_i)_{i=1,\cdots,m}$ that lead to the robustness level wanted by decision-makers, we proposed a methodology based on the following three iterative modules:

Module 1: ***The Operations Research Module (OR Module).*** According to the current value of $(\Omega_i)_{i=1,\cdots,m}$, the Robust Mixed Integer Programming Model using Bertsimas and Sim's framework is designed (Bertsimas and Sim 2004). Then, this is input into a Solver to get the optimal solution taking into account the robustness parameters. The obtained solution X is then sent to the second Module.

Module 2: ***The Discrete Event Systems Module (DES Module).*** Considering the system-of-interest and the solution proposed by the Operations Research module, a Stochastic Timed Automata model is first designed. Then, the different robustness levels Γ_i (as instantiations of the service level $\mathcal{SL}$) to be assessed are defined as properties to be checked on the resulting

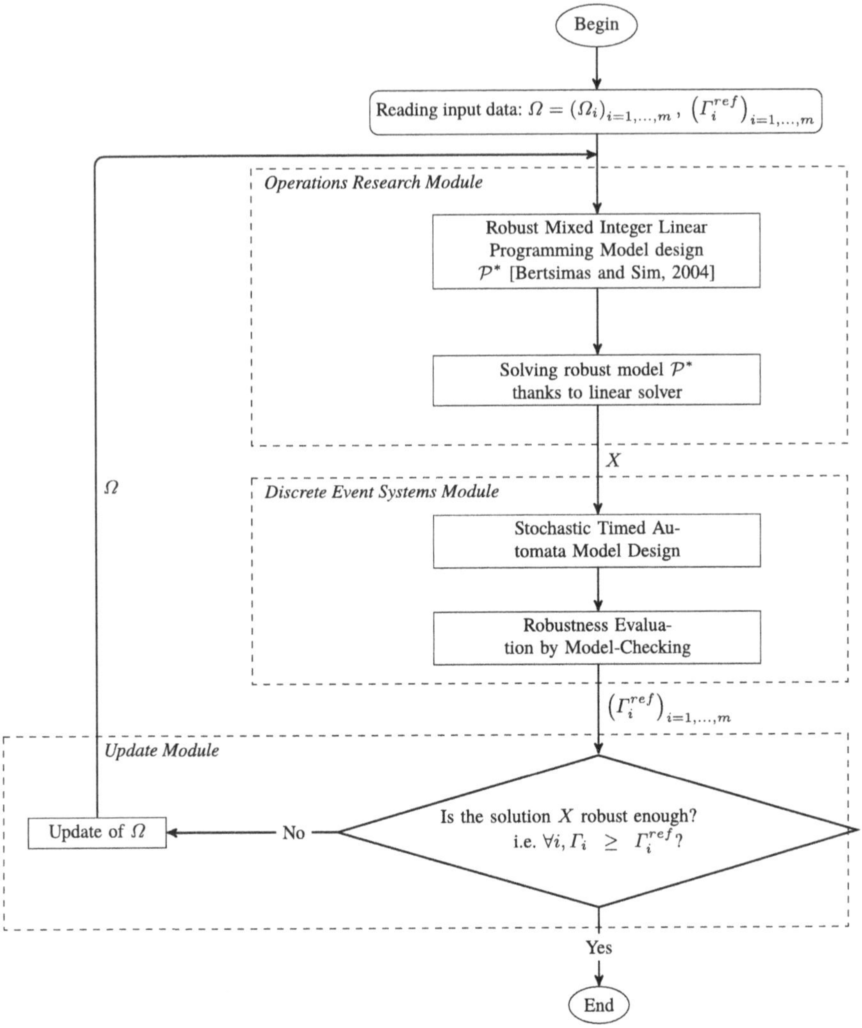

Fig. 6.2 The proposed methodology

model by a model-checker. The resulting $(\Gamma_i)_{i=1,\dots,m}$ are sent to the third module.

Module 3: ***Update Module***. Depending on the robustness levels $(\Gamma_i)_{i=1,\dots,m}$ assessed by the Discrete Event Systems Module, if the robustness levels required by the decision-maker are reached, then the process is stopped and the solution is given. Otherwise, $(\Omega_i)_{i=1,\cdots,m}$ are updated and sent back to the Operations Research Module for a new iteration.

6.3 Application to the Problem of Scheduling Under Perturbations

6.3.1 Scheduling Under Perturbations

The issue of production scheduling is an important decision-making problem in industrial processes. Actually, to guarantee the production performances, the decision-maker has to find an adapted schedule to its production system and the associated constraints. A production scheduling problem consists usually in (1) allocating the workshop resources to operations needed to make the jobs, (2) sequencing the operations on resources (defining the execution order of operations on resources), and (3) eventually defining the starting and ending dates of each operation. The schedule obtained should satisfy the workshop constraints (precedence constraints, non-preemption of operations, etc.). Indeed, each type of workshop has its own constraints in order to satisfy the production objective (like minimizing the total completion time of operations, number of late jobs, production cost, etc.).

Monostori et al. (2016) consider robust scheduling as one of the six main challenges in Research and Development for Cyber-Physical Production Systems. Others like Zhong et al. (2017) prefer to talk about a need of intelligent scheduling able to generate, from captured data, a reliable schedule in real time.

6.3.2 Instantiation of the Approach to Scheduling Under Perturbations

Here, we present an illustration of our methodology applied to a scheduling problem in a production area composed of two non-identical parallel machines: this means that the machines can perform the same operations but with different processing times. Then scheduling problem in this production cell involves both machine allocation and sequencing, rather than simply sequencing (Mokotoff 2001). Figure 6.3 shows the considered production cell.

In our case, we seek to minimize the completion time of the last scheduled job: this criterion is usually called the Makespan and is denoted as C_{max}. The main assumptions of our problem are the following:

- All jobs are available at time 0,
- The two machines are always available (no breakdown, ...),
- Processing times for the jobs are independents,
- A machine cannot process more than one product at any time.

This problem which is referred to as $R2||C_{max}$ has been shown to be NP-hard in the weak sense (Lenstra et al. 1977). Here, we also assume that processing times are not deterministic.

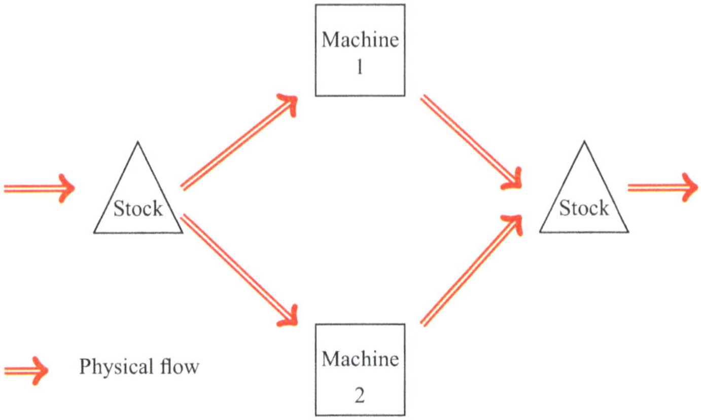

Fig. 6.3 The production area

6.3.2.1 Instantiation of the Methodology

The methodology presented in Fig. 6.2 can be instantiated as in Fig. 6.4. The MILP formulation of the problem $R2||C_{max}$ under uncertainties is presented in Sect. 6.3.2.2. The DES Module is presented in Sect. 6.3.2.3. The Update Module is presented in Sect. 6.3.2.4.

6.3.2.2 Operations Research Module

First, we give the MILP formulation of this scheduling problem when all the processing times are deterministics.

The parameters of the model are given in Table 6.1.

The decision variables are summarized in Table 6.2.

The $R2||C_{max}$ problem can be formulated as follows:

$$\text{Minimize } C_{max} \tag{6.18}$$

s.t.

$$\sum_{k=1}^{2} x_{jk} = 1 \quad \forall j \in \{1, \ldots, N\} \tag{6.19}$$

$$C_{max} - \sum_{j=1}^{N} t_{jk} x_{jk} \geq 0 \quad \forall k \in \{1, 2\} \tag{6.20}$$

$$C_{max} \geq 0 \tag{6.21}$$

$$x_{jk} \in \{0, 1\} \quad \forall (j, k) \in \{1, \ldots, N\} \times \{1, 2\} \tag{6.22}$$

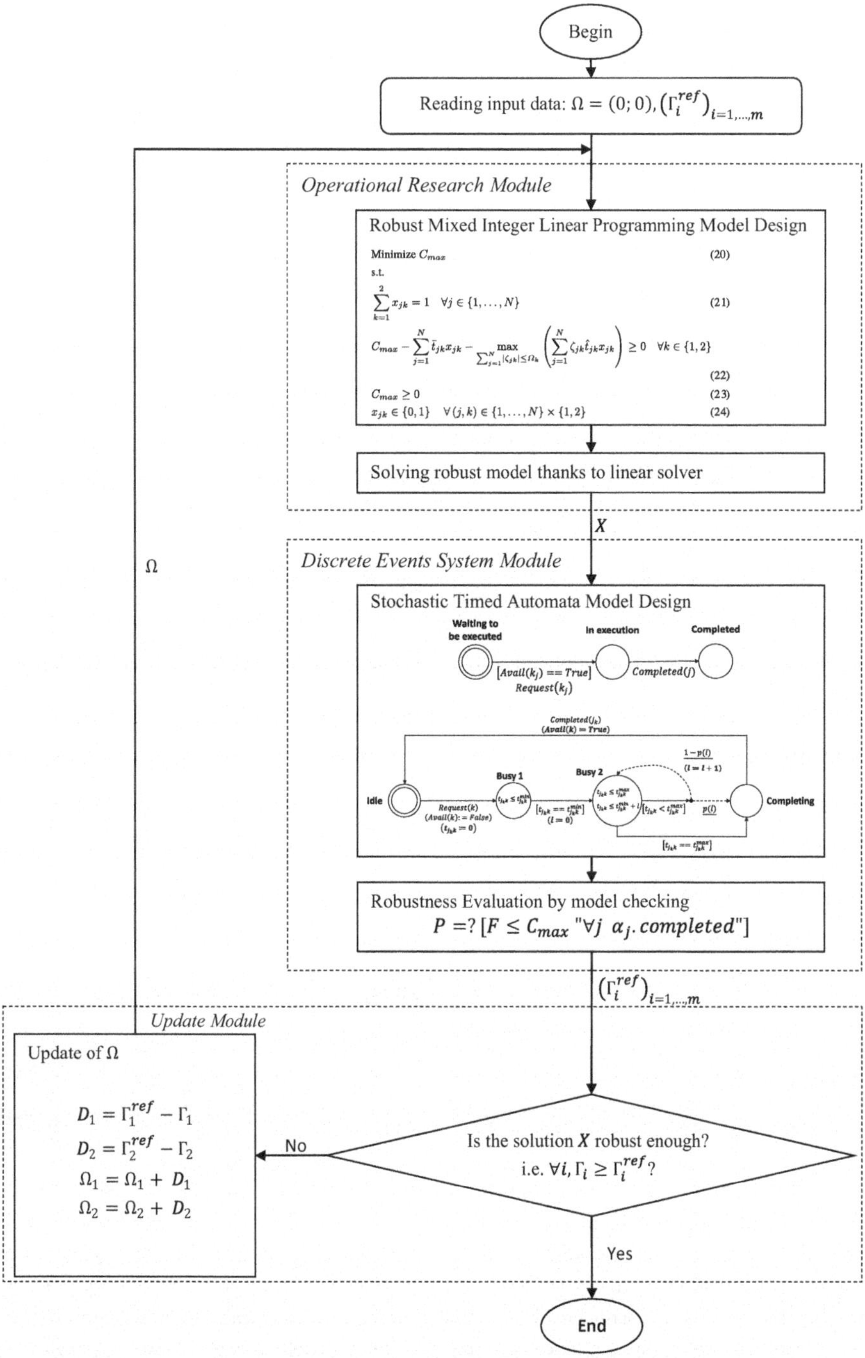

Fig. 6.4 Instantiated methodology to $R2||C_{max}$

Table 6.1 Parameters of the model

N: Number of jobs we have to schedule	
t_{jk}: Processing time for job j on the machine k, $(j,k) \in \{1,\dots,N\} \times \{1,2\}$	

Table 6.2 Decision variables of the model

$X = \left(x_{jk}\right)_{jk}$:	$\begin{cases} x_{jk} = 1 & \text{if machine } k \text{ is allocated to job } j \\ x_{jk} = 0 & \text{otherwise} \end{cases}$ $(j,k) \in \{1,\dots,N\} \times \{1,2\}$
C_{max}	: is the makespan value

Equation (6.18) is the objective function we seek to minimize. Equation (6.19) ensures that every job is executed by a single machine. Constraint (6.20) requires that total completion time C_{max} is higher than the completion time on each machine. Equations (6.21) and (6.22) are positivity and integrity constraints.

Now, we suppose that there are some uncertainties related to the jobs' processing times. As presented in the robustness section, there is a random variable ζ_{jk} which takes its values in $[-1, 1]$ such that

$$t_{jk} = \bar{t}_{jk} + \zeta_{jk}\hat{t}_{jk}$$

According to the Bertsimas and Sim (2004) approach, we can formulate the robust model as follows:

$$\text{Minimize } C_{max} \tag{6.23}$$

s.t.

$$\sum_{k=1}^{2} x_{jk} = 1 \quad \forall j \in \{1,\dots,N\} \tag{6.24}$$

$$C_{max} - \sum_{j=1}^{N} \bar{t}_{jk}x_{jk} - \max_{\sum_{j=1}^{N}|\zeta_{jk}| \le \Omega_k} \left(\sum_{j=1}^{N} \zeta_{jk}\hat{t}_{jk}x_{jk} \right) \ge 0 \quad \forall k \in \{1,2\} \tag{6.25}$$

$$C_{max} \ge 0 \tag{6.26}$$

$$x_{jk} \in \{0,1\} \quad \forall (j,k) \in \{1,\dots,N\} \times \{1,2\} \tag{6.27}$$

In this context, Ω_k represents the maximal deviation (using the $\mathcal{L}^1$-Norm) that is taking into account in the model for each machine k. If $\Omega_k = 0$, which means that no uncertainties are taken into account. In fact, the constraints (6.25) become equivalent to the constraints (6.20) and the robust formulation becomes equivalent to the deterministic formulation. On the contrary, if we want to consider all the uncertainties, Ω_k must be chosen as equal to N. If it is the case, the most conservative solution will be obtained. In fact, the constraints (6.25) become equivalent to the following:

$$C_{max} - \sum_{j=1}^{N} \bar{t}_{jk}x_{jk} - \sum_{j=1}^{N} \hat{t}_{jk}x_{jk} \geq 0, \ \forall k \in \{1, 2\} \tag{6.28}$$

Thus, this corresponds to the worst-case formulation: i.e. the most conservative, considering that the worst case (all the ζ_{jk} are equal to 1) is more important than the other cases.

Here, the idea is to fix Ω_k (as a "maximum amount of deviation" on the operation durations) but with guaranteeing that the desirable robustness levels Γ_k^{ref} are reached.

6.3.2.3 Discrete Event Systems Module

This section presents the DES formulation such that the allocation $X = \left[x_{jk}\right]_{jk}$ resulting from the solving of the robust MILP formulation given in the previous section can be evaluated regarding its robustness level and the result is sent to the Update Module for updating accordingly $(\Omega_k)_{k=1,2}$ (and a new iteration is launched) or not. The DES module contains two steps:

Step 1: Stochastic Timed Automata Model design: using STA for modeling the behavior of jobs and machines when executing the allocation X.

Step 2: Robustness evaluation by Model-Checking: evaluating the robustness levels Γ_k associated with each machine k.

Stochastic Timed Automata Model Design

First we propose to model the behavior of the jobs and the machines when they are not subjected to uncertainties. We define a job pattern that will be instantiated for each job j and a machine pattern that will be instantiated for each machine k.

In the following, we denote as k_j the machine that is allocated to the job j (defined by X coming from the Operations Research Module). Formally $k_j = \sum_k x_{jk}.k$.

The job pattern (named α_j) is presented in Fig. 6.5a. First, in the **Waiting to be executed** location, the job j waits the availability of its allocated machine k_j (through the guard $\left[Avail\left(k_j\right) == True\right]$). When the guard is satisfied, the job pattern sends a request to the machine pattern (by the synchronizing event $Request\left(k_j\right)$) and reaches the **In execution** location waiting for its completion (the reception of the synchronizing event $Completed\,(j)$). After its completion, the job reaches the **Completed** location.

The machine pattern is represented in Fig. 6.5b. In this model, the job that is going to be executed is denoted as j_k. First, in the **Idle** location, the machine waits a request from a job (the synchronizing event $Request\,(k)$) and then reaches the **Busy** location after updating its availability status ($Avail(k) := False$) and initializing the local clock $t_{j_k k}$ to 0. In the **Busy** location, the machine executes the job until the

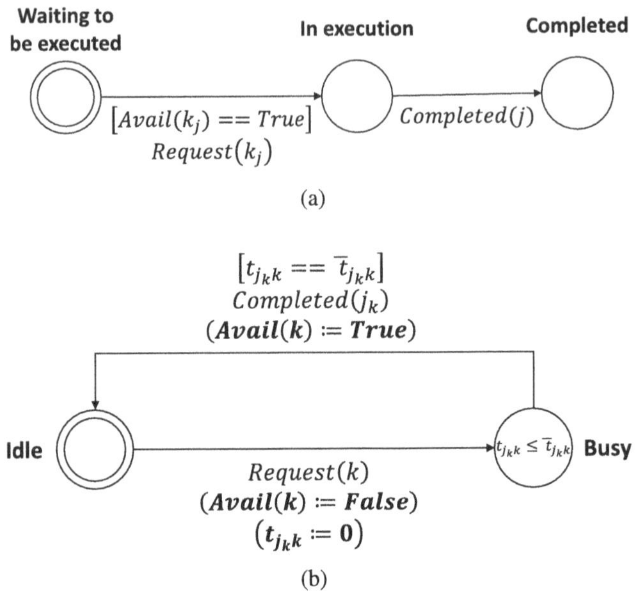

Fig. 6.5 STA models of job and machine without considering perturbations. (**a**) Job STA: α_j. (**b**) Deterministic machine STA

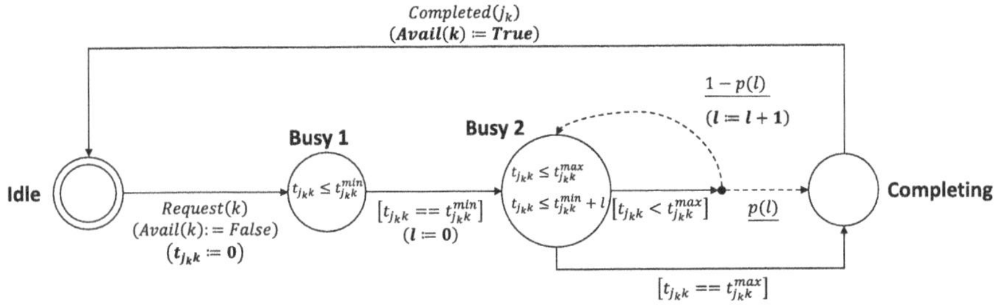

Fig. 6.6 Perturbed machine STA

local clock reaches the deterministic duration $\bar{t}_{j_k k}$. When the duration is reached, the job can be informed of its completion (by the synchronizing event $Completed\,(j_k)$) and the machine updates its availability status ($Avail(k) := True$). The machines then go back to the **Idle** location.

Now, we integrate the uncertainties on the job duration into the machine pattern. The resulting updated machine pattern is represented in Fig. 6.6. In the following, we denote the upper value of the job duration as $t_{j_k k}^{max} = \bar{t}_{j_k k} + \hat{t}_{j_k k}$ and the lower value of the job duration as $t_{j_k k}^{min} = \bar{t}_{j_k k} - \hat{t}_{j_k k}$.

In the **Busy 1** location, the machine waits to reach the minimal duration $t_{j_k k}^{min}$ (through the guard $\left[t_{j_k k} == t_{j_k k}^{min}\right]$). Moreover, the iteration counter l is initiated to

0. The idea is to let the duration increase according to a discrete probability $p(l)$ that is evolving depending on the iterations number l. In the **Busy 2** location, if the maximal duration is reached ($\left[t_{j_k k} == t_{jk}^{max}\right]$), then the machine reaches the **completing** location. If it is not the case ($\left[t_{j_k k} < t_{jk}^{max}\right]$), there are two possible probabilistic choices: (1) with the probability $1 - p(l)$, the duration can increase and the iteration counter is updated ($l := l + 1$) or (2) with the probability $p(l)$, the current duration is the final duration.

Finally, the probability that $t_{j_k k} = t_{j_k k}^{min} + l$ is the probability to loop into the **Busy 2** location $l - 1$ times and to get out from the loop in the lth iteration.

Actually, $p(l)$ is a probabilistic parameter that can be calculated from the probability distribution followed by $t_{j_k k}$.

Modeling the execution of the job as previously presented allows to not be restricted to any kind of probability distribution (symmetric or not, discrete or not, etc.). We could even imagine to cut the interval of the job duration in several sub-intervals in which the probability distributions could be different. That makes this approach a good complement to the robust linear programming of the Operations Research module.

Robustness Evaluation by Model-Checking

In the second step, model-checking tools are used to assess the robustness level of X.

In a scheduling problem, we can instantiate the service level presented in Eq. (6.17) as follows: z is the total completion time despite the considered uncertainties and z_{max} is the referential completion time C_{max} associated with X given by the Operations Research module. So we define the robustness level as the probability that the executed makespan is smaller or equal than the referential completion time C_{max}. Formally, this metric is given by Eq. (6.29):

$$\mathcal{SL} = \mathbb{P}\left(C_{max}\left(X, U\right) \leq C_{max}\right) \tag{6.29}$$

where $C_{max}\left(X, U\right)$ is the executed makespan of an allocation X subjected to uncertainties U.

So to assess the value of $\mathcal{SL}$ using DES models and associated Model-Checking, the property to check is: "*What is the probability that all the paths lead to a global state where all the job models α_j are in the marked location **Completed** in a time that is less or equal to C_{max}?*"
Using PCTL, this property can be expressed as follows:

$$P =? \left[F \leq C_{max} \text{``}\forall j\ \alpha_j.Completed''\right] \tag{6.30}$$

where $P = ?$ means that we want to assess the probability that the property that is inside the brackets [ ] is reached. This property can be translated as follows:

- "$F \leq C_{max}$" means "There exists in the future in a time that is less or equal to C_{max}."
- "$\forall j\ \alpha_j.Completed$" means "a state where, for all j, all the stochastic timed automata α_j are in the marked location **Completed**."

That means that the formula $\left[F \leq C_{max}\ \text{“}\forall j\ \alpha_j.Completed\text{”}\right]$ is a PCTL expression for: $C_{max}(X, U) \leq C_{max}$.

Here, two robustness levels associated, respectively, with each machine can be defined. They consist to consider only the uncertainties are only taken into account on machine 1 or machine 2. So, we can evaluate which machine is more sensitive than the other. These two robustness levels are defined as follows:

$$\begin{aligned} \Gamma_1 &= \mathbb{P}\left(C_{max}\left(X, \left(\hat{t}_{j1}\right)_{j1}\right) \leq C_{max}\right) \\ \Gamma_2 &= \mathbb{P}\left(C_{max}\left(X, \left(\hat{t}_{j2}\right)_{j2}\right) \leq C_{max}\right) \end{aligned} \tag{6.31}$$

where $\left(\hat{t}_{j1}\right)_{j1}$ (resp. $\left(\hat{t}_{j2}\right)_{j2}$) are the uncertainties on the operation durations when considering that there are no uncertainties on the machine 2 (resp. 1). Finally, these robustness levels assess whether the inequation (6.9) is satisfied or not. This result is used in the Update Module for updating or not Ω.

Moreover, we are able to evaluate a general robustness level considering the global uncertainties $\left(\hat{t}_{jk}\right)_{jk}$ as follows:

$$\Gamma = \mathbb{P}\left(C_{max}\left(X, \left(\hat{t}_{jk}\right)_{jk}\right) \leq C_{max}\right) \tag{6.32}$$

As the two machines are independent, we have $\Gamma = \Gamma_1 \times \Gamma_2$.

6.3.2.4 Update Module

Update of Ω

Here we assumed that the decision-maker is able to fix a robustness level Γ^{ref} he would like to be achieved by the system. This robustness level assessed the minimal acceptable probability that the executed makespan is smaller than the reference makespan C_{max}. Moreover, we considered that:

- the machine are independent: $\Gamma^{ref} = \Gamma_1^{ref} \times \Gamma_2^{ref}$
- the contributions of each machine to the global robustness level are equivalent (no machine is more critical than the other).

Thus, Γ_k^{ref} (defined in the inequation (6.9)) can be fixed as follows:

$$\forall k \in \{1, 2\}, \Gamma_k^{ref} = \sqrt{\Gamma^{ref}}$$

Following the assessment of Γ, Γ_1, and Γ_2 by the Discrete Event Systems Module, the following algebraic distances to the required minimal robustness level Γ^{ref}, Γ_1^{ref}, and Γ_2^{ref} can thus be calculated:

$$D = \Gamma^{ref} - \Gamma \tag{6.33}$$

$$D1 = \Gamma_1^{ref} - \Gamma_1 \tag{6.34}$$

$$D2 = \Gamma_2^{ref} - \Gamma_2 \tag{6.35}$$

If $D1 > 0$ or $D_2 > 0$, which means that the required robustness levels are not reached and the parameters Ω_1 and Ω_2 have to be updated. We propose to do it as follows:

$$\Omega_1 = \Omega_1 + D_1 \tag{6.36}$$

$$\Omega_2 = \Omega_2 + D_2 \tag{6.37}$$

We can note that these update formulas are arbitrarily defined. However, they express the fact that the further away from the objective (the bigger D_k), the more the parameters Ω_k must be amplified.

6.4 Application

In the application, 10 jobs are considered, with execution times having the uncertainties defined in Table 6.3. Moreover, Γ^{ref} is fixed to 0.90: meaning that the probability that the executed makespan will be effectively less than or equal to the optimal value is at least equal to 0.90.

Table 6.3 Characteristics of jobs

	t_{1k}^{min}	t_{1k}^{max}	t_{2k}^{min}	t_{2k}^{max}	t_{3k}^{min}	t_{3k}^{max}	t_{4k}^{min}	t_{4k}^{max}	t_{5k}^{min}	t_{5k}^{max}
machine k = 1	1	1	1	3	2	8	1	5	3	13
machine k = 2	1	3	1	1	1	3	2	6	3	9
	t_{6k}^{min}	t_{6k}^{max}	t_{7k}^{min}	t_{7k}^{max}	t_{8k}^{min}	t_{8k}^{max}	t_{9k}^{min}	t_{9k}^{max}	t_{10k}^{min}	p_{10k}^{max}
machine k = 1	1	3	2	6	1	3	3	5	1	1
machine k = 2	4	6	1	5	1	7	1	3	1	1

Table 6.4 Iterations for the application

Iteration k	Input Ω	OR module		DES module		Update module
		Solution	C_{max}	$\{\Gamma, \Gamma_1, \Gamma_2\}$	$\{D, D_1, D_2\}$	Output Ω
0	$[0, 0]$	X_1	12	$\{0.65, 0.85, 0.77\}$	$\{0.25, 0.10, 0.18\}$	$[0.10, 0.18]$
1	$[0.10, 0.18]$	X_1	12.54	$\{0.65, 0.85, 0.77\}$	$\{0.25, 0.10, 0.18\}$	$[0.21, 0.35]$
2	$[0.21, 0.35]$	X_1	13.05	$\{0.80, 0.93, 0.88\}$	$\{0.10, 0.02, 0.07\}$	$[0.23, 0.42]$
3	$[0.23, 0.42]$	X_1	13.26	$\{0.80, 0.93, 0.88\}$	$\{0.10, 0.02, 0.07\}$	$[0.25, 0.49]$
4	$[0.25, 0.49]$	X_1	13.47	$\{0.80, 0.93, 0.88\}$	$\{0.10, 0.02, 0.07\}$	$[0.27, 0.56]$
5	$[0.27, 0.56]$	X_1	13.54	$\{0.80, 0.93, 0.88\}$	$\{0.10, 0.02, 0.07\}$	$[0.29, 0.63]$
6	$[0.29, 0.63]$	X_2	13.58	$\{0.80, 0.85, 0.95\}$	$\{0.10, 0.10, 0.00\}$	$[0.39, 0.63]$
7	$[0.39, 0.63]$	X_2	13.78	$\{0.80, 0.85, 0.95\}$	$\{0.10, 0.10, 0.00\}$	$[0.49, 0.63]$
8	$[0.49, 0.63]$	X_1	13.89	$\{0.80, 0.92, 0.87\}$	$\{0.10, 0.03, 0.07\}$	$[0.51, 0.70]$
9	$[0.51, 0.70]$	X_2	14.02	$\{0.90, 0.93, 0.99\}$	$\{0.00, 0.02, -0.04\}$	$[0.53, 0.66]$
10	$[0.53, 0.66]$	X_1	13.98	$\{0.92, 0.96, 0.95\}$	$\{-0.02, -0.01, 0.00\}$	$\emptyset$

Table 6.4 gives the different iterations of the combined approach. We started with $\Omega = [0, 0]$ (meaning that no uncertainty is considered). Two solutions are explored during different iterations. The solution X_1 allocates the first machine to jobs 1, 4, 6, 7, 8 and the second machine to jobs 2, 3, 5, 9, 10. The solution X_2 allocates the first machine to jobs 1, 4, 6, 7, 8, 10 and the second machine to jobs 2, 3, 5, 9.

This application shows that combining the two approaches allows to converge to a solution with a good robustness level without degrading too much the makespan. Initially (without perturbations), the makespan was of 12 and the associated robustness level was of 0.65. If the decision-maker accepts to degrade this makespan of around 20% (increasing the makespan to 14), then the robustness level reaches 0.90. Moreover, this approach is a good means for tuning the Ω parameters even if the probability distribution associated with the uncertainties is not symmetrical. We can note here that the makespan for a robustness level of 1 is $C_{max} = 18$ (namely the most conservative solution). Thus, the couple ($C_{max} = 13.98$, $\Gamma = 0.9$) is a good compromise between optimality and robustness.

6.5 Conclusions

Among the major issues related to *Industry 4.0*, risk management and, consequently, robust decision support play a significant role in the concerns of decision-makers.

In order to provide an efficient answer to this problem, we have proposed a generic method combining robust mathematical programming and Discrete Event Systems models. This allows to reach the level of robustness desired by the decision-maker by finely assessing the degree of robustness of the solutions provided by the optimization module, regardless of the probability distributions that follow the uncertainties on the model input data. We have illustrated the latter on the case of a scheduling problem with parallel machines.

However, as far as our methodology is generic, it will have to be adapted to the context of use. In particular, the mechanism for updating robustness coefficients Ω can be designed more efficiently to increase the rate of convergence of the methodology towards a solution with the required robustness level. In addition, instead of considering an equidistribution of the levels of robustness to be obtained over all the constraints of the model, a more specific distribution of these can be considered, taking into account, for example, the configuration of the production system, the criticality of certain machines (for instance, requiring greater robustness for bottleneck machines, etc.).

With the development of *Industry 4.0* and, more particularly, the increasing use of digital twins, these hybridization between decision support and performance evaluation models are likely to develop. The advent of Big Data and its consequences in terms of model calibration (and in particular through a more realistic estimation of probability laws modeling data uncertainties) combined with ever-increasing computing power will make it possible to implement this type of methodology in decision support tools in an industrial context.

Acknowledgments This chapter is the result of a collaboration between the two working groups Bermudes[1] and SED[2] from the French National Centre for Scientific Research (CNRS).

The Bermudes group has been created in 1996. It is labeled by the GDR RO[3] and the GDR MACS [4], which are two research groups depending on the French CNRS. Focused at the beginning on classical scheduling issues such as classical scheduling workshop problems (Job Shop, Flow Shop, Generalized Job Shop, Hybrid Flow Shop, etc.), scheduling problems in manufacturing systems (Flexible Manufacturing Systems, Hoist Scheduling Problems), and Resource Constrained Project Scheduling Problems, its research topics have evolved following Industry 4.0 context and include now integrated scheduling problems taking into account both several related activities (maintenance, transport, etc.) and different constraints (environmental, human, energetic, etc.). Determination of robust or reactive scheduling has become an important issue.

The Discrete Event Systems Working Group (DES) has been created in 2014. It is a French working group from the GDR MACS. Its objectives are to promote exchanges between the various specialists, whether they come from the world of automation, computer science, or mathematics, and thus to provide a better knowledge of the problems related to DES and the solutions that can be provided. The topics covered include (1) the study of the syntax and semantics of DES formalisms; (2) the application of these formalisms for modeling based on specifications, system performance analysis, simulation, property verification, system control, supervision, observation, detection, diagnosis, decision support, architecture selection, reconfiguration, etc.

[1] http://www.gt-bermudes.fr/.

[2] https://sites.google.com/site/gtsedmacs/.

[3] http://gdrro.lip6.fr/.

[4] https://gdr-macs.cnrs.fr/.

References

Alur, R., & Dill, D. L. (1994). A theory of timed automata. *Theoretical Computer Science, 126*(2), 183–235.

Baier, C., & Kwiatkowska, M. (1998). Model checking for a probabilistic branching time logic with fairness. *Distributed Computing, 11*(3), 125–155.

Ballarini, P., Djafri, H., Duflot, M., Haddad, S., & Pekergin, N. (2011). Petri nets compositional modeling and verification of Flexible Manufacturing Systems. In *IEEE International Conference on Automation Science and Engineering*, pp. 588–593. New York: IEEE.

Bertsimas, D., & Sim, M. (2004). The price of robustness. *Operations Research, 52*(1), 35–53.

Billaut, J.-C., Moukrim, A., & Sanlaville, E. (2013). *Flexibility and robustness in scheduling*. London: Wiley.

Cassandras, C. G., & Lafortune, S. (2009). *Introduction to discrete event systems*. New York: Springer Science & Business Media.

Chiola, G., Marsan, M. A., Balbo, G., & Conte, G. (1993). Generalized stochastic petri nets: A definition at the net level and its implications. *IEEE Transactions on Software Engineering, 19*(2), 89–107.

Chvátal, V. (1983). *Linear programming. A series of books in the mathematical sciences*. New York: W. H. Freeman.

Dauzères-Pérès, S., Castagliola, P., & Lahlou, C. (2010). Service level in scheduling. In J.-C. Billaut, A. Moukrim, & E. Sanlaville (Eds.), *Flexibility and robustness in scheduling* (chapter 5, pp. 99–121). Hoboken: Wiley-ISTE.

Ierapetritou, M. G., & Jia, Z. (2007). Short-term scheduling of chemical process including uncertainty. *Control Engineering Practice, 15*(10), 1207–1221. Special Issue - International Symposium on Advanced Control of Chemical Processes (ADCHEM).

Larsen, K. G., Pettersson, P., & Yi, W. (1997). Uppaal in a nutshell. *International Journal on Software Tools for Technology Transfer, 1*(1–2), 134–152.

Lenstra, J., Kan, A. R., & Brucker, P. (1977). Complexity of machine scheduling problems. In P. Hammer, E. Johnson, B. Korte, & G. Nemhauser (Eds.), *Studies in integer programming. Annals of discrete mathematics* (Vol. 1, pp. 343–362). Amsterdam: Elsevier.

Mokotoff, E. (2001). Parallel machine scheduling problems: A survey. *Asia-Pacific Journal of Operational Research, 18*(2), pp. 193–242.

Monostori, L., Kádár, T., Bauerhansl, T., Kondoh, S., Kumara, S., Reinhart, G., et al. (2016). Cyber-physical systems in manufacturing. *CIRP Annals, 65*(2), 621–641.

Morel, G., Pétin, J.-F., & Johnson, T. L. (2009). *Reliability, maintainability, and safety* (pp. 735–747). Berlin/Heidelberg: Springer.

Nemhauser, G. L., & Wolsey, L. A. (1999). *Integer and combinatorial optimization. Wiley-Interscience series in discrete mathematics and optimization*. New York, NY: Wiley.

Panetto, H., Iung, B., Ivanov, D., Weichhart, G., & Wang, X. (2019). Challenges for the cyber-physical manufacturing enterprises of the future. *Annual Reviews in Control, 47*, 200–213.

Papadimitriou, C. H., & Steiglitz, K. (1998). *Combinatorial optimization: Algorithms and complexity*. Mineola, NY: Dover Publications.

Plateau, B., & Atif, K. (1991). Stochastic automata network of modeling parallel systems. *IEEE Transactions on Software Engineering* **17**(10), 1093–1108.

Wolsey, L. A. (1998). *Integer programming. Wiley-Interscience series in discrete mathematics and optimization*. New York: Wiley.

Zhong, R. Y., Xu, X., Klotz, E., & Newman, S. T. (2017). Intelligent manufacturing in the context of industry 4.0: A review. *Engineering, 3*(5), 616–630.

Chapter 7
Integrated Scheduling of Information Services and Logistics Flows in the Omnichannel System

Dmitry Ivanov and Boris Sokolov

Abstract This chapter develops a model for dynamic integrated scheduling of information services and logistics flows in the omnichannel system. The proposed service-oriented description makes it possible to coordinate the information services and material process schedules simultaneously. It also becomes possible to determine the volume of information services needed for physical supply processes. In addition, impact of disruptions in information services on the schedule execution in the physical structure is analyzed. The results provide a base for information service scheduling according to actual physical process execution.

7.1 Background

Nowadays, companies start adopting the decentralized distributed information services (ISs). One of these concepts is omnichannel that is commonly understood as a multichannel promotion actions (in-store, social media, and mobile applications) to improve the customer experience (Ailawadi and Farris 2017). Combination of traditional retail stores and online sales is the core idea in the omnichannel concept.

Omnichannel implementation in practice is challenged by cross-channel logistics coordination, the resulting increase in coordination complexity. In addition, it requires extensions to traditional functionality in enterprise resource planning (ERP) and warehouse management system (WMS) systems (Pagani and Pardo 2017). At the same time, omnichannel is expected to increase the reaction flexibility to demand fluctuations as well as to positively influence the lead times and capacity utilization.

D. Ivanov (✉)
Berlin School of Economics and Law, Berlin, Germany
e-mail: divanov@hwr-berlin.de

B. Sokolov
Saint Petersburg Institute for Informatics and Automation of the RAS (SPIIRAS), St. Petersburg, Russia
e-mail: sokol@iias.spb.su

B. Sokolov et al. (eds.), *Scheduling in Industry 4.0 and Cloud Manufacturing*,
International Series in Operations Research & Management Science 289,
https://doi.org/10.1007/978-3-030-43177-8_7

Moreover, the literature on supply chain resilience and first analyses of COVID-19 pandemic impacts on the supply chains and production systems show that firms with omnichannel distribution systems were able to sustain the disruptions along with the ripple effect control (Lee et al. 2020; Ivanov and Dolgui 2019; Hosseini et al. 2019; Ivanov et al. 2019; Panetto et al. 2019; Ivanov 2020; Ivanov and Dolgui 2020a, b; Ivanov and Das 2020).

Omnichannel concept is based on the IS usage. Due to the increasing role of IS in different forms, for example, cloud computing requires service-based approaches to integrated scheduling of both material and information flows (Bardhan et al. 2010; Li et al. 2010). The impact of information technology (IT) on the material processes in supply network (SN) became crucial in recent years (Choi et al. 2002; Giard and Mendy 2008; Camarinha-Matos and Macedo 2010; Cannella et al. 2014). Recent research indicated that an aligning of business processes and information systems may potentially provide new quality of decision-making support and an increased performance (Surana et al. 2005; Dedrick et al. 2008; Jain et al. 2009; Ivanov et al. 2014).

Most of the new IT share attributes of intelligence. Examples include data mining, cloud computing, physical internet, pattern recognition, knowledge discovery, to name a few. In addition, the beginning era of Internet of Things and explicit inclusion of wireless sensor networks, machine-to-machine systems, and mobile apps into the management require the data-driven business models instead of static information architectures. Elements of physical processes are supported by information services. In addition, such systems evolve through adaptation and reconfiguration of their structures, that is, through *structural dynamics* (Ivanov et al. 2004, 2007, 2010, 2015; Ivanov and Sokolov 2012a, b, 2013). Such SCs are common not only in manufacturing but also in different cyber-physical systems, for example, in networks of emergency response units, energy supply, city traffic control, and security control systems.

It can be observed that current concepts and models for schedule integration do not provide adequate decision support from intelligent IS; we regard this shortcoming as an opportunity for research and development, which could significantly improve the practice of logistics management. On one hand, the alignment of new intelligent elements of IS infrastructures with real material flows can be achieved. On the other hand, investments into IS can be estimated regarding real schedules.

This chapter faces these two decision domains on the basis of structural dynamics control approach that is built upon tools from optimal program control theory (Ivanov et al. 2005, 2010). Although recent research has extensively dealt with supply chain (SC) scheduling (Chen 2010) and IS scheduling (see, e.g., works on scheduling in telecommunications) in isolation, the integrated scheduling of both material and information flows still represents a research gap (Dolgui et al. 2019; Ivanov et al. 2016, 2018, 2020; Panetto et al. 2019).

In this chapter, the problem of coordinated dynamic scheduling of IS and material flows in the context of cyber-physical systems is stated and solved with the help of optimal control approach. In addition, specific research contributions are the

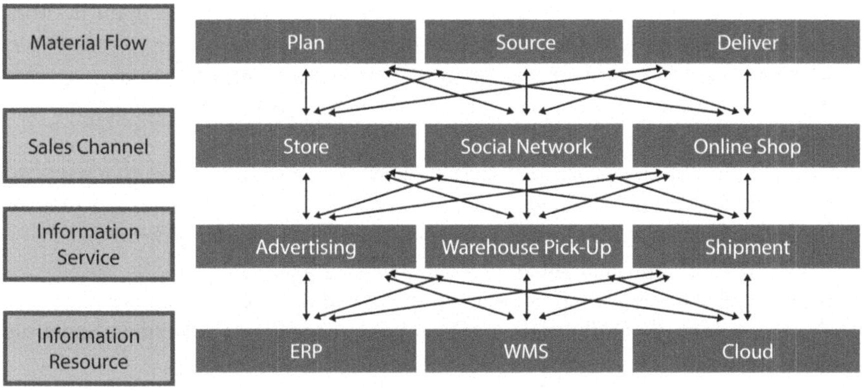

Fig. 7.1 Interrelations among material flows, sales channels, IS, functions, and IR

considerations of IS reconfiguration in a real execution stage and monetary estimation of investments into IS.

7.2 Problem Statement

Consider a simple example of the interrelations among physical processes, ISs, information functions, and information resources (IRs) that is presented in Fig. 7.1.

Such a framework is based on recent developments in cloud computing, see for example, studies by Wang et al. (2010) and Jiang et al. (2012). For material flow scheduling and control, some ISs are needed. They should be available when material flow is scheduled and executed. The ISs are provided by some distributed IRs which may be subject to full or partial unavailability due to planned upgrades or unpredicted disruptions. Therefore, such network needs to be considered as a dynamic system (Ivanov & Sokolov 2010; Ivanov et al. 2012).

7.2.1 General Assumptions

Let us define a formal scheduling problem for this framework. The majority of the technical part of this chapter is based upon the study by Ivanov et al. (2014).

- The jobs in material flows are independent and available for processing at time zero. Each of the jobs has a release date that is known in advance through the schedule coordination.
- Precedence constraints exist, that is, the operations are logically arranged in jobs.

- The material flow operations are executed at one of the enterprises in the network and are supported by ISs from different IRs.
- Machines and IRs have unequal information processing rates which may also differ for various operations and therefore influence the processing time and processing volume.
- Each IS may be composed of functions from different IRs and is characterized by availability time windows, productivity, that is, the processed volume of operations at an instant of time, and costs (fixed cost and operation cost).
- Setup times are independent and included in the processing time.
- Initial state and the desired end state of the dynamic system are known.
- Transition from the initial state to the end state depends on selection of controls in material flow, IS, and IR reconfiguration scheduling models.

7.2.2 Notations

- Denote $A = \{A_\nu; \nu = 1, \ldots, n\}$ as jobs in a material flow.
- Each of the jobs A_ν is composed of the operations $D^{(\nu)} = \left\{D_i^{(\nu)}; i = 1, \ldots, k_\nu\right\}$.
- a_i is the planned processing volume (e.g., lot-size) of the operation $D_i^{(\nu)}$.
- Consider a set of enterprises (machines) $B = \{B_j; j = 1, \ldots, m\}$.
- Denote $B^{(\nu,i)} = \left\{B_r^{(\nu,i)}; r = 1, \ldots, \rho_\nu\right\}$ as a set of IRs.
- Denote a_i as processing volume of the operation $D_i^{(\nu)}$.
- Denote $e_r^{(i)}$, $V_r^{(i)}$, $\Phi_r^{(i)}$ as maximal processing intensity of the operation $D_i^{(\nu)}$ at the IR $B_r^{(\nu,i)}$, maximal capacity of the IR $B_r^{(\nu,i)}$, and maximal productivity of the IR $B_r^{(\nu,i)}$ before the reconfiguration correspondingly; $\overline{e}_r^{(i)}$, $\overline{V}_r^{(i)}$, $\overline{\Phi}_r^{(i)}$ are given variables characterizing the same domains but after a disruption-based reconfiguration.
- Let t be current instant of time, $T = (t_0, t_f]$ the scheduling horizon, and t_0 (t_f) the start (end) instant of time for the scheduling horizon, respectively.
- Denote $\varepsilon(t)$ as an element of the matrix of time-spatial constraints ($\varepsilon(t) = 1$, if $t_0^k < t \le t_f^k$, $\varepsilon(t) = 0$ otherwise), where k are the numbers of time windows available for operation execution (e.g., subject to maintenance).
- Denote $S^{(\nu)} = \left\{S_l^{(\nu)}; l = 1, \ldots, d_j\right\}$ as a set of IT services to execute operations $D^{(\nu)}$.
- Denote $F^{(\nu,l)} = \left\{F_\chi^{(\nu,l)}; \chi = 1, \ldots, S_l\right\}$ as a set of functions of IR to implement the service.
- Denote fixed cost as $c_{il}^{(\nu,1)}(t)$ and operation cost as $c_{il}^{(\nu,2)}(t)$.
- Denote $g_l^{(\nu)}$ as a number of operations $D_i^{(\nu)}$ which may be processed by a service $S_l^{(\nu)}$.
- Denote $h_i^{(\nu)}$ as a given number of services $S_l^{(\nu)}$ which may be simultaneously used by execution of the operation $D_i^{(\nu)}$.

- Denote $D_l^{(\nu,i)} = \left\{ D_{<l,\chi>}^{(\nu,i)}; l = 1, \ldots, d_j, \chi = 1, \ldots, S_l \right\}$ as operations of IR (e.g., information processing, storage, transmission, and protection).
- Denote $D_r^{(p,i)} = \left\{ D_{<r,k>}^{(p,i)}; p = 1, \ldots, P^{(r)}; k = 1, \ldots, \pi_i^{(r)} \right\}$ as operations in the jobs for reconfiguration of the IR $B_r^{(\nu,i)}$.
- Denote $V_\chi^{(\nu)}$ as the online storage capacity of the IR $B_r^{(\nu,i)}$ to execute the operation $D_{<l,\chi>}^{(\nu,i)}$ and $\delta_{\chi\ r}^{(\nu,l)}(\tau)$ as a quality function to estimate the execution results.
- Denote $c_{\chi\ r}^{(l,1)}(\tau), c_{\chi\ r}^{(l,2)}(\tau)$ as given time functions of fixed and operation costs of an IR $B_r^{(\nu,i)}$ used for the operation $D_{<l,\chi>}^{(\nu,i)}$ by realization of the function $F_\chi^{(\nu,l)}$.
- Denote $\eta_{il}^{(\nu)}(t)$ as a given time function which characterizes the costs of idle time of services for the operation $D_i^{(\nu)}$.
- $y_{il}^{(\nu)}$ denotes the value of current idle cost due to a backlog in the operation $D_i^{(\nu)}$ caused by unavailability of the service $S_l^{(\nu)}$.

In order to describe the execution of operations, let us introduce the *state variables*:

- $x_{il}^{(\nu)}(t)$ characterizes the execution of the operation $D_i^{(\nu)}$ with the use of the service $S_l^{(\nu)}$;
- $x_{il}^{(\nu,1)}(t)$ is an auxiliary variable characterizing the current state of the operation $D_i^{(\nu)}$. Its value is numerically equal to the time interval that has elapsed since the beginning of the scheduling interval and the execution start of the operation $D_i^{(\nu)}$;
- $x_{il}^{(\nu,2)}(t)$ is an auxiliary variable characterizing the current state of the processing operation. Its value is numerically equal to the time interval that has elapsed since the end of the execution of the operation $D_i^{(\nu)}$ and the end of the scheduling interval;
- $x_r^{(\nu,l)}$ is an auxiliary variable characterizing the employment time of the IR $B_r^{(\nu,j)}$;
- $x_\chi^{(\nu,l)}$ is an auxiliary variable which characterizes the execution of the operation $D_{<l,\chi>}^{(\nu,j)}$;
- $x_{rS_l}^{(\nu,l)}(t)$ is an auxiliary variable characterizing the current state of the information processing operation. Its value is numerically equal to the time interval that has elapsed since the end of the execution of the operation $D_{<l,\chi>}^{(\nu,j)}$ and the instant of time t;
- $u_{il}^{(\nu)}(t)$ is a control that is equal to 1 if the operation $D_i^{(\nu)}$ is assigned to the service $S_l^{(\nu)}$ at the moment t; otherwise $u_{il}^{(\nu)}(t) = 0$;
- $\vartheta_{il}^{(\nu,1)}(t)\left(\vartheta_{il}^{(\nu,2)}(t)\right)$ are auxiliary control variables that are equal to 1 if the operation $D_i^{(\nu)}$ has not started and is equal 0 otherwise;
- $w_{\chi\ r}^{(\nu,l)}$ is a control that is equal to 1 if the operation $D_{<l,\chi>}^{(\nu,j)}$ is assigned to the IR $B_r^{(\nu,i)}$ and is equal 0 otherwise;
- $\omega_{rS_l}^{(\nu,l)}(t)$ is auxiliary control that is equal to 1 if all the operations $D_{<l,\chi>}^{(\nu,j)}$ in the function $F_\chi^{(\nu,l)}$ are completed and is equal 0 otherwise;

- $\vartheta_r^{(p,2)}(t)$ is auxiliary control that is equal to 1 if the reconfiguration from old parameters $e_r^{(i)}, V_r^{(i)}, \Phi_r^{(i)}$ to new ones $\overline{e}_r^{(i)}, \overline{V}_r^{(i)}, \overline{\Phi}_r^{(i)}$ is completed and is 0 otherwise.

The *problem* is to find a joint schedule for dynamic execution of information services and physical flows, that is, two schedules should be generated in a coordinated manner, that is,

- an optimal program control (OPC) (schedule) for the integrated execution of material flows and information services (model M1),
- an OPC (schedule) for the execution of information services within the IRs (model M2).

Jobs are to be scheduled subject to maximal customer service level (i.e., minimal lateness), minimal backlogs, minimal idle time of services, and minimal costs of IT (e.g., fixed, operation, and idle costs).

7.3 Methodology

In this section, we describe both general methodology and method for formulation of the integrated scheduling model in particular.

7.3.1 Structure Dynamics Control Methodology

The logistics network dynamic characteristics are distributed upon different structures, that is:

- organizational structure dynamics (i.e., agile supply structure),
- functional structure dynamics (i.e., flexible competencies),
- information structure dynamics (i.e., fluctuating information availability), and
- financial structure dynamics (i.e., cost and profit sharing).

This multidimensional dynamic space along with the coordinated and distributed decision-making leads us to the understanding of the logistics network as *multistructural systems with structure dynamics*. The main idea of the proposed method is the dynamic interpretation of planning in accordance with the natural logic of time with the help of OPC. The solution procedure is transferred to mathematical programming (MP). In this setting, the solution procedure becomes undependable from the continuous optimization and can be of a discrete nature, for example, an integer linear program (Ivanov et al. 2020).

The modeling procedure is based on the dynamic representation where the scheduling decisions are taken for certain intervals of structural constancy and regarding problems of significantly smaller dimensionality. For each interval, a static optimization problem of a smaller dimensionality can be solved with the help of MP. The transitions between the intervals are modeled in the dynamic OPC model. The computational time decreases considerably.

Besides, *a priori* knowledge of the logistics network structure, and moreover, structure dynamics, is no more necessary—the structures and corresponding functions are optimized simultaneously as the control becomes a function of both states and structures. The splitting of the planning period into the intervals occurs according to the natural logic of time and events. As the proposed method is based on control theory, it is a convenient approach to describe intangible services due to abstract nature of state variables which can be interpreted as abstract service volumes.

7.3.2 Formulation of the Integrated Scheduling Model

The basic *conceptual idea* of this approach is that the operations and machine availability are dynamically distributed in time on the scheduling horizon. As such, not all operations and machines are involved in decision-making at the same time. Therefore, it becomes quite natural to transit from large-size allocation matrices with a high number of binary variables to a scheduling problem that is dynamically decomposed.

In following an approach to decompose the solution space and to use exact methods over its restricted subspaces, we propose to use the OPC theory for the dynamic scheduling problem decomposition. Computational procedure will be based on modified maximum principle in continuous form blended with MP.

That is why the basic *technical idea* of our approach, which extends the previous application of maximum principle to production and logistics, is to apply the methods of discrete optimization for combinatorial tasks within certain time intervals and to use the OPC with all its advantages (i.e., accuracy of continuous time, integration of planning and control, and the operation execution parameters as time functions) for (1) flow control within the operations and (2) interlinking the partial (decomposed) solutions into the optimal schedule.

The SN is modeled as a networked control system described through a *dynamic* interpretation of the operations' execution. The execution of operations is characterized by (1) results (e.g., processed volume and completion time), (2) intensity consumption of the machines, and (3) supply and information flows resulting from the schedule execution. The operations control model (M1) is first used to assign and sequence ISs to operations in material flows, and then a flow control model (M2) is employed to assign and schedule jobs at IRs subject to the requirements on the ISs availability. The basic interaction of these two models is that after the solving M1, the found control variables are used in the constraints of M2. Note that in the calculation procedure, the models M1 and M2 will be solved simultaneously, that is, the scheduling problems in all the structures (i.e., material flows, ISs, and IRs) will be integrated.

7.4 Mathematical Model

7.4.1 Mathematical Model M1

The *model of operation execution dynamics* can be expressed as (7.1)–(7.3):

$$\frac{dx_i^{(\nu,l)}}{dt} = \varepsilon_{il}(t) \cdot u_{il}^{(\nu)}(t) \tag{7.1}$$

$$\frac{dy_{il}^{(\nu)}}{dt} = \eta_{il}(t)\left[1 - \vartheta_{il}^{(\nu,1)} - u_{il}^{(\nu)} - \vartheta_{il}^{(\nu,2)}\right] \tag{7.2}$$

$$\frac{dx_{il}^{(\nu,1)}}{dt} = \vartheta_{il}^{(\nu,1)}; \quad \frac{dx_{il}^{(\nu,2)}}{dt} = \vartheta_{il}^{(\nu,2)} \tag{7.3}$$

Equation (7.1) describes operation execution dynamics subject to availability of IS described in the matrix function $\varepsilon_{il}(t)$. $u_{il}^{(\nu)}(t) = 1$ if service $S_l^{(\nu)}$ is assigned to the operation $D_i^{(\nu)}$, $u_{il}^{(\nu)}(t) = 0$ otherwise. Equation (7.2) represents idle time in the material flow caused by unavailability of the IS $S_l^{(\nu)}$. Equation (7.3) represents the dynamics of operation's execution according to precedence constraints.

The control actions are *constrained* as follows:

$$\sum_{i=1}^{k_j} u_{il}^{(\nu)}(t) \leq g_l^{(\nu)}; \forall l; \sum_{l=1}^{d_j} u_{il}^{(\nu)}(t) \leq h_i^{(\nu)}; \forall i \tag{7.4}$$

$$\sum_{l=1}^{d_j} u_{il}^{(\nu)}\left[\sum_{\alpha\in\Gamma_{\nu 1}}\left(a_\alpha^{(\nu,l)} - x_\alpha^{(\nu,l)}\right) + \prod_{\beta\in\Gamma_{\nu 2}}\left(a_\beta^{(\nu,l)} - x_\beta^{(\nu,l)}\right)\right] = 0; \forall \nu \tag{7.5}$$

$$\vartheta_{il}^{(\nu,1)} \cdot x_{il}^{(\nu,l)} = 0; \quad \vartheta_{il}^{(\nu,2)}\left(a_{il}^{(\nu,l)} - x_{il}^{(\nu,l)}\right) = 0; \forall i; \forall l \tag{7.6}$$

$$u_{il}^{(\nu)}(t) \in \{0, 1\}; \vartheta_{il}^{(\nu)}(t) \in \{0, 1\} \tag{7.7}$$

Constraints (7.4) are assignment problem constraints. They define possibilities of parallel use of many services for one operation and for parallel processing of many operations at one service. Constraints (7.5) determine the precedence relations. Constraints (7.6) interconnect main and auxiliary controls. Equation (7.7) constraints control to be Boolean variables.

Remark 7.1 Note that constraints (7.4)–(7.7) are identical to those in MP models. However, at each t-point of time, the number of variables is determined by the operations which are currently in the "scheduling window." Therefore, the tendency will be to have small-size instances and to apply known methods for the solution of MP models (e.g., Hungarian or Branch & Bound methods) subject to the problems (7.1)–(7.12).

The *boundary conditions* are defined as shown in Eqs. (7.8) and (7.9):

$$t = t_0^{(j)} : x_i^{(\nu)}\left(t_0^{(j)}\right) = y_{il}^{(\nu)}\left(t_0^{(j)}\right) = x_{il}^{(\nu)}\left(t_0^{(j)}\right) = 0 \tag{7.8}$$

$$t = t_f^{(j)} : x_i^{(\nu)}\left(t_f^{(j)}\right) = a_i^{(\nu)};\; y_l^{(\nu)}\left(t_f^{(j)}\right); x_i^{(\nu)}\left(t_f^{(j)}\right) \in \mathbf{R}^1 \tag{7.9}$$

Equations (7.8) and (7.9) define initial and end values of the variables $x_i^{(\nu)}(t)$, $y_{il}^{(\nu)}(t)$, $x_{il}^{(\nu)}(t)$ at the moments $t_0^{(j)}$ and $t_f^{(j)}$.

Remark 7.2 End conditions in OPC models play the role of demand variables in MP models. Conditions (7.9) reflect the desired end state. The right parts of equations are predetermined at the planning stage subject to the planned demand for each job.

The *goals* are defined as shown in Eqs. (7.10)–(7.12):

$$\min J_1^{(\nu)} = \sum_{i=1}^{k_\nu}\sum_{l=1}^{d_j} y_{il}^{(\nu)}\left(t_f^{(j)}\right) \tag{7.10}$$

$$\max J_2 = \sum_{i=1}^{k_\nu}\sum_{l=1}^{d_j} \frac{1}{x_{il}^{(\nu,2)}\left(t_f^{(j)}\right)} \int_{t_0^{(j)}}^{t_f^{(j)}} \vartheta_{il}^{(\nu,2)}(\tau)\,d\tau \tag{7.11}$$

$$\min J_3 = \sum_{i=1}^{k_\nu}\sum_{l=1}^{d_j} \int_{t_0^{(j)}}^{t_f^{(j)}} \left[c_{il}^{(\nu,1)}(\tau) + c_{il}^{(\nu,2)}(\tau)\right] \cdot u_{il}^{(\nu)}(\tau)\,d\tau \tag{7.12}$$

Equation (7.10) minimizes losses from the idle time of services. Equation (7.11) estimates the service level by the volume of on-time completed jobs in the material flow. Equation (7.12) minimizes total costs of IS.

7.4.2 Mathematical Model M2

The *model of operation execution dynamics* in the IRs can be expressed as (7.13):

$$\frac{dx_{\chi}^{(\nu,l)}}{dt}=\sum_{r=1}^{\rho_{\nu}}u_{\chi\ r}^{(\nu,l)};\ \frac{dx_{r}^{(\nu,l)}}{dt}=\sum_{\chi=1}^{S_l}w_{\chi\ r}^{(\nu,l)};\ \frac{dx_{rS_l}^{(\nu,l)}}{dt}=\omega_{rS_l}^{(\nu,l)} \tag{7.13}$$

Equation (7.13) describes operation's execution dynamic in the IR subject to operation of the IRs and recovery operations in the case of disruptions in the information structure.

The control actions are *constrained* as shown in Eqs. (7.14)–(7.20):

$$0\le u_{\chi\ r}^{(\nu,l)}\le\left[e_{\chi\ r}^{(j)}\left(1-\vartheta_r^{(p,2)}(t)\right)+\overline{e}_{\chi\ r}^{(j)}\vartheta_r^{(p,2)}(t)\right]w_{\chi\ r}^{(\nu,l)}, \tag{7.14}$$

$$\sum_{\nu=1}^{n_j}\sum_{\chi=1}^{S_\nu}V_{\chi}^{(\nu)}\cdot w_{\chi\ r}^{(\nu,l)}\le\left[V_r^{(j)}\left(1-\vartheta_r^{(p,2)}(t)\right)+\overline{V}_r^{(j)}\vartheta_r^{(p,2)}(t)\right]\xi_r^{(j,1)}, \tag{7.15}$$

$$\sum_{\nu=1}^{n_j}\sum_{\chi=1}^{S_\nu}u_{\chi\ r}^{(\nu,l)}(t)\le\left[\Phi_r^{(j)}\left(1-\vartheta_r^{(p,2)}(t)\right)+\overline{\Phi}_r^{(j)}\vartheta_r^{(p,2)}(t)\right]\xi_r^{(j,2)}, \tag{7.16}$$

$$\sum_{r=1}^{\rho_\nu}w_{\chi\ r}^{(\nu,l)}\left[\sum_{\pi\in\Gamma_{\chi 3}}\left(a_{\pi}^{(\nu,l)}-x_{\pi}^{(\nu,l)}\right)+\prod_{\mu\in\Gamma_{\mu 4}}\left(a_{\mu}^{(\nu,l)}-x_{\mu}^{(\nu,l)}\right)\right]=0, \tag{7.17}$$

$$\sum_{r=1}^{\rho_\nu}w_{\chi r}^{(\nu,l)}(t)\le\psi_{\chi};\forall\chi;\ \sum_{\chi=1}^{s_l}w_{\chi r}^{(\nu,l)}(t)\le\phi_r;\forall r, \tag{7.18}$$

$$\omega_{rS_l}^{(\nu,l)}\left(a_{S_l}^{(\nu,l)}-x_{S_l}^{(\nu,l)}\right)=0, \tag{7.19}$$

$$w_{\chi r}^{(\nu,l)}\in\left\{0,u_{il}^{(\nu)}\right\};\ \vartheta_r^{(p,2)}(t),\omega_{rS_l}^{(\nu,l)}\in\{0,1\};\ \xi_r^{(j,1)}(t);\xi_r^{(j,2)}(t)\in[0,1]. \tag{7.20}$$

With the help of functions $0 \le \xi_r^{(j,1)}(t) \le 1$ and $0 \le \xi_r^{(j,2)}(t) \le 1$, perturbation impacts on the IR $B_r^{(\upsilon,j)}$ can be modeled. Equations (7.14)–(7.16) constraint information processing at $B_r^{(\upsilon,j)}$ before and after the reconfiguration. Constraints (7.17) set precedence relations on information processing operation similar to Eq. (7.5). Constraints (7.18) are related to assignment problem and are similar to (7.4). Equation (7.19) determines the conditions of processing completion.

The *boundary conditions* are defined as shown in Eqs. (7.21) and (7.22):

$$t = t_0^{(j)} : x_\chi^{(\upsilon,l)}\left(t_0^{(j)}\right) = x_r^{(\upsilon,l)}\left(t_0^{(j)}\right) = x_{rS_l}^{(\upsilon,l)}\left(t_0^{(j)}\right) = 0, \tag{7.21}$$

$$t = t_f^{(j)} : x_\chi^{(\upsilon,l)}\left(t_f^{(j)}\right) = a_\chi^{(\upsilon,l)};\ x_r^{(\upsilon,l)}\left(t_f^{(j)}\right); x_{rS_l}^{(\upsilon,l)}\left(t_f^{(j)}\right) \in \mathbf{R}^1. \tag{7.22}$$

The *goals* are defined as shown in Eqs. (7.23)–(7.26):

$$J_4 = \sum_{r=1}^{\rho_\upsilon - 1} \sum_{r_1=r+1}^{\rho_\upsilon} \int_{t_0^{(j)}}^{t_f^{(j)}} \left(x_r^{(\upsilon,l)}(\tau) - x_{r_1}^{(\upsilon,l)}(\tau)\right) d\tau, \tag{7.23}$$

$$J_5 = \sum_{r=1}^{\rho_\upsilon} \sum_{\chi=1}^{S_l} \int_{t_0^{(j)}}^{t_f^{(j)}} \delta_{\chi r}^{(\upsilon,l)}(\tau) \cdot w_{\chi r}^{(\upsilon,l)}(\tau)\, d\tau, \tag{7.24}$$

$$J_6 = \frac{1}{2} \sum_{\chi=1}^{S_l} \left(a_\chi^{(\upsilon,l)} - a_\chi^{(\upsilon,l)}\left(t_f^{(j)}\right)\right)^2, \tag{7.25}$$

$$J_7 = \sum_{\chi=1}^{S_l} \sum_{r=1}^{\rho_\upsilon} \int_{t_0^{(j)}}^{t_f^{(j)}} \left[c_{\chi r}^{(l,1)}(\tau) + c_{\chi r}^{(l,2)}(\tau)\right] w_{\chi r}^{(\upsilon,l)}(\tau)\, d\tau. \tag{7.26}$$

Equation (7.23) estimates uniformity of the use of the IRs $B_r^{(\upsilon,j)}$ and $B_{r_1}^{(\upsilon,j)}$; r, $r_1 \in \{1, \ldots, \rho_\upsilon\}$. Equation (7.24) estimates amount of completed operations $D_{\langle l,\chi \rangle}^{(\upsilon,j)}$. Equation (7.25) takes into account losses from nonfulfilled operations. Equation (7.26) assesses total cost of ownership (TCO) for the IR $B_r^{(\upsilon,j)}$.

7.4.3 Model Integration

The developed modeling complex is composed of dynamic models of IS and IR control subject to execution of material flows. It also includes elements of IR reconfiguration [e.g., in Eqs. (7.14)–(7.16) and (7.20)]. Full consideration of the reconfiguration model can be found in (Ivanov and Sokolov 2013).

The presented models M1 and M2 are interconnected with the help of Eq. (7.6) where elements from M2 are used in M1. In its turn, M1 influences M2 through Eqs. (7.14) and (7.20).

The proposed models and algorithms have been validated in a developed prototype based on C++ and XML. The OPC calculation is based on the *Hamiltonian* function. In integrating the main and the conjunctive equation systems, the values of variables in both of the systems can be obtained at each point of time. The maximum principle guarantees that the optimal solutions (i.e., the solution with maximal values) of the instantaneous problems (i.e., at each point of time) give the optimal solution to the overall problem. For these subproblems, optimal solutions can be found, for example, with the help of MP. Then these solutions are linked into an OPC.

7.4.4 Model Analysis

Let us discuss optimality and sufficiency properties that have been proved theoretically and experimentally. The formulated scheduling model satisfies the conditions of the existence theorem in Lee and Markus (1967, Theorem 4, Corollary 2), which allows us to assert the existence of the optimal solution in the appropriate class of admissible controls. The formulated scheduling problem is the standard problem of OPC with mixed constraints and its optimal solution and relaxed system can be obtained with the help of local cut method-based modification of the continuous maximum principle. An analysis of constraints in M1 and M2 shows that both state and control variables are constrained (i.e., the mixed state-control constraints exist Boltyanskiy 1973) and form therefore a dynamic system with a variable control domain. To obtain necessary conditions of control optimality, Boltyansky's method of local sections can be used. Then the necessary conditions can be formulated in the form of the Boltyanskiy's theorem (maximum principle) (1973).

Corollary 7.1 Analysis of Boltyanskiy (1973) and Moiseev (1974) shows that for the linear nonstationary finite-dimensional systems (models M1 and M2) with the convex area of admissible control $Q(\boldsymbol{x})$ and performance indicators (7.10)–(7.12) and (7.23)–(7.26), the stated necessary conditions of optimality are also the conditions of sufficiency.

According to study (Ivanov and Sokolov 2012a), the initial problem of nonclassical calculus of variations can be transformed to the two-point boundary problem help of local cut method. The assignment and routing at each instant of time are performed on the basis of the "dynamic priority" of the operations. The dynamic priority includes both the values of conjunctive variables and the current values of the goal functions (7.10)–(7.12) and (7.23)–(7.26).

7.5 Algorithmic Realization

Theorem 7.1 Let Γ be a relaxed problem for the basic OPC problem. Then

(a) If the problem Γ does not have allowable solutions, then this is true for the problem *PS* as well.
(b) If the OPC of the problem Γ is allowable, then it is the OPC for the problem *PS* as well.

Proof

(a) If the problem Γ does not have allowable solutions, then a control $\mathbf{u}(t)$ transferring dynamic system (7.1)–(7.3) and (7.13) $\dot{\mathbf{x}} = \mathbf{f}(\mathbf{x}, \mathbf{u}, t)$ from a given initial state to a given final state does not exist. The same end conditions are violated in the OPC problem.
(b) Let $\mathbf{u}^*(t)$, $\forall\ t \in (T_0, T_f]$ be an OPC in Γ and $\mathbf{x}(t)$ be a solution to models M1 and M2 subject to $\mathbf{u}(t) = \mathbf{u}^*(t)$. Then $\mathbf{u}^*(t)$ meets the requirements of the local cut method and maximizes Hamiltonian for the OPC problem. Hence, vector $\mathbf{u}^*(t)$ and $\mathbf{x}^*(t)$ return minimum to performance indicators (7.10)–(7.12) and (7.23)–(7.26). The proof is complete. □

As the dynamics of state and conjunctive variables is described by differential equations, it becomes possible to calculate these variables at any instant of time subject to given initial conditions. Therefore, the Hamiltonian becomes the function of only one variable $\mathbf{u}$ that can be calculated at any t subject to allowable control from $u \in G_u$. Therefore, the *OPC problem* can be reduced to a boundary problem with the help of the local cut method.

Let us consider the algorithmic realization of the above-described modified maximum principle. After transforming to the boundary problem, a relaxed problem can be solved to receive OPC for the schedule of the model M1, for computation of which the main and conjunctive systems are integrated, that is, the OPC vector $\mathbf{u}^*(t)$ and the state trajectory $\mathbf{x}^*(t)$ are obtained. The OPC vector at time $t = T_0$ and for the given value of $\boldsymbol{\psi}(t)$ should return maximum criteria indicators (7.10)–(7.12) and (7.23)–(7.26) have been transformed to a general performance index and expressed in a scalar form J_G.

The basic peculiarity of the considered boundary problem is that the initial conditions for the conjunctive variables $\psi(t_0)$ are not given. At the same time, an OPC should be calculated subject to end conditions (7.8) and (7.9) and (7.21) and (7.22). To obtain the conjunctive system vector, we use the Krylov-Chernousko method for OPC problem with free right end that is based on joint use of modified successive approximations method and branch-and-bound method. We propose to use a heuristics schedule $\overline{\mathbf{u}}(t)$ to obtain the initial conditions for $\psi(t_0)$. Then, the algorithm can be stated as follows:

- **Step 1**: An initial solution $\overline{\mathbf{u}}(t), t \in (T_0, T_f]$ is calculated and iteration step $r = 0$.
- **Step 2**: The parameters of the gained schedule $\overline{\mathbf{u}}(t), t \in (T_0, T_f]$ are put into Eqs. (7.1)–(7.3) and (7.13) and integrated. As a result of the dynamic model run, a new trajectory of operation states $\mathbf{x}^{(r)}(t)$ is received. Besides, if $t = T_f$ then the record value $J_G = J_G^{(r)}$ can be calculated.
- **Step 3**: Then, the transversality conditions are evaluated. The conjugate system is integrated subject to $\mathbf{u}(t) = \overline{\mathbf{u}}(t)$ and over the interval from $t = T_f$ to $t = T_0$. For $t = T_0$, the first approximation $\psi_l^{(r)}(T_0)$ is received as a result. Here, the iteration number $r = 0$ is completed.
- **Step 4**: The control $\mathbf{u}^{(r)}(t)$ being searched for subject to maximization of the Hamiltonian function. The iterative process of the optimal schedule search is terminated as follows: either the allowable solution is determined, or at the fourth step no significant improvement is achieved.

Analogously, the OPC for the schedule of the model M2 can be obtained through the integration of corresponding conjunctive systems. Subsequently, through the reverse integration of the main equation systems, the mutual interrelating of the models M1 and M2 is realized.

Acknowledgments This research has received funding from the European Union's Horizon 2020 research and innovation program under the Marie Sklodowska-Curie grant agreement No. 873077 (MAIA-H2020-MSCA-RISE 2019) (Sect. 7.3). The research described in this chapter is partially supported by the Russian Foundation for Basic Research grants: 18-07-01272 (Sect. 7.4), state research 0073-2019-0004 (Sect. 7.5).

References

Ailawadi, K. L., & Farris, P. W. (2017). Managing multi- and omni-channel distribution: Metrics and research directions. *Journal of Retailing, 93*(1), 120–135.

Bardhan, I. R., Demirkan, H., Kannan, P. K., Kauffman, R., & Sougstad, R. (2010). An interdisciplinary perspective on IT services management and service science. *Journal of Manufacturing Information Systems, 26*(4), 13–64.

Boltyanskiy, B. (1973). *Optimal control of discrete systems*. Moscow: Nauka.

Camarinha-Matos, L. M., & Macedo, P. (2010). A conceptual model of value systems in collaborative networks. *Journal of Intelligent Manufacturing, 21*(3), 287–299.

Cannella, S., Framinan, J. M., & Barbosa-Póvoa, A. (2014). An IT-enabled supply chain model: A simulation study. *International Journal of Systems Science, 45*(11), 2327–2341.

Chen, Z.-L. (2010). Integrated production and outbound distribution scheduling: Review and extensions. *Operations Research, 58*(1), 130–148.

Choi, J., Kim, Y., Park, Y. T., & Kang, S. H. (2002). Agent-based product-support logistics system using XML and RDF. *International Journal of Systems Science, 33*(6), 467–484.

Dedrick, J., Xu, S., & Zhu, K. (2008). How does information technology shape supply-chain structure? Evidence on the number of suppliers. *Journal of Manufacturing Information Systems, 25*(2), 41–72.

Dolgui, A., Ivanov, D., Potryasaev, S., Sokolov, B., Ivanova, M., & Werner, F. (2019). Blockchain-oriented dynamic modelling of smart contract design and execution control in the supply chain. *International Journal of Production Research*. https://doi.org/10.1080/00207543.2019.1627439.

Giard, V., & Mendy, G. (2008). Scheduling coordination in a supply chain using advance demand information. *Production Planning and Control, 19*(7), 655–667.

Hosseini, S., Ivanov, D., & Dolgui, A. (2019). Review of quantitative methods for supply chain resilience analysis. *Transportation Research: Part E, 125*, 285–307.

Ivanov D. (2020) Predicting the impact of epidemic outbreaks on the global supply chains: A simulation-based analysis on the example of coronavirus (COVID-19 / SARS-CoV-2) case. Transportation research – Part E, 136., https://doi.org/10.1016/j.tre.2020.101922.

Ivanov D., Das A. (2020). Coronavirus (COVID-19 / SARS-CoV-2) and supply chain resilience: A research note. International journal of integrated supply management, forthcoming.

Ivanov, D., & Dolgui, A. (2019). New disruption risk management perspectives in supply chains: Digital twins, the ripple effect, and resileanness. 9th IFAC conference on manufacturing modelling, management and control (MIM 2019), edited by Dmitry Ivanov, Alexandre Dolgui, Farouk Yalaoui. *IFAC-PapersOnLine, 52*(13), 337–342.

Ivanov, D., & Dolgui, A. (2020a). Viability of intertwined supply networks: Extending the supply chain resilience angles towards survivability. In *A position paper motivated by COVID-19 outbreak*. International Journal of Production: Research. https://doi.org/10.1080/00207543.2020.1750727.

Ivanov D., Dolgui A. (2020b). A digital supply chain twin for managing the disruptions risks and resilience in the era of industry 4.0. Production planning and control, forthcoming.

Ivanov, D., & Sokolov, B. (2010). *Adaptive supply chain management*. London: Springer.

Ivanov, D., & Sokolov, B. (2012a). Dynamic supply chain scheduling. *Journal of Scheduling, 15*(2), 201–216.

Ivanov, D., & Sokolov, B. (2012b). Structure dynamics control approach to supply chain planning and adaptation. *International Journal of Production Research, 50*(21), 6133–6149.

Ivanov, D., & Sokolov, B. (2013). Dynamic coordinated scheduling in the supply chain under a process modernization. *International Journal of Production Research, 51*(9), 2680–2697.

Ivanov, D., Arkhipov, A., & Sokolov, B. (2004). Intelligent supply chain planning in virtual enterprises. In L. Camarihna-Matos (Ed.), *Virtual Enterprises and Collaborative Networks, Proceedings of the IFIP Conference on Virtual Enterprises PRO-VE 2004* (pp. 215–223). Boston, MA: Kluwer Academic Publishers.

Ivanov, D., Dolgui, A., & Sokolov, B. (2019). The impact of digital technology and industry 4.0 on the ripple effect and supply chain risk analytics. *International Journal of Production Research, 57*(3), 829–846.

Ivanov, D., Käschel, J., Arkhipov, A., Sokolov, B., & Zschorn, L. (2005). Quantitative models of collaborative networks. In L. Camarihna-Matos, H. Afsarmanesh, & A. Ortiz (Eds.), *Collaborative Networks and Their Breeding Environments, Proceedings of the IFIP Conference on Virtual Enterprises PRO-VE 2005* (pp. 387–394). Valencia: Springer.

Ivanov, D., Kaeschel, J., & Sokolov, B. (2007). Integrated modeling of agile enterprise networks. *International Journal of Agile Systems and Management, 2*(1), 23–49.

Ivanov, D., Sokolov, B., & Kaeschel, J. (2010). A multi-structural framework for adaptive supply chain planning and operations with structure dynamics considerations. *European Journal of Operational Research, 200*(2), 409–420.

Ivanov, D., Dolgui, A., & Sokolov, B. (2012). On applicability of optimal control theory to adaptive supply chain planning and scheduling. *Annual Reviews in Control, 36*(1), 73–84.

Ivanov, D., Sokolov, B., & Dilou Raguinia, E. A. (2014). Integrated dynamic scheduling of material flows and distributed information services in collaborative cyber-physical supply networks. *International Journal of Systems Science: Operations & Logistics, 1*(1), 18–26.

Ivanov, D., Sokolov, B., Hartl, R., Dolgui, A., Pavlov, A., & Solovyeva, I. (2015). Integration of aggregate distribution and dynamic transportation planning in a supply chain with capacity disruptions and ripple effect considerations. *International Journal of Production Research, 53*(23), 6963–6979.

Ivanov, D., Sokolov, B., Dolgui, A., Werner, F., & Ivanova, M. (2016). A dynamic model and an algorithm for short-term supply chain scheduling in the smart factory Industry 4.0. *International Journal of Production Research, 54*(2), 386–402.

Ivanov, D., Sethi, S., Dolgui, A., & Sokolov, B. (2018). A survey on the control theory applications to operational systems, supply chain management and Industry 4.0. *Annual Reviews in Control, 46*, 134–147.

Ivanov, D., Sokolov, B., Chen, W., Dolgui, A., Werner, F., & Potryasaev, S. (2020). A control approach to scheduling flexibly configurable jobs with dynamic structural-logical constraints. *IISE Transactions*. https://doi.org/10.1080/24725854.2020.1739787.

Jain, V., Wadhwa, S., & Deshmukh, S. G. (2009). Revisiting information systems to support a dynamic supply chain: Issues and perspectives. *Production Planning and Control, 20*(1), 17–29.

Jiang, N., Xu, L., Vrieze, P. D., Lim, M. G., & Jarabo, O. (2012). A cloud based data integration framework. In L. Camarinha-Matos, L. Xu, & H. Afsarmanesh (Eds.), *Proceedings of the IFIP Conference on Virtual Enterprises PRO-VE 2012 IFIP AICT* (Vol. 380, pp. 177–185). Heidelberg: Springer.

Lee, E. B., & Markus, L. (1967). *Foundations of optimal control theory*. New York: Wiley & Sons.

Lee, H.L., Zhao X., Li X., Voss C. (2020). The Effects of COVID-19 on Global Supply Chains:Responsiveness, Resilience, and Restoration (3Rs). Available at https://onlinelibrary.wiley.com/pb-assets/assets/18731317/JOM%20CFP%20-%20COVID-19%20and%20Global%20Supply%20Chains-1585601180383.pdf., accessed on April 4, 2020.

Li, Q., Zhou, J., Peng, Q.-R., Li, C.-Q., Wang, C., Wu, J., & Shao, B.-E. (2010). Business processes oriented heterogeneous systems integration platform for networked enterprises. *Computers in Industry, 61*(2), 127–144.

Moiseev, N. N. (1974). *Element of the optimal systems theory [in Russian]*. Moscow: Nauka.

Pagani, M., & Pardo, C. (2017). The impact of digital technology on relationships in a business network. *Industrial Marketing Management, 67*, 185–192.

Panetto, H., Iung, B., Ivanov, D., Weichhart, G., & Wang, X. (2019). Challenges for the cyber-physical manufacturing enterprises of the future. *Annual Reviews in Control, 47*, 200–213.

Panetto, H., Iung, B., Ivanov, D., Weichhart, G., & Wang, X. (2019b). Challenges for the cyber-physical manufacturing enterprises of the future. *Annual Reviews in Control, 47*, 200–213.

Surana, A., Kumara, S., Greaves, M., & Raghavan, U. N. (2005). Supply-chain networks: A complex adaptive systems perspective. *International Journal of Production Research, 43*(20), 4235–4265.

Wang, L., von Laszewski, G., Younge, A., He, X., Kunze, M., Tao, J., & Cheng, F. (2010). Cloud computing: A perspective study. *New Generation Computing, 28*(2), 137–146.

Chapter 8
Human-Oriented Assembly Line Balancing and Sequencing Model in the Industry 4.0 Era

Daria Battini, Serena Finco, and Fabio Sgarbossa

Abstract Ergonomics plays a crucial role in the design process of manual assembly systems, since a poorly ergonomic workplace leads to injuries, accidents, and musculoskeletal disorders. Using Industry 4.0 solutions, smart technologies, and cloud platforms, the well-being of workers can be improved more easily than in the past. In this context, smartwatches can be used to monitor workers' health and to collect data about the physical efforts of each worker during the working day, in relation to energy expenditure or heart rate monitoring. Managers can use data collected via these smart solutions to improve sequencing and scheduling processes in terms of both ergonomics and time, achieving a trade-off between ergonomics and productivity. Using real-time monitoring, a dynamic scheduling and sequencing approach can be implemented to guarantee the right safety level for each worker. In this chapter, we give a general overview of smart tools for measuring and quantifying the ergonomics level. Based on the data from smartwatches, we propose a multi-objective assembly line balancing model and an ergo-sequencing model, and demonstrate the benefits of using smart solutions and Industry 4.0 tools. The limitations are discussed using a real case application. Our conclusions can guide managers and practitioners during the design phase.

D. Battini (✉) · S. Finco
Department of Management and Engineering, University of Padova, Vicenza, Italy
e-mail: daria.battini@unipd.it; serena.finco@unipd.it

F. Sgarbossa
Department of Mechanical and Industrial Engineering Valgrinda, NTNU—Norwegian University of Science and Technology, Trondheim, Norway
e-mail: fabio.sgarbossa@ntnu.no

B. Sokolov et al. (eds.), *Scheduling in Industry 4.0 and Cloud Manufacturing*, International Series in Operations Research & Management Science 289,
https://doi.org/10.1007/978-3-030-43177-8_8

8.1 Introduction

Two significant movements have engaged manufacturing systems over the past 10 years. One is the Industry 4.0 revolution, entailing the digitalization of products and processes across manufacturing sectors and supply chains (i.e., Ivanov et al. 2016a, b; Panetto et al. 2019). The second development concerns social sustainability and in particular human-centered design (HCD), workplace safety and ergonomics (i.e., Battini et al. 2011).

The main objective of ergonomics is to achieve an optimal relationship between people and their work environment. However, to reach this optimal point, two main conflicting factors must be addressed. On one hand, managers and companies require the maximum efficiency level and productivity, while on the other, workers need comfortable and safe workplaces to guarantee their health and physical well-being. The workers' needs become even more critical when an ageing workforce must be employed and motivated to work until 65 or 70 years old in industrial system shopfloors (e.g., Calzavara et al. 2020; Finco et al. 2019). For this reason, several studies have been carried out over recent decades to achieve the right trade-off between the needs of the workers and the companies, with the common goal of avoiding work-related musculoskeletal disorders (WMSDs), diseases and accidents. WMSDs represent a significant concern globally, not only from a workers' point of view but also due to their economic impact. According to some estimations for the manufacturing sector, 12.5% of the workforce missed days of work due to illness or injury in 2015. In the European Union, more than 40 million workers are affected by musculoskeletal disorders (MSDs) (about one in seven people), while in the United States, MSDs represent about 30% of occupational injuries. In the United States, the median number of days absent from work due to WMSDs was 10 in 2012, while in the European Union this figure was about 12 days.

Consequently, both companies and states must allocate extra financial resources to deal with this crucial problem, since these costs negatively impact both companies' earnings and the GDPs of countries. The decrease in the gross national product of the European Union due to WMSDs was estimated at up to 2% in 2010, while in Canada (resp. the United States), the impact was estimated at up to 3.4% (resp. 2.5%) of the GDP.

However, according to several studies, MSDs can be avoided through ergonomic improvements to the workplace, and may have a pay-back period of less than 1 year. Starting from these assumptions, academics and experts have focused their attention on this problem in recent years, and research works have been published on strategies, approaches, and methods for improving workers' well-being and safety, including the ergonomic features of manufacturing systems with an emphasis on manual assembly systems (Otto and Battaïa 2017). Several studies have been performed to include classical ergonomic indexes such as OCRA, NIOSH, OWAS, or RULA in assembly line balancing, scheduling, or sequencing problems in the form of multi-objective functions or additional constraints.

More recently, other strategies for including ergonomic and human factors into the design of assembly systems have focused on general and local physical fatigue from performing single tasks or a set of activities (Battini et al. 2015). Moreover, new smart technologies and Industry 4.0 solutions can provide useful data concerning the workers' physical state, general health conditions, or anthropological data. Wearables can also provide a wide range of sensors for measuring acceleration, motion and stress (e.g., number of steps, times of day when the operator is standing/sitting, and pace of work), which can be associated with the operators' physical workload.

Several benefits can be linked to the use of these new and innovative technologies. First, the same device can be used to collect several types of data during the execution of tasks. These are also noninvasive solutions, since smart devices are light and easy to wear, and do not interfere with the working environment. All data collected with smart devices can be shared between managers or staff coordinators via cloud platforms. In this way, suggestions or warnings can be provided to workers within a few seconds, and dynamic scheduling or sequencing of the tasks to be performed can be done based on the workers' condition. In this way, the risk of injury among workers is reduced, and a correct balance can be found not only for working time but also ergonomic effort.

Moreover, using the cloud platform, all data can be used to implement new assembly workstations. The collected data can provide valuable measures of the ergonomics effort required to perform specific tasks or general activities, and thus can be used to correctly balance new workstations, not only from the point of view of time but also from an ergonomics perspective. In these circumstances, multi-objective approaches can be used to find a good trade-off between productivity and ergonomics.

In this chapter, we provide an overview of wearable 4.0 devices that can be used to evaluate ergonomics conditions. Energy expenditure will be used as an ergonomics constraint in a mixed-assembly line balancing and sequencing problem. In Sect. 8.2, general guidelines for direct ergonomics measurements and wearable tools are given, while in Sect. 8.3, general considerations about assembly systems are discussed. In Sects. 8.4 and 8.5, the mixed-assembly line balancing problem and the sequencing problem will be detailed and a numerical example will be provided. Finally, conclusions and general guidelines are provided in Sect. 8.6.

8.2 Methods and Tools to Measure Fatigue

In this section, the main methods and tools used to quantify the physical effort required to execute a set of activities are described, and the pros and cons of each are listed.

When workers execute assembly tasks, they are subjected to a physical effort that may involve the whole body or only certain parts. When the whole body is used, general fatigue arises, while local muscle fatigue arises if only certain parts of the body are involved in a strenuous effort. In both cases, several methods and tools can be used to evaluate this fatigue. However, only some of the available tools can be considered wearable devices, and only a few can be connected to the cloud, allowing the possibility of evaluating workers' conditions in real time (Battini et al. 2018).

Firstly, these methods can be categorized into qualitative and quantitative approaches (Abdous et al. 2018). From within the quantitative approaches, we identify direct measurement tools or observational methods, and these will be discussed below.

Qualitative methods consist of subjective evaluations, based on verbal estimations made by the operators during execution of the task. The advantages of using these techniques are related to their low cost in comparison to other methodologies, which require high levels of investment to buy the required equipment and significant amounts of time to understand how to use and test it in the specific industrial context. Moreover, subjective evaluations can give feedback not only on the stress on the muscles and joints during the activity but also on the central nervous system.

Despite these advantages, they are influenced by subjectivity, and this leads to difficulty in assessing the accuracy and variability of a given measure between different operators. Evaluations by operators for the same load may be different according to their physical capacity and general health. In addition, the precision of a measure may be different if the operator has previous exposure to the benchmark. For this reason, qualitative methods can be used as a practical tool to involve workers in some ergonomic decisions, giving them the opportunity to evaluate their working environment in a straightforward way.

The other approach involves quantitative methods, which are related to the real measurement of the load using existing devices. Observational methods fall into the category of quantitative approaches.

With regard to general muscular fatigue, the most widely used observation methods are those proposed by Garg et al. (1978) and the predeterminate motion energy system (PMSE) proposed in 2016 by Battini et al. Both are based on a measure of energy expenditure. The positive aspect of both methods is that they can take into account the differences between one person and another in terms of age, body weight, and height. However, even if these approaches can provide accurate values, they cannot be put into practice quickly since they are based on an evaluation of every individual movement of the operator performing a task. Thus, these methods are very time-consuming approaches, and cannot be used to monitor workers' well-being in real-time. For this reason, wearable devices are preferable in order to collect data that can be compared and then used to make appropriate changes.

8.2.1 Wearable Devices

From the point of view of muscular fatigue, two main types of tools can be used to obtain direct, quantitative values of the local effort expended by a worker executing a task.

An electromyography (EMG) sensor (Fig. 8.1) is the first type of device that made it possible to evaluate muscular fatigue, and is used to detect electrical activity

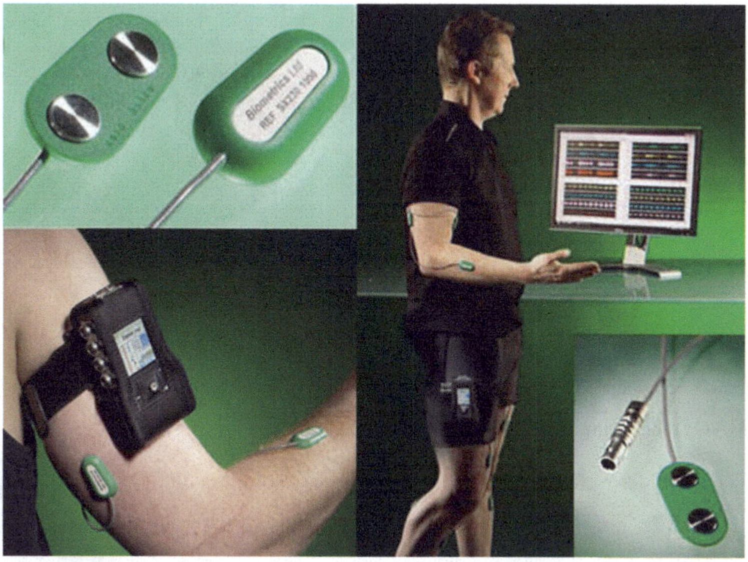

Fig. 8.1 EMG sensors. (Source: NexGen Ergonomics.com)

in the muscles. It involves the placement of electrodes on the skin surface above the muscle, and the contraction is monitored in order to evaluate the percentage maximum voluntary contraction (MVC) of the muscle during performance of the activity. The disadvantages of the EMG are the influence of other muscle movements, interference from the electrical supply, and mechanical problems with the recorded measurements of MVC.

Moreover, it is associated with certain problems related to the application, since different individuals may use different groups of muscles for the same task, and it is difficult to interpret the measure of MVC for multiple muscle groups.

This technology is complicated and costly for application in an industrial context. Moreover, the equipment used can affect the usual way of executing a task, since the electrodes are connected to the main hardware with wires.

There are also dynamometers and grip force sensors (Fig. 8.2), which are tools that are able to measure the peak and average force in kilograms during carrying, pushing, and pulling activities. They are fixed to the object to be carried, pushed, or pulled, and slipping must be prevented. Before they are used, it is essential to understand the direction of the forces representing the path of motion of the operator. These devices are easy to use, and the output data can reveal the kind of movement that a given operator performed in addition to the influence of the height and weight of the item. Based on the force level, it is then possible to estimate the local fatigue.

General or global fatigue is measured using two main tools: oxygen consumption (VO_2) monitoring systems or heart rate (HR) monitoring devices (available in the latest generation of smartwatches).

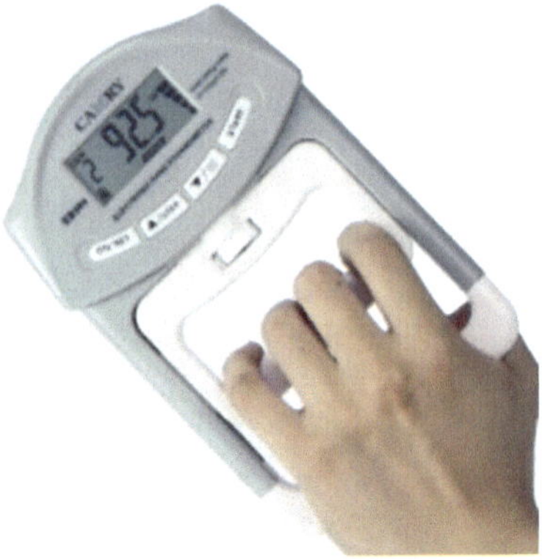

Fig. 8.2 Hand dynamometer and hand-grip force sensors

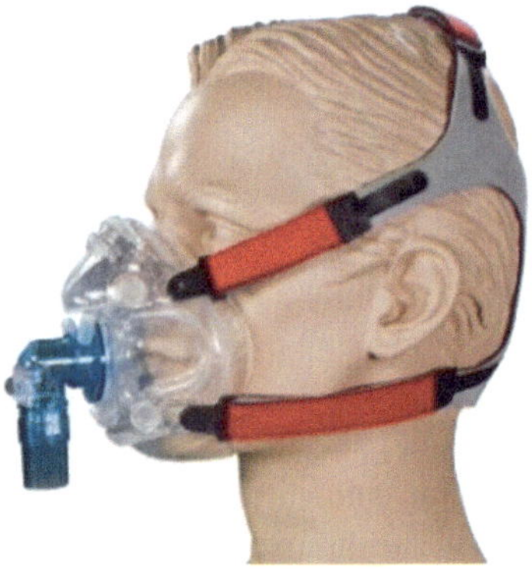

Fig. 8.3 VO_2 monitoring system

VO_2 monitoring (Fig. 8.3) is the most widely validated method in the literature, and its relationship with the activity performed has been demonstrated. However, it cannot be easily applied in an industrial context, since the investment required is considerable, and a certain level of preparation is needed for the use of the equipment. In addition, the most significant limitations are the size of the equipment and the inconvenience of using a mask for taking measurements, as it can influence the operator's performance due to stress and difficulties in breathing.

In recent years, the technology related to VO_2 measurements has developed a great deal; wearable wireless equipment is now available, and data can now be processed in real time. However, a laboratory testing phase is needed in order to evaluate whether additional effort is linked to wearing this technology, and only if positive laboratory results are achieved, can this technology be implemented in companies.

Another type of tool that can be used to evaluate general fatigue is the HR monitoring system. Nowadays, this tool is included in all smartwatches, and can provide and predict also some other important data such as the stress level of each workers, the number of steps perform during the day.

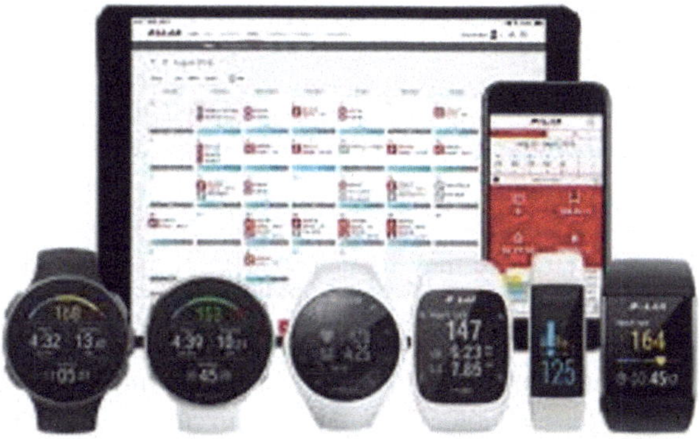

Fig. 8.4 HR monitoring systems. (Source: polar.com)

The traditional HR monitor (Fig. 8.4) is based on a Bluetooth HR sensor connected to a watch, where the trend of the HR and the duration of the activity are visualized. It is commonly used to obtain feedback regarding training status and to improve the physical fitness of a person through accurate planning of the next training activities. It does not require specific knowledge, and does not interfere with the operator's activity. It allows real-time feedback to be given to the operator, who can be conscious of his or her physical condition, and if appropriate, can speed up or slow down the rate of the activity. Moreover, it recognizes the effect that personal characteristics such as age, weight, VO_2, HR at rest, and training status can have on the accumulation of fatigue. The use of HR as a measure of energy expenditure has been analyzed both in the past and in more recent literature, and several works have demonstrated its applicability in evaluating general fatigue in terms of energy expenditure (Calzavara et al. 2018).

Measurements taken with an HR monitor are easier than those with a VO_2 mask. Everyone can use the HR monitor without difficulty, and the measurement of HR to monitor fatigue levels can be carried out for several activities without disturbing the operators. It can also be connected to a cloud platform, and real-time values can be used to evaluate the workers' physical state.

Due to these advantages, an HR device may be the best device to carry out a fatigue analysis and to evaluate the energy expenditure required to perform a task or a set of activities.

Using these devices, sequencing solutions can be analyzed and continuously changed with the aim of finding a solution that can assign more strenuous and heavier activities first and lighter ones afterward. However, these solutions may change according to the type of worker and his/her features, and for this reason, HR devices play a crucial role during this phase. If used correctly, they can provide continuous information on the health status of workers.

8.3 Manual Assembly Systems: A General Overview

Manual assembly systems, also known as manual assembly lines, are used in several industrial contexts since they allow workers to collect and fit together various parts or components to create a final product (Boysen et al. 2007). They were initially introduced to increase efficiency in the mass production of standardized products, but more recently they have gained importance in the low-volume production of customized products.

Assembly systems or lines consist of several workstations at which a set of operations, generally called tasks, are performed by one or more workers (Fig. 8.5).

Due to technological and organizational restrictions, certain tasks can be performed only after the execution of others, and it is therefore necessary to define a so-called precedence graph (Fig. 8.6).

Depending on the variety of products assembled on the same line, three types of assembly lines (Fig. 8.7) may be used:

1. Single-model lines: The same product is manufactured in massive quantities in the same line. In this case, the tasks executed at each workstation are always the same, as is the workload.

Fig. 8.5 A simple assembly line

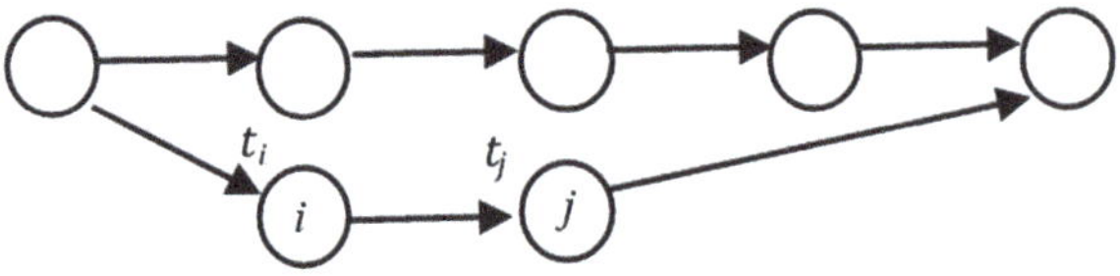

Fig. 8.6 Precedence graph

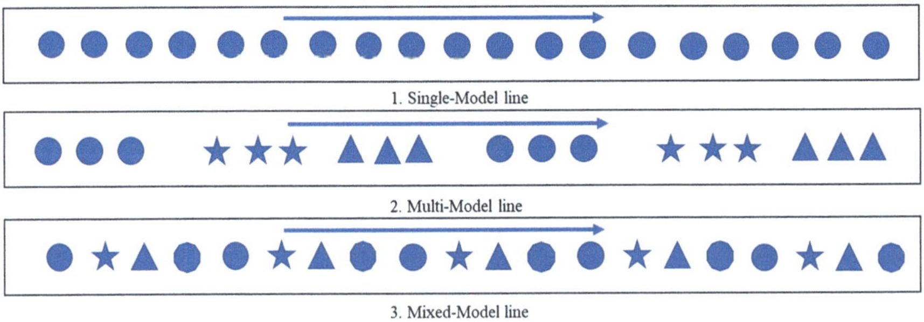

Fig. 8.7 Examples of assembly lines

2. Multimodel lines: Several similar products are constructed on one or more lines. In this case, there are significant differences in the manufacturing processes of different products, and setups are generally required.
3. Mixed-model lines: Several versions of the same family of an item are produced on the same line. In this case, models differ in terms of certain attributes or features. Some products may or may not require certain tasks, and a given task may require a variable process time depending on the variety of products. In this case, since products are very similar to each other, no or short setup times are required when the product changes.

The execution of the tasks required to obtain the final product is called manual assembly, and this represents one of the most critical phases of the production systems due to its high added value, its contribution to the final product quality, and its direct connection with the final market. For these reasons, practitioners and academics are continuously developing new approaches and improving the existing ones to increase efficiency and productivity, and to guarantee the required flexibility.

The problem of defining which tasks must be executed at each workstation is called the assembly line balancing problem (ALBP). The first attempt to create a balancing model was made by Salveson (1955), who suggested a linear program to describe all possible task assignments for an assembly line.

Three different methodologies to address the ALBP are described in the literature: the single-model assembly line balancing problem (SALBP), the mixed-model assembly line balancing problem (MALBP), and the batch-model (or multi-model) assembly line balancing problem (BMALBP).

In an ALBP, the objective or goal function may be the minimization of the number of workstations (ALBP-1), minimization of the cycle time (ALBP-2), or maximization of efficiency (ALBP-E). In each case, the aim is to evaluate the quality of a feasible solution based on the final goal and the constraints.

There also the feasibility problem (ALBP-F), in which the number of stations and the cycle time are known, and the problem involves evaluating whether or not a line with these characteristics can be operated.

To solve ALBPs, it is necessary to take into account different kinds of constraints (such as assignment or cycle time constraints). The solution methods might be exact, simple heuristics or metaheuristics, and a compromise is required between two or more conflicting objectives (Becker and Scholl 2006). This is known as a multi-criteria approach, and includes lexicographic resolution, fuzzy goal programming, and Pareto-based ranking.

As discussed in Sect. 8.1, during recent decades, many studies have highlighted the relevance of integrating ergonomics aspects into the design of an assembly system, and for this reason, several studies have been carried out in the ALBP field.

Most of the proposed methods integrate ergonomics using multi-criteria approaches, and single-model assembly lines are generally analyzed. Finco et al. (2018) proposed a heuristic approach to integrate human energy expenditure into SALBP-1. Moreover, to improve workers' well-being Finco et al. (2020) developed a mathematical model able to minimize smoothness index by integrating also workers' recovery time. However, the single-model assembly line represents a strong restriction for many companies, since the customization level is very high nowadays, and thus mixed-model assembly lines are preferable over single-model alternatives.

8.3.1 Mixed-Model Assembly Line

In this section, the characteristics of mixed-model assembly lines are described as part of our discussion of approaches that are able to integrate ergonomics into mixed-model assembly systems.

As stated above, in a mixed-model assembly line, different models of the same product family are processed with no or minor setups or rearrangements required between the different workpieces. In this case, the production process is similar for each product, and the differences are mainly due to customized features.

For this type of assembly system, several decision problems may arise depending on the planning horizon. The first decision concerns the design of the assembly line in which the number of workstations, production rates, and workload must be defined. All of these decisions are linked to the MALBP. A practical example of mixed model line balancing and sequencing is provided by Azzi et al. (2012a, b). In addition to the SALBP, the MALBP-1, MALBP-2, MALBP-E, and MALBP-F consist of defining the number of workstations, cycle time, line efficiency, and the feasibility problem, respectively. However, in these cases, the problem is more complicated, since each workstation must be balanced in each model. Smart solutions and cloud platforms could help managers and practitioners in this regard, since historical data on energy expenditure or the physical effort by workers in similar tasks could be used and adapted to the context under analysis. In this context, multi-objective approaches are preferable over mono-objective ones, since ergo-time solutions can achieve higher efficiency levels and improvements in the health status of workers.

Results obtained over a long period represent a basis for the division of the labor force and the production rate in the short term. The short-term decision problem is known as a mixed-sequencing problem, and consists of finding a sequence of model units for assembly, based on the short-term production program of maximizing or minimizing an objective function. Smart and Industry 4.0 solutions can also lead to improvements in the working environment. In this phase, real-time measures can be taken, and real-time sequencing adjustments can be made based on physical effort data that are continuously monitored through smartwatches.

The two problems described above are closely connected, since results obtained over the long term form the input data for a problem in the short term. This means that the quality and efficiency of sequencing decisions and planning are strongly correlated with workload balancing. On the other hand, the quality of balancing solutions depends on the model mix and the possible sequences. However, these data are generally not available before line balancing, since demand cannot be forecast exactly, and inefficiency can therefore occur.

Moreover, the planning horizons are different, and problems must be solved separately. For this reason, a hierarchical planning system is generally used.

Based on these assumptions, a MALPB and a sequencing model will be developed in the following sections. In both cases, ergonomics will be integrated according to the energy expenditure required to execute each task. A hierarchical approach is followed.

8.4 Ergo-Mixed-Model Assembly Line Balancing Problem

In this section, an ergo-mixed-model assembly line balancing problem is discussed. The ergonomic level of each assembly task is defined using the energy expenditure rate, which can be measured with an HR monitoring device (Sgarbossa et al. 2016).

A multi-objective MALB model that evaluates both the energy and time required for each task is developed to evaluate the effects of moving from a time-optimal solution to an energy-time optimal one.

First, the MALBP is converted into an SALBP using the joint precedence graph. Then, the virtual average model (VAM) is considered, since this can simulate a set of various products, and the SALBP model is then solved. It is not easy to evaluate the behavior and efficiency of a mixed-model assembly line, and the use of a VAM can help make the balancing problem easier; however, for a multi-objective model that optimizes time and energy, this approach can lead to optimal solutions in terms of time and energy that are different from those found by considering the entire mix instead of the VAM.

In the following, the steps required to solve the MALBP will be described.

8.4.1 Virtual Average Model with Time and Ergonomics Approaches

Depending on the product mix and demand, it may be easier to turn the mixed-model assembly line into a single-model case by joining the precedence graph of each model into a join precedence graph. A join precedence diagram implies that:

- there is a common subset of tasks among the considered models;
- some tasks may be required for one model but not for others;
- the same task may have different operating times for different models, implying that it must be performed at the same station.

The balancing of a mixed-model assembly line requires not only a joint precedence graph but also the concept of the VAM, which consists of a dummy average model representing all the products assembled on this line. The time for each activity of the VAM can be calculated based on the following points:

- the maximum time required for this activity, considering all products;
- the average time required for this activity, considering all products;
- the weighted average time required for this activity, considering the mix of products.

In this chapter, the third approach to evaluating the VAM is applied, and it is integrated with the multi-objective problem based on energy expenditure.

In the multi-objective approach, we consider both the solution that optimizes time and the one that optimizes energy. The analysis of the Pareto frontiers allows us to evaluate the trade-off from one solution to the other. This gives the set of nondominated solutions for a multi-objective system, where the solutions optimize one of the objectives of the problem.

In order to apply this kind of approach, it is necessary to know the task time and the energy expenditure for each task in each model. We denote t_{jm} as the time for task j in model m, e_{jm} as the relative energy expenditure, and d_m as the percentage demand of model m in the considered mix. The formulae for t_j and e_j for a VAM are as follows:

$$t_j = \sum_m t_{jm} d_m \tag{8.1}$$

$$e_j = \sum_m e_{jm} d_m \tag{8.2}$$

8.4.2 *Time-SALBP and Energy-SALBP with VAM*

The definition of t_j allows us to balance the mixed-model assembly line with the SALBP-2, which is the SALBP model that minimizes the cycle time with a predefined number of stations, in order to increase the productivity. Based on the binary linear model in the single-model assembly line, the binary variable x_{jk} is used to indicate the assignment of task j to station k, and B_k is the set of tasks assignable to station k (within a set of workstations from 1 to K). To solve the SALBP-2, the following constraints need to be considered:

- Occurrence constraint:

$$\sum_k x_{jk} = 1 \qquad \forall j = 1, \ldots, n \tag{8.3}$$

- Cycle time constraint:

$$\sum_j x_{jk} t_j \le c \qquad \forall k = 1, \ldots, K \tag{8.4}$$

- Precedence constraint:

$$\sum_{k} kx_{hk} \leq \sum_{i} ix_{ji} \qquad \forall (h, j) \in A \tag{8.5}$$

There may be many different balancing solutions, and each one needs to be evaluated. In this case, unlike traditional approaches, the objective functions considered are the time smoothness index (SX-T) and the energy smoothness index (SX-E), which measure the equality of workload distribution among the stations and the physical load on workers at different stations, respectively. They are defined as follows:

$$\min\ SX - T = \min \sqrt{\sum_{k=1}^{K} \left(c_r - \sum_{j} x_{jk} t_j\right)^2} \tag{8.6}$$

$$\min\ SX - E = \min \sqrt{\sum_{k=1}^{K} \left(e_r - \sum_{j} x_{jk} e_j\right)^2} \tag{8.7}$$

where c_r (resp. e_r) is the maximum station time (resp. energy expenditure) for all stations.

To evaluate the trade-off between the time-based and energy-based optimal solutions, the Pareto frontier (the set of nondominated solutions of a multi-objective system) is defined using the following function:

$$\begin{aligned} &\min \{SX - T; SX - E\} \\ &= \min \left\{ \sqrt{\sum_{k=1}^{K} \left(c_r - \sum_{j} x_{jk} t_j\right)^2}; \sqrt{\sum_{k=1}^{K} \left(e_r - \sum_{j} x_{jk} e_j\right)^2} \right\} \end{aligned} \tag{8.8}$$

8.4.3 *Time- and Energy MALBPs*

In a mixed-model assembly line, the time-SALBP and energy-SALBP refer to the concept of VAM might not be enough to evaluate the changing properly in the Pareto frontier. When the same balancing solutions are evaluated knowing the number of stations and the cycle time (MALBP-F), the idle time and the workloads may be different depending on the distribution of work. To evaluate a balancing solution properly, it is necessary to include objective functions for the analysis of smoothed station loads. In a mixed-model assembly line, we can define T_{mk} and E_{mk} as the processing time and energy expenditure per unit of model m at station k. S_k is the set of tasks assigned to station k, and T_{mk} and E_{mk} can be defined as follows:

$$T_{mk} = \sum_{j \in S_k} t_{jm} \tag{8.9}$$

$$E_{mk} = \sum_{j \in S_k} e_{jm} \tag{8.10}$$

Knowing the values of T_{mk} and E_{mk}, we can define c_{mr} and e_{mr} for the model m as

$$c_{mr} = \max \{T_{mk} | m = 1, \ldots, M\} \tag{8.11}$$

$$e_{mr} = \max \{E_{mk} | m = 1, \ldots, M\} \tag{8.12}$$

Having defined these instances, the balancing solutions can be evaluated not only in terms of the operation time but also the operation energy.

The following functions express the maximal deviation of the operation time/operation energy of a model from the maximum station time/energy weighed on the demand of each model:

$$\Psi_t = \sum_{m=1}^{M} \sum_{k=1}^{K} | T_{mk} - c_{mr} | d_m \tag{8.13}$$

$$\Psi_e = \sum_{m=1}^{M} \sum_{k=1}^{K} | E_{mk} - e_{mr} | d_m \tag{8.14}$$

The multi-objective functions of this second approach are introduced to allow us to compare them with a time-based and energy-based approach applied to a virtual average product, in order to demonstrate how the Pareto frontier changes when one approach or the other is used for a mixed-model assembly line.

8.4.4 Numerical Application

A numerical example is provided here to evaluate the effects that energy expenditure can have on the solution to the MALBP.

In the case study described here, there are three models, and the joint precedence graph contains 17 tasks that are denoted by A, ..., Q (Fig. 8.8). For each model and each task, the time and the energy expenditure are given in Table 8.1.

In this case, smartwatches are used to evaluate HR values while executing the tasks, and energy expenditure values are then obtained using regression models and general formulae.

Different kinds of VAM are also considered, and these are obtained by considering a different demand for each model weighted in the mix. The correlation index R between the time and energy expenditure for each model and VAM is known.

This analysis aims to apply a multi-objective approach through the utilization of the Pareto frontier in a MALBP, and to discover whether or not the traditional practice of evaluating a mixed model assembly line by approximating all the considerations to the concept of a VAM has a meaning, and in which cases.

This numerical example is calculated by computing different balancing solutions for each kind of VAM using the Patterson and Albracht algorithm, and evaluating each possible assignment of the task to each station in order to respect the occurrence, cycle time, and precedence constraints.

For each feasible balancing solution for each VAM (whose values of t_j and e_j depend on the particular mix considered), SX-T (resp. SX-E) is calculated based on the concept that a mixed-model line can be transformed to a single-model one by producing a VAM. We then calculate Ψ_t and Ψ_e, based on which we derive the line efficiency in terms of time and energy, considering the effect that a particular mix might have.

Using these functions, it is possible to define two different kinds of frontiers. The aim of obtaining the nondominant solutions in terms of time and energy is the same, but the first considers SX-E and SX-T, and the second Ψ_e and Ψ_t.

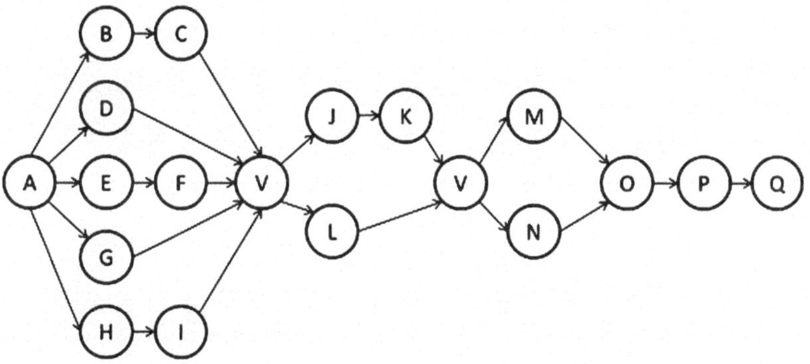

Fig. 8.8 The joint precedence graph

Table 8.1 Input data for the numerical example

							VAM sceneries							
							Mix A		Mix B		Mix C		Mix D	
	Model 1		Model 2		Model 3		33.3%,33.3%, 33.3%		5.56%, 22.2%, 72.2%		22.2%, 72.2%, 5.56%		72.2%, 5.56%, 22.2%	
R	0.559		0.729		0.678		0.662		0.684		0.696		0.598	
Task	t_j (s)	e_j (kcal)	t_j (s)	e_j (kcal)	t_j (s)	e_j (kcal)	t_j (s)	e_j (kcal)	t_j (s)	e_j (kcal)	t_j (s)	e_j (kcal)	t_j (s)	e_j (kcal)
A	19.20	0.76	24.00	0.95	28.80	1.14	24.00	0.95	27.20	1.07	23.20	0.92	21.60	0.85
B	46.00	2.69	55.20	3.23	36.80	2.15	46.00	2.69	41.40	2.42	52.13	3.05	44.47	2.60
C	15.60	1.08	13.00	0.90	10.40	0.72	13.00	0.90	11.27	0.78	13.43	0.93	14.30	0.99
D	7.00	0.18	5.60	0.14	8.40	0.21	7.00	0.18	7.70	0.19	6.07	0.15	7.23	0.18
E	20.00	0.95	25.00	1.19	30.00	1.43	25.00	1.19	28.33	1.35	24.17	1.15	22.50	1.07
F	15.00	0.88	12.00	0.70	18.00	1.06	15.00	0.88	16.50	0.97	13.00	0.76	15.50	0.91
G	4.00	0.12	5.00	0.15	6.00	0.18	5.00	0.15	5.67	0.17	4.83	0.15	4.50	0.14
H	45.60	1.47	30.40	0.98	38.00	1.23	38.00	1.23	36.73	1.19	34.20	1.11	43.07	1.39
I	11.00	0.96	8.80	0.77	13.20	1.16	11.00	0.96	12.10	1.06	9.53	0.84	11.37	1.00
J	64.00	3.65	96.00	5.47	80.00	4.56	80.00	4.56	82.67	4.71	88.00	5.02	69.33	3.95
K	68.00	4.06	85.00	5.07	102.00	6.09	85.00	5.07	96.33	5.75	82.17	4.90	76.50	4.56
L	30.00	1.20	25.00	1.00	20.00	0.80	25.00	1.00	21.67	0.87	25.83	1.03	27.50	1.10
M	60.00	1.08	72.00	1.30	48.00	0.87	60.00	1.08	54.00	0.97	68.00	1.23	58.00	1.05
N	65.00	1.17	52.00	0.94	78.00	1.41	65.00	1.17	71.50	1.29	56.33	1.02	67.17	1.21
O	36.00	0.74	45.00	0.92	54.00	1.10	45.00	0.92	51.00	1.04	43.50	0.89	40.50	0.83
P	20.00	0.40	25.00	0.50	30.00	0.60	25.00	0.50	28.33	0.57	24.17	0.48	22.50	0.45
Q	19.20	0.76	12.80	0.50	16.00	0.63	16.00	0.63	15.47	0.61	14.40	0.57	18.13	0.72

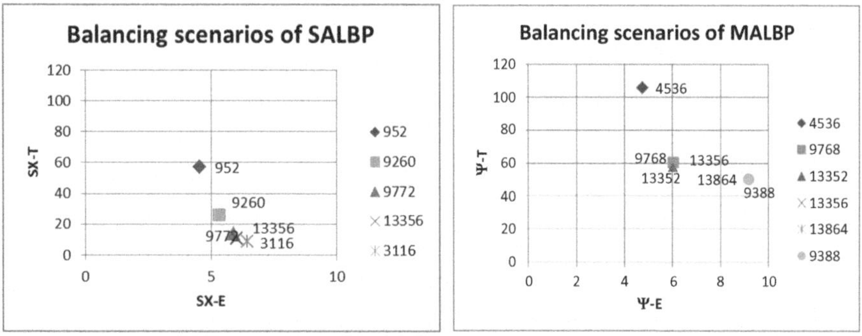

Fig. 8.9 SALBP and MALBP frontiers for mix A

These two approaches cannot have the same frontier, as can be seen in the example below, which involves a mix containing 33.3% of M1, 33.3% of M2, and 33.3% of M3. The numbers and types of scenarios are different (where each one implies a specific balancing solution within the frontier) (Fig. 8.9).

If the scenarios for the Pareto frontier are different between the two approaches, we need to evaluate whether the difference is more evident for some mixes than others. If this difference is substantial, the choice of the SALBP approximation is not the right way to evaluate a mixed-model line, or if used, the choice of one of the SALBP solutions would not correspond appropriately to the MALBP solution.

In Fig. 8.10, the efficient frontier for the MALBP is compared with the frontier for the MALBP for each mix, the scenarios for the SALBP frontier are considered and the scenarios of the time-optimal solution and energy-optimal solution are highlighted.

The difference between the two frontiers is not the same in all mixes, but the frontier for the SALBP is always to the right of the real frontier for each mix considered. A frontier can move from left to right if the correlation index R (the relation between t_j and e_j) decreases; the more it increases, the more the frontier moves to the left and is reduced to a point.

In order to evaluate the deviation between the two frontiers ΔT_{opt} and ΔE_{opt}, it is necessary to consider the two-point of time-optimal and energy-optimal solutions of the two frontiers and to analyze the subsequent measures. We define Ψ_{T-VAM} and Ψ_{E-VAM} as the optimum points of the SALBP frontier for time and energy and Ψ_T^*, Ψ_E^* the optimum points for time and energy for the other frontier in the MALBP:

$$\Delta T_{opt} = \frac{\Psi_{T-VAM} - \Psi_T^*}{\Psi_T^*} \tag{8.15}$$

$$\Delta E_{opt} = \frac{\Psi_{E-VAM} - \Psi_E^*}{\Psi_E^*} \tag{8.16}$$

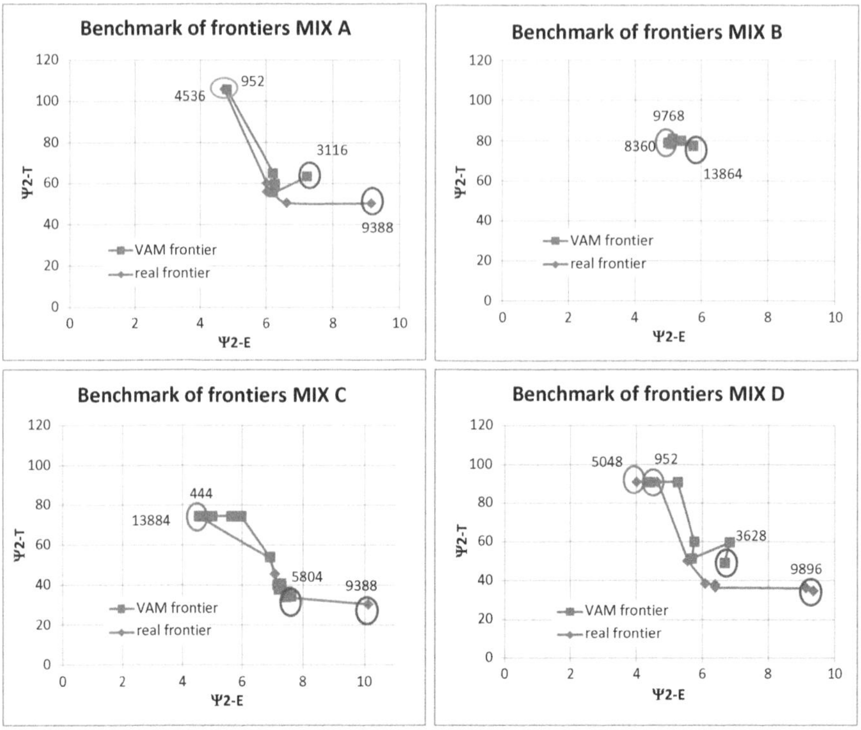

Fig. 8.10 Comparison of SALBP and MALBP frontiers for all mixes

Table 8.2 Results for the different mixes

Mix	ΔT_{opt} (%)	ΔE_{opt} (%)
A	20.6	1.7
B	–	2.6
C	11.9	9.7
D	29.7	9.2

The results for the mixes considered here are given in Table 8.2.

As it can be seen from the table above, the error in the points of the energy-optimal solutions is lower, since the feasible balancing solutions take into account only the time (cycle time). Conversely, there is a higher value of error for the time-optimal solutions.

8.5 Ergo-Sequencing Problem

As stated in Sect. 8.3, for mixed-model assembly systems, the balancing phase represents the long-term decision process. During the balancing phase, important decisions concerning the assembly system design are taken. Moreover, the assignment of tasks to stations is conducted based on the long-term demand for items and the VAM.

However, in the short-term, for mixed-model assembly systems, the main issue is to define the sequence of products to launch down the line in order to respect the short-term demand for products and to minimize certain objective functions such as the total work overload, the total idle time or the labor cost. In the short term, the demand mix may be slightly different from the long term one due to problems with material suppliers. Companies must therefore schedule the assembly process according to the availability of material, and the model mix may therefore change. Proper assembly sequencing can be used to cover inefficiencies.

In recent years, companies have solved the sequencing problem in several ways, but in most cases, ergonomics and working conditions have been neglected. However, in the same way as the balancing process, wearable devices, Industry 4.0 solutions, IoT, or cloud platforms can be used to achieve good product sequencing, including workers' physical conditions. Dynamic scheduling can therefore be conducted, and real-time changes can be made based on the operators' fatigue level. In this way, a reduction in productivity can be avoided by assigning light tasks to more fatigued workers and heavier tasks to less fatigued ones.

Starting from balancing solutions provided with a multi-objective approach defined in the previous section, a sequencing model that evaluates energy expenditure inefficiencies is presented. Since the balancing methodology used above generates several solutions, the sequencing model is applied to each balancing scenario associated with the Pareto frontier. Finally, the model sequence that provides the best results in terms of both work overload and energy overload is chosen.

8.5.1 Sequencing Model

Starting from the line balancing phase and considering the features of each model in terms of operating times than energy expenditure, Eq. (8.9) (resp. Eq. (8.10) can be used to evaluate the processing time (resp. energy expenditure) in model m at station k. The set of data acquired through Eqs. (8.9) and (8.10) forms the input data for the sequencing model. The following data are required to solve the problem. For each model, the short-term demand is known, and this is set to d'_m, while the sequence length I is equivalent to the sum of the demand in each model $I = \sum_m d'_m$.

The sequencing decision variable is introduced as follows:

$$y_{mi} = \begin{cases} 1, if\ m\ is\ assigned\ to\ the\ ith\ position\ of\ the\ sequence \\ 0, otherwise \end{cases} \tag{8.17}$$

The other variables that must be included are as follows:

- ΔE^*_{ki} $\forall\, k, i$, defined as $\max\{0; \sum_m E_{mk} y_{m,\,i+1} - \sum_m E_{mk} y_{m,\,i}\}$. This represents the energy expenditure overload between two consecutive units processed. Only positive gaps are considered, since models that require higher ergonomic effort should be assessed before the lighter ones.

- s_{ki} $\forall$ k, i represents the operator start position at station k for the ith unit;
- wo_{ki} $\forall$ k, i represents the work overload at station k for the ith unit. This is defined as $\max\{0; s_{ki} - \sum_m T_{mk} y_{mi} - c_r\}$

The objective function of the sequencing model is defined as follows:

$$\min\ EO = \min \sum_k \sum_i \Delta E^*{}_{ki} \tag{8.18}$$

This minimizes $\Delta E^*{}_{ki}$ considering all stations and the sequence of the product. The following constraints then need to be considered:

- Only one model unit must be assigned to each position of the sequence, according to the following formula:

$$\sum_m y_{mi} = 1 \quad \forall i \tag{8.19}$$

- For each model, the short-term demand d'_m must be met, according to following equation:

$$\sum_i y_{mi} = d'_m \qquad \forall m \tag{8.20}$$

- The processing of a model unit must start only when the previous unit has been completed, as defined by

$$s_{k,i+1} \geq s_{ki} + \sum_m T_{mk} y_{mi} - c_r - wo_{ki} \quad \forall k, i \tag{8.21}$$

- The line must be in the initial state before and after unit production:

$$s_{k1} = s_{k,I+1} = 0 \qquad \forall k \tag{8.22}$$

- The objective function $\Delta E^*{}_{mi}$ is defined as the maximum value, and is nonlinear. To linearize this variable, the following additional constraints and an additional Boolean variable must be included in the final model:

$$\Delta E^*{}_{ki} \geq \sum_m E_{mk} y_{m,i+1} - \sum_m E_{mk} y_{m,i} \qquad \forall k; \forall i = 1, .., I-1 \tag{8.23}$$

$$\Delta E^*{}_{ki} \geq 0 \qquad \forall k; \forall i = 1, .., I-1 \tag{8.24}$$

$$\Delta E^*{}_{ki} \leq \sum_m E_{mk} y_{m,i+1} - \sum_m E_{mk} y_{m,i} + UB\,(1 - z_{ik}) \qquad \forall k;\forall i = 1,..,I-1 \tag{8.25}$$

$$\Delta E^*{}_{ki} \leq 0 + UB z_{ik} \qquad \forall k;\forall i = 1,..,I-1 \tag{8.26}$$

- While the following constraints set the type of variables:

$$s_{ki} \geq 0 \quad \forall k, i \tag{8.27}$$

$$wo_{ki} \geq 0 \quad \forall k, i \tag{8.28}$$

$$y_{mi} \in \{0;\,1\} \qquad \forall m, i \tag{8.29}$$

$$z_{ki} \in \{0;\,1\} \qquad \forall k;\forall i = 1,..,I-1 \tag{8.30}$$

The sequencing model proposed here assigns a model unit in a point of the sequencing length to minimize the total energy overload of the assembly systems, based on the difference in energy expenditure between two consecutive product units. Moreover, the processing of each unit model starts only after the previous one has been completed. In this way, both the time and ergonomics aspects are considered simultaneously.

8.5.2 Numerical Example

A numerical example is used to test the balancing approach and to evaluate the sequencing methodology. The balancing solutions obtained for mix A are used. In the short-term, the demand for each model and among models can change, and six scenarios are therefore analyzed, as shown in Table 8.3.

The sequencing model is then applied for each mix, and each balancing solution is related to each point of the Pareto frontier (see Fig. 8.9).

It is interesting to note from Table 8.4 that for the same point on the Pareto frontier, and thus the same balancing solution, the total energy overload is assumed

Table 8.3 Short-term demand mix

Model	Mix 1	Mix 2	Mix 3	Mix 4	Mix 5	Mix 6
Model 1	6	4	7	5	7	7
Model 2	6	7	7	6	6	3
Model 3	6	7	4	7	5	8

Table 8.4 Energy overload results

		Energy overload					
SX-E	SX-T	Mix 1	Mix 2	Mix 3	Mix 4	Mix 5	Mix 6
4.56	57.16	0.86	0.86	0.86	0.86	0.86	0.86
5.31	25.79	0.96	0.96	0.96	0.96	0.96	0.96
5.91	13.96	0.93	0.93	0.93	0.93	0.93	0.93
6.01	10.82	0.89	0.89	0.89	0.89	0.89	0.89
6.45	8.77	1.02	1.02	1.02	1.02	1.02	1.02

Table 8.5 Work overload results

		Work overload					
SX-E	SX-T	Mix 1	Mix 2	Mix 3	Mix 4	Mix 5	Mix 6
4.56	57.16	120	140	116	128	112	100
5.31	25.79	180	210	120	210	150	240
5.91	13.96	228	266	164	262	194	284
6.01	10.82	240	280	181	237	207	285
6.45	8.77	258	301	211	288	228	279

to always have the same value. Conversely, Table 8.5 shows that the total work overload, defined as $WO = \sum_k \sum_i wo_{ki}$, varies between mixes for the same balancing solution. Moreover, the same model is used continuously until the short-term demand is achieved. In this way, the sequencing approach is closely linked to the differences in energy expenditure at each station between models.

Figure 8.11 provides information that is valuable for identifying which point on the Pareto frontier is preferable compared to the others, and this can help in the selection of the mix. For each point on the Pareto front, both the *EO* and *WO* for each mix are illustrated.

The point on the Pareto front that provides acceptable results in terms of *EO* that *WO* is the one at which the value of SX-E is minimized, while the point that minimizes SX-T provides a higher *EO* and *WO*.

It is interesting to compare the ergo-sequencing results with those of the traditional sequencing model ($WO*$ and $WO*$) that minimizes the work overload. We define $\Delta EO_{mix(i)}$ and $\Delta WO_{mix(i)}$ for a generic mix *i*th as

$$\Delta EO_{mix(i)} = \frac{EO_{\mathrm{mix}(i)} - EO^*{}_{mix(i)}}{EO^*{}_{mix(i)}} \tag{8.31}$$

$$\Delta WO_{mix(i)} = \frac{WO_{\mathrm{mix}(i)} - WO^*{}_{mix(i)}}{WO^*{}_{mix(i)}} \tag{8.32}$$

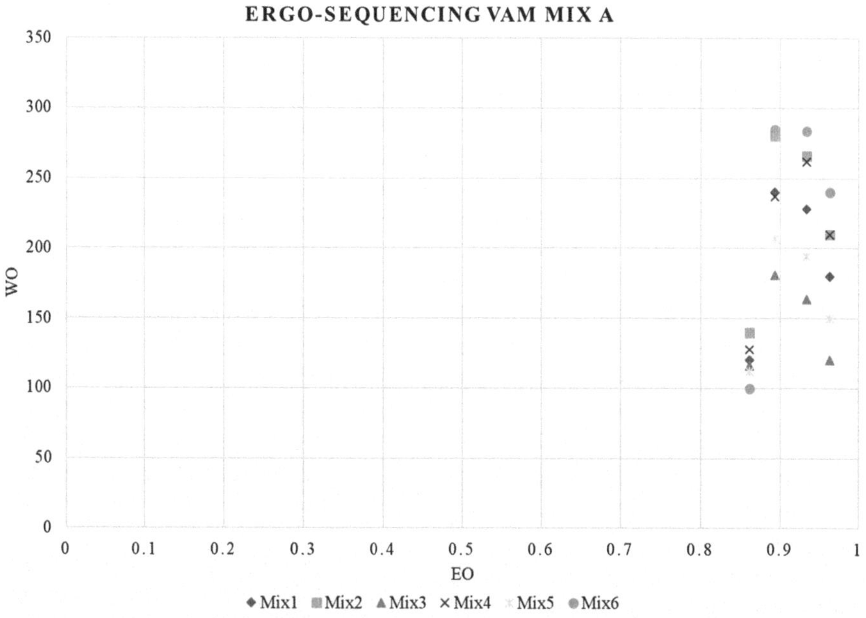

Fig. 8.11 Graph of energy and work overload

Table 8.6 Results for energy and work overload

SX-E	SX-T	Δ*EO* (%)	Δ*WO* (%)
4.56	57.16	−80.21	11.33
5.31	25.79	−76.71	24.31
5.91	13.96	−78.92	28.47
6.01	10.82	−79.64	16.34
6.45	8.77	−79.82	13.87

Table 8.6 shows the deviation between the results of the ergo and traditional sequencing models. The mean value of the mix is shown.

These results are very interesting, and confirm that the point on the Pareto front that minimizes SX-E is preferable over the others for two main reasons.

Firstly, *WO* is higher than WO^*, but it is closer to WO^* than the other points. Its ΔEO is −80.21%, meaning that the solution provided by the ergo-sequencing model can achieve the minimum *EO* and at the same time can provide correct solutions in terms of *WO*.

Moreover, the higher the value of SX-T, the lower the value of *WO*, since the work load is not well balanced between workstations, and idle time can occur in support of work-overload. Good sequencing results can therefore be achieved.

8.6 Conclusion

The design of efficient assembly systems requires the integration of ergonomics aspects, since the well-being and safety of operators implies an improvement in the final product quality and a reduction in costs related to absenteeism and employee turnover caused by accidents or injuries. Moreover, an ergonomics evaluation can be quickly conducted using smart solutions such as smartwatches, which can provide several forms of information about a worker's physical condition. Since many types of data can be collected with wearable devices, the cloud platform represents the best solution for collecting all of these data, which will be used to modify or improve the assignment of tasks to each workstation.

In this chapter, a balancing and sequencing model for a mixed-model assembly line has been described and discussed. The ergonomics level related to each task is defined based on the energy expenditure, since this can be easily quantified with smartwatches or HR monitoring systems. Using a multi-objective balancing model, SX-E and SX-T are minimized, and an in-depth analysis has been performed using a numerical example.

For each balancing solution belonging to the Pareto front, the ergo-sequencing model was applied to minimize the total energy overload. In this phase and the balancing phase, smartwatches can be used to monitor and quantify the physical effort required to execute the set of tasks assigned to each station according to the mix used.

Positive results are obtained since the optimal ergo-sequencing solution means that both energy and time overload can be minimized. Future researches in the field could try to develop similar approaches by considering other kind of human activities, as material storage, material handling, parts feeding to assembly systems, machines loading, and unloading (e.g., Zennaro et al. 2019).

Acknowledgments This research has received funding from the European Union's Horizon 2020 research and innovation program under the Marie Sklodowska-Curie grant agreement no. 873077 (MAIA-H2020-MSCA-RISE 2019).

References

Abdous, M. A., Finco, S., & Visentin, V. (2018, September). Workload evaluation of industrial work: Existing methods and practical applications.

Azzi, A., Battini, D., Faccio, M., & Persona, A. (2012a). Mixed model assembly system with multiple secondary feeder lines: Layout design and balancing procedure for ATO environment. *International Journal of Production Research, 50*(18), 5132–5151.

Azzi, A., Battini, D., Faccio, M., & Persona, A. (2012b). Sequencing procedure for balancing the workloads variations in case of mixed model assembly system with multiple secondary feeder lines. *International Journal of Production Research, 50*(21), 6081–6098. https://doi.org/10.1080/00207543.2011.653456.

Battini, D., Faccio, M., Persona, A., & Sgarbossa, F. (2011). New methodological framework to improve productivity and ergonomics in assembly system design. *International Journal of Industrial Ergonomics, 41*, 30–42.

Battini, D., Delorme, X., Dolgui, A., & Sgarbossa, F. (2015). Assembly line balancing with ergonomics paradigms: Two alternative methods. *IFAC-PapersOnLine, 48*(3), 586–591.

Battini, D., Calzavara, C., Persona, A., Sgarossa, S., Visentin, V., & Zennaro, I. (2018). Integrating mocap system and immersive reality for efficient human-centred workstation design. *IFAC-PapersOnLine, 51*(11), 188–193.

Becker, C., & Scholl, A. (2006). A survey on problems and methods in generalized assembly line balancing. *European Journal of Operational Research, 168*(3), 694–715.

Boysen, N., Fliedner, M., & Scholl, A. (2007). A classification of assembly line balancing problems. *European Journal of Operational Research, 183*(2), 674–693.

Calzavara, M., Persona, A., Sgarbossa, F., & Visentin, V. (2018). A device to monitor fatigue level in order-picking. *Industrial Management & Data Systems, 118*(4), 714–727.

Calzavara, M., Battini, D., Bogataj, D., Sgarbossa, F., & Zennaro, I. (2020). Ageing workforce management in manufacturing systems: State of the art and future research agenda. *International Journal of Production Research, 58*(3), 729–747.

Finco, S., Battini, D., Delorme, X., Persona, A., & Sgarbossa, F. (2018). Heuristic methods to consider rest allowance into assembly balancing problem. *IFAC-PapersOnLine, 51*(11), 669–674.

Finco, S., Zennaro, I., Battini, D., & Persona, A. (2019). Workers' availability definition through the energy expenditure evaluation. In *Proceedings—25th ISSAT International Conference on Reliability and Quality in Design* (pp. 29–33).

Finco, S., Battini, D., Delorme, X., Persona, A., & Sgarbossa, F. (2020). Workers' rest allowance and smoothing of the workload in assembly lines. *International Journal of Production Research, 58*(4), 1255–1270.

Garg, A., Chaffin, D. B., & Herrin, G. D. (1978). Prediction of metabolic rates for manual materials handling jobs. *The American Industrial Hygiene Association Journal, 39*(8), 661–674.

Ivanov, D., Sokolov, B., Dolgui, A., Werner, F., & Ivanova, M. (2016a). A dynamic model and an algorithm for short-term supply chain scheduling in the smart factory Industry 4.0. *International Journal of Production Research, 54*(2), 386–402.

Ivanov, D., Sokolov, B., & Ivanova, M. (2016b). Schedule coordination in cyber-physical supply networks Industry 4.0. *IFAC-PapersOnLine, 49*(12), 839–844.

Otto, A., & Battaïa, O. (2017). Reducing physical ergonomic risks at assembly lines by line balancing and job rotation: A survey. *Computers & Industrial Engineering, 111*, 467–480.

Panetto, H., Iung, B., Ivanov, D., Weichhart, G., & Wang, X. (2019). Challenges for the cyber-physical manufacturing enterprises of the future. *Annual Reviews in Control, 47*, 200–213.

Salveson, M.E. (1955). The assembly line balancing problem. *Journal of Industrial Engineering, 6*(3), 18–25.

Sgarbossa, F., Battini, D., Persona, A., & Visentin, V. (2016). Including ergonomics aspects into mixed-model assembly line balancing problem. In *Advances in physical ergonomics and human factors* (pp. 991–1001). Cham: Springer.

Zennaro, I., Finco, S., Battini, D., & Persona, A. (2019). Big size highly customised product manufacturing systems: A literature review and future research agenda. *International Journal of Production Research, 57*(15–16), 5362–5385.

Chapter 9
A Generic Decision Support Tool to Planning and Assignment Problems: Industrial Applications and Industry 4.0

Nathalie Klement and Cristóvão Silva

Abstract Decision support tools are essential to help the management of industrial systems at different levels: strategic to size the system; tactical to plan activities or assign resources; operational to schedule activities. We present a generic and modular decision support tool to solve different problems of planning, assignment, scheduling, or lot-sizing. Our tool uses a hybridization between a metaheuristic and a list algorithm. The specification of the considered problem is taken into account in the list algorithm. Several tactical and operational problems have been solved with our tool: a problem of planning activities with resources assignment for hospital systems, a lot-sizing and scheduling problem taking into account the setup time for a textile application and for a plastic injection problem, and a scheduling problem with precedence constraints. At the strategic level, this tool can also be used as part of Industry 4.0 to design reconfigurable manufacturing systems. This paper summarizes some problems solved with the proposed tool and presents its evolution.

9.1 Introduction

Industry 4.0 is the main international program which aims at improving the operational system in companies. More companies are concerned by this approach. Thanks to Internet of Things, "things" are connected to ease the communication, but other improvements may be envisaged. An integration of the whole production system is needed. Managers need to fully control the production, the actual one and the future one. The concept of Decision Support System (DSS) is known for, at least four decades, and it can be defined as a computer-based information system that

N. Klement (✉)
Arts et Métiers Institute of Technology, LISPEN, HESAM Université, Lille, France
e-mail: nathalie.klement@ensam.eu

C. Silva
Department of Mechanical Engineering, University of Coimbra, CEMMPRE, Coimbra, Portugal
e-mail: cristovao.silva@dem.uc.pt

B. Sokolov et al. (eds.), *Scheduling in Industry 4.0 and Cloud Manufacturing*,
International Series in Operations Research & Management Science 289,
https://doi.org/10.1007/978-3-030-43177-8_9

supports business or organizational decision-making activities. They are especially valuable in situations in which the amount of available information is prohibitive for the intuition of an unaided human decision-maker, and in which precision and optimality are of importance. DSS can aid human cognitive deficiencies by integrating various sources of information, providing intelligent access to relevant knowledge, and aiding the process of structuring decisions (Druzdzel and Flynn 2010). DSS can be used to help the decider to solve many problems such as sizing of the shop floor, sizing of resources, planning of activities, assignment of resources, scheduling of activities. For now, most of the DSS are developed to solve specific problems. For instance, McKay and Wiers (2003) developed an integrated decision support for planning, scheduling, and dispatching tasks in a focused factory. Recently, after discussing with the stakeholders, some industrialists from small and medium size enterprises or big companies, we realized that no generic DSS was actually proposed. Given our conclusion from a literature review, no generic DSS found, and our discussion with the stakeholders, we decided to propose a generic and modular decision support tool. In the twenty-first century, we also have to consider the new available technologies.

Because of Industry 4.0, more data are available, more variability can be considered, so DSS is more important than ever. How can the future demand be integrated without major changes in the actual layout? The production system needs to be flexible to face the variety of products and the quantities of products. To propose some decision support tools to help the manager, new trends need to be considered such as continuous improvement, Big Data, or collaborative robotics (Guérin et al. 2016). For instance, it would be necessary to use the data in the shop floor to treat them in real time to adapt the schedule and the planning to the hazards, a link with the used Enterprise Resource Planning by the company needs to be done. Thanks to collaborative robotics, flexible production means can be used in the future flexible and agile production system, which will be reconfigurable. Many companies are actually thinking about converting their actual system into a reconfigurable manufacturing system.

This paper proposes a decision support tool which can be used to design reconfigurable manufacturing system. At the beginning, four static problems have been studied, in which the demands are already known. We focused on many problems: planning, scheduling, resources assignment. These problems can be summarized: they represent a system in which activities have to be done over a horizon planning. Each activity has some characteristics such as processing time and needed resources. Some resources are available to process the activities. The system is ruled by some constraints. Different objectives can be achieved such as optimizing the productivity of the system. Once this tool has been validated for the static problems, we can focus on the dynamic ones: the demands are not completely known, they can vary in quantity and/or variety of products. Information system has to be considered to integrate results from the tool to the shop floor.

Our research proposal may be considered as part of Information Technology (March and Smith 1995). Section 9.2 presents the developed generic and modular decision support tool. Some examples of problems that have been identified and solved are presented in Sect. 9.3. The future work will focus on the reconfigurable manufacturing systems and Information systems provided by Industry 4.0, presented in Sect. 9.4. This paper ends with a conclusion in Sect. 9.5.

9.2 Generic and Modular Decision Support Tool

9.2.1 Genericity

The proposed tool, illustrated by Fig. 9.1, uses a hybridization of a metaheuristic and a heuristic, specifically a list algorithm. A single solution based metaheuristic or a population based metaheuristic can be used. The encoding used by the metaheuristic is a list Y of activities. List algorithm L considers the activities according to their order in list Y, to plan and assign them to the required resources, considering the problem constraints. This builds solution X. Objective function H evaluates solution X. According to this evaluation, the solution is chosen or not by the metaheuristic. At the end of the computation, the solution given by the hybridization is the best list Y^* of activities: the one which optimizes the objective function by applying the list algorithm. This hybridization can be used to solve many problems: only the list algorithm and the objective function need to be specific to the considered problem by integrating the different constraints which rule the system and the objectives to achieve.

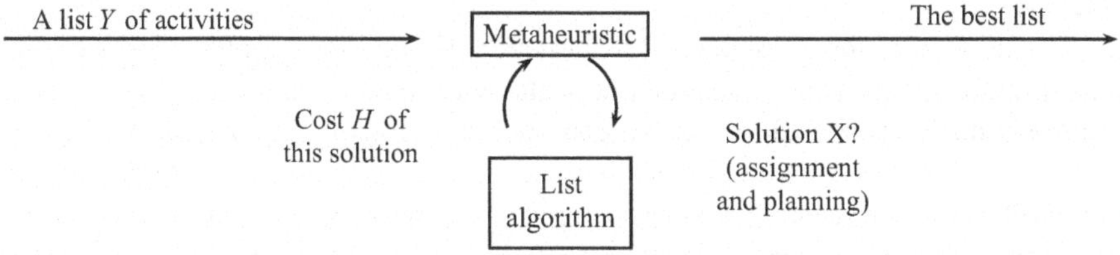

Fig. 9.1 Hybridization metaheuristic—list algorithm

9.2.2 A List Y of Activities

The general scheme of the encoding is given by Eq. (9.1), with Ω the set of all lists Y and S the set of all admissible solutions X built by the list algorithm L. Ω is the set of all permutations of activities. Cardinal of Ω is $N!$ with N the number of activities.

$Y \in \Omega$ is a list of activities. More details about the encoding are given in Gourgand et al. (2014).

$$Y \in \Omega \underset{Heuristic\ L}{\longrightarrow} L(Y) = X \in S \underset{Criterion\ H}{\longrightarrow} H(X) \tag{9.1}$$

9.2.3 *Metaheuristic*

The metaheuristic performs in Ω. An initial solution is randomly computed: a list of activities randomly sorted between one and the number of activities. A neighborhood system is used to visit the set of solutions; it allows to go from one solution to another one. Neighborhood system V is a permutation of two activities in list Y: the activity at position i permutes with the one at position j, with i and j being two different random numbers. V satisfies the accessibility and reversibility properties. Several metaheuristics have been used: some single solution based metaheuristics such as iterated local search and simulated annealing.

9.2.3.1 Iterated Local Search

A simple descent stays in the first found local minimum. Lourenço et al. (2002) showed that an iterated local search allows to go out from this local minimum. After having applied a local search, the current solution is disrupted to go out from the local minimum. Then, a new local search is applied to the disrupted solution.

Kangaroo algorithm is an iterated local search. It consists of applying a stochastic descent, but if there is no improvement of the current solution during A iterations, a jump is made. The used formula to compute the number of iterations A is given by Eqs. (9.2) and (9.3). To make this jump, a solution is chosen in a neighborhood system W, different from V. Kangaroo algorithm converges in probability to the set of optimal solutions if neighborhood system W satisfies the accessibility property. We choose W as the consecutive application five times of V.

$$A \geq card(V) \times \ln(2) \tag{9.2}$$

$$card(V) = \frac{N \times (N-1)}{2} \tag{9.3}$$

9.2.3.2 Simulated Annealing

Simulated annealing is inspired by a process used in metallurgy which consists of alternating cycles of slow cooling and heating. Inhomogeneous simulated annealing was used by Metropolis et al. (1953) to simulate the physical cooling in metallurgy.

Applied to the optimization field, it consists of executing a descent with a non-zero probability to choose a worst solution than the current one. This probability decreases while the number of iterations increases. Aarts and Laarhoven (1987) proved that simulated annealing converges in probability to the set of optimal solutions if neighborhood system V satisfies the accessibility and reversibility properties.

Two parameters are used:

- The initial temperature T_0 is chosen such that all the transitions are accepted at the beginning, defined by Eq. (9.4).

$$e^{-\frac{H(Y')-H(Y)}{T_0}} \simeq 1, \forall(Y, Y') \tag{9.4}$$

- The decreasing factor α is chosen such that the final temperature T_a is close to zero, computed as Eq. (9.5), with $IterMax$ the maximum number of iterations.

$$\alpha = \sqrt[IterMax]{(\frac{T_a}{T_0})} \tag{9.5}$$

9.2.4 List Algorithm

A list algorithm is used to build solution X from list Y: it assigns the activities to resources over the horizon planning according to the problem constraints.

List scheduling algorithms are one-pass heuristics that are widely used to make schedules. Zhu and Wilhelm (2006) defined standard list scheduling algorithm as the construction of a schedule by assigning each activity in the listed order to the first resource that becomes idle. It is important to work with a list algorithm because the metaheuristic browses the set of lists Y.So the used algorithm needs to consider the order of the list to assign activities to resources over the horizon planning.

The developed list algorithms will be detailed according to the considered problem in Sect. 9.3.

9.2.5 Objective Function

Solutions are compared according to an objective function which characterizes the quality of the solution. The aim of our tool is to find the solution X^* defined by Eq. (9.6).

$$X^* = \min_{\forall X \in S} H(X) \tag{9.6}$$

Depending on the considered problem and the objectives to achieve, the objective function can be computed differently. The used objective function for each application will be detailed in Sect. 9.3. Most of the time, the weighed criteria method

Algorithm 1: Hybridization of simulated annealing and a list algorithm

Data: Initial solution Y; Temperature T_0; Decreasing factor α

$T := T_0$; $X := L(Y)$; $X^* := X$

while *necessary* **do**

  Choose uniformly and randomly $Y' \in V(Y)$

  $X' := L(Y')$

  if $H(X') < H(X^*)$ **then**

    $X^* := X'$; $Y^* := Y'$

  if $H(X') \leq H(X)$ **then**

    $X := X'$; $Y := Y'$

  else if $rand[0, 1) \leq e^{-\frac{H(Y')-H(Y)}{T}}$ **then**

    $X := X'$; $Y := Y'$

  Compute the new temperature $T := \alpha \times T$

return Y^*

defined by Coello (2000) is used. The objective function is a weighed sum between some criteria. An example with two criteria H_1 and H_2 is defined by Eq. (9.7). ω_2 is defined by Eq. (9.8), so both criteria can be easily readable. This method can be used with more than two criteria. The function H has to be minimized.

$$H(X) = 10^{\omega_2} \times H_2(X) + H_1(X) \tag{9.7}$$

$$10^{\omega_2} > \max_{\forall X \in S} H_1(X) \tag{9.8}$$

9.2.6 *The Best List Y**

Algorithm 1 describes the whole method with the example of simulated annealing as the used metaheuristic. Set Ω of the lists of activities is browsed, thanks to the metaheuristic using neighborhood system V. Lists are compared, thanks to list algorithm L and objective function H. According to an acceptance criterion, some lists are selected. At the end, the metaheuristic gives the best found list Y^*.

9.3 Applications

In the industrial engineering field, many problems can be solved according to the used decision level. Some of these problems are summarized in Table 9.1. Some identified real problems have already been solved with our tool. Each of these problems has been classified, thanks to a literature review. The already identified applications are about tactical or operational level. Section 9.3.1 presents the first application: a

Table 9.1 Decision levels and problems

Decision level	Horizon planning	Problems	Use case	Literature
Strategic	Years	Sizing of the system	∅	∅
Tactical	Months	Activities planning	Hospital case	BPP
		Resources assignment	Textile	DLSP
Operational	Weeks	Activities scheduling	Injection	CLSP
	Days	Response to hazards	Fridge	JSSP

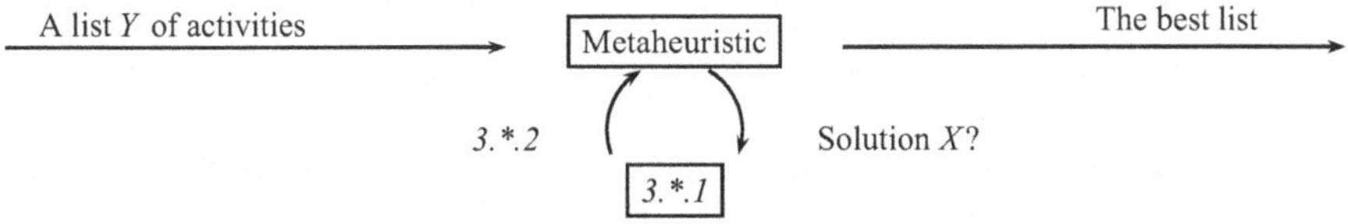

Fig. 9.2 Presentation of the applications

problem of activities planning and resources assignment in the hospital system. The second application is about a lot-sizing and scheduling problem in a textile company, presented in Sect. 9.3.2. The third one is also a lot-sizing and scheduling problem but about an injection plastic case, presented in Sect. 9.3.3. The difference between these two problems is the fabrication order policy: in the textile case, only one product may be made per period, in the injection plastic case, several products may be produced in one period. The last one, presented in Sect. 9.3.4, takes into account precedence between activities. For each problem, an analysis of the problem is made. Then, the description of the proposed list algorithm and the objective function, which are the only two parts of our tool which need to be specific to the problem, is given and is explained in Fig. 9.2. Then, the results of the use of our tool are presented.

9.3.1 Activities Planning and Resources Assignment

This method has firstly been used to solve an activities planning and resources assignment problem in a multi-place hospital context (Klement et al. 2017). The application has been made for the medical imaging department. The planning horizon is divided into periods. Each period represents one half-day. Activities are exams.

The hospital system, called Hospital Territorial Grouping (HTG), is made up of several places. Some material resources belong to each place. Each material resource has an opening schedule. All material resources cannot treat all the exams, there are some incompatibilities between exams and material resource. Each exam has a given

Algorithm 2: List algorithm for a hospital system

```
Data: List of exams (Y_i)_{i∈{1,N}}; Opening schedule of all resources during all periods;
      Processing time of all exams
Occupied time := 0 for all resources and all periods
forall the i do
    First resource, first period, assigned := false
    while (assigned = false) AND current period ≤ max of periods do
        while (assigned = false) AND current resource ≤ max of resources do
            if Exam Y_i is compatible with current resource then
                if Exam Y_i fits in couple (resource, period) then
                    Assign exam Y_i to couple (resource, period)
                    Update occupied time of couple (resource, period)
                    assigned := true
            Next resource
        Next period
```

processing time and a due date before it has to be done. The objective is to assign each exam to one material resource during one period, respecting the time constraints and the compatibility constraints.

9.3.1.1 List Algorithm

This problem has been analyzed as a bin packing problem Gourgand et al. (2014). The list algorithm used to assign each exam to one resource and one period is an extension of an existing list algorithm to solve the bin packing problem, presented by Algorithm 2.

N exams have to be assigned. The proposed list algorithm consists in assigning the exams to the first available couple (resource, period). Compatibility between exam and couple must be respected. Exams have to be done as soon as possible so all resources from one period are tested before going to the next period. Current exam Y_i is assigned to couple (resource, period) if available time in this couple is bigger than the processing time of exam Y_i and if exam Y_i is compatible with the resource. A new period is considered if there is no couple in the current one with enough time to receive the exam.

9.3.1.2 Objective Function

The objective function represents the timing aspect of our problem. Exams have to be done as soon as possible, thus the makespan, the period assigned to the last exam, should be considered. Because many solutions may have the same makespan, we choose instead the sum of assigned periods to all exams, so the solutions can be dissociated. This criterion is written H_S. Another criterion is considered to ensure that most of the exams are assigned before their due date. This criterion, written H_D, is computed as the number of exams assigned after their due date. The weighed criteria method presented in Sect. 9.2.5 is used.

9.3.1.3 Results

The proposed tool provides a good planning of exams, with the assignment of one resource and one period, minimizing the objective function.

The data are randomly generated but the characteristics and the size of the data represent real instances. The HTG is made up of three places. The planning horizon is made by 8–40 periods. As a reminder, one period represents one half-day, thus the planning horizon is between 4 and 20 days. 4–8 resources are available. 50–500 exams need to be planned and assigned. Incompatibilities between exams and resources are randomly generated. Each processing time is between 5 and 100 min. Each material resource has an opening schedule equal to 300 min.

The results are detailed in Table 9.2. The host machine is powered by an Intel Xeon X5687 quad-core CPU running at 3.6 GHz. The computation has been stopped after thirty minutes. Each reported result is the value of the objective function for the best solution found in less than 30 min, but most of the time, the best solution is found in a few minutes. The results are presented as a couple of values $(H_D; H_S)$ with H_D the number of exams assigned after their due dates, and H_S the sum of assigned periods to all the exams.

The results compare three methods:

- The resolution of the mathematical model with an exact method by using the solver CPLEX. If no optimal solution has been found in less than thirty minutes by the solver, no result is written.

Table 9.2 Results (number of exams assigned after their due dates; sum of assigned periods to all the exams; best found solution in bold)

Number of exams	CPLEX	Gourgand et al. (2014)	ILS*	SA*
50	**(0; 51)**	(0; 51)	(0; 51)	**(0; 51)**
50	**(1; 150)**	(10; 147)	(1; 151)	**(1; 150)**
100	**(0; 131)**	(0; 131)	(0; 131)	**(0; 131)**
100	**(0; 517)**	(2; 535)	(1; 516)	(0; 518)
200	**(0; 266)**	(0; 266)	(0; 266)	**(0; 266)**
200	–	(3; 1197)	(0; 1154)	**(0; 1135)**
300	–	(0; 548)	(0; 537)	**(0; 534)**
400	**(0; 830)**	(0; 890)	(0; 841)	(0; 835)
500	–	(0; 1350)	(0; 1241)	**(0; 1234)**
500	–	(194; 8218)	(19; 6382)	**(18; 6659)**

- Results from the method previously published (Gourgand et al. 2014), using two single solution based metaheuristics (iterated local search and simulated annealing) in a classical way: the best value found by all these methods is reported.
- Results from our proposed method detailed in the current paper. The used metaheuristics are distinguished: iterated local search and simulated annealing, written ILS* and SA*.

The results are promising. Firstly, this problem has been solved by CPLEX, thanks to our mathematical model previously proposed (Gourgand et al. 2015). The solver finds an optimal solution only for small size of problems (less than 200 exams over 4 days). The solver does not find any solutions when the size of the problem increases. Then, it has been solved with two approximate methods: in a classical way and with a hybridization. Both methods find an optimal solution for the small instances. For biggest instances, the hybridization between a metaheuristic and a list algorithm outperforms the previous method. Simulated annealing seems to work better than iterated local search. Recently, the results have been improved, using the particle swarm optimization as a metaheuristic (Laurent and Klement 2019).

The assignment to human resources could also be added. Human resources can move over the different places belonging to the hospital system. Each human resource has a planning defining its availability. Each human resource has some skills to use some material resource. The new objective could be to assign each exam to one material resource and one human resource during one period. Some additional constraints have to be considered for each assigned exam: the capability of the human resource to work on the material resource, the availability of the human resource, and the location of the human resource at the place where the material resource is.

9.3.2 Discrete Lot-Sizing and Scheduling

The second application has been identified in a company (Klement et al. 2017). The horizon planning is 1 month, divided by periods of 1 day. Activities are jobs to produce.

The considered company is producing acrylic fiber used in the textile industry. These fibers are made following three steps. Dope preparation: a polymer is dissolved in an organic solvent to make the dope. Spinning: dope is going through a spinneret (a tool which is a metal plate with holes from different diameters) to obtain the synthetic filaments. After this step the fiber is called tow. Cutting and packing: the tow obtained in the previous step can suffer two different types of operation: it can simply be packed before being sent to the warehouse or, it can be cut in small segments, originating a new kind of fiber called raw, which is also packed before expedition.

Our study focuses on the spinning area. There are 10 non-identical spinning machines on the shop floor. Machines are dedicated, all products cannot be produced on all machines. A compatibility machine-product is defined. All machines do not have the same production rate: some of them can produce more tons of product per hour than others.

Fibers can be made from different diameters, within three colors: shiny, mat, and black. There are two types of fibers: tow and raw. In total, the company can make 60 different products. A change of color may induce the stop of the machine for its cleaning. Transition shiny–shiny and shiny–mat/black does not induce a stop, but transition mat/black–shiny does. If there is a setup (change of tool), change of color is made in the meantime. A tool can produce fibers from different diameters by changing the used tensity during the production. Moreover, each tool has a lifetime, from 8 to 45 days. When its lifetime is over, a new tool has to be used. A setup lasts 2 h. Lifetime of tools depends on the type and the color of the product and depends on the used machine. Constraints have to be respected:

- All or nothing: if a product is planned, production lasts 24 h even if the needed quantity is less. Over-quantity is stocked and used next month, it will be deduced from the next orders.
- Possible setup between two products:
  - Change of tool between two products,
  - Cleaning of the machine (transition mat–shiny or black–shiny),
  - Setup if lifetime is over.

The company wants to plan products within a planning horizon of 1 month, by periods of 24 h. Customers' orders are analyzed to make a production plan. Our objective is to schedule these fabrication orders on the different machines and the different periods during the considered month. The result will be a schedule over 28 days by periods of 24 h.

This problem has been identified as a Discrete Lot-Sizing and Scheduling Problem (Klement et al. 2017).

9.3.2.1 List Algorithm

As a reminder, the algorithm is considering jobs one by one respecting the order defined by the list. The used list algorithm is summarized in Algorithm 3. The list algorithm needs to be efficient to find a solution (good or not) in a small computation time. It does not need to be effective because a good solution will be found after having tried some neighbors, thanks to the metaheuristic.

The hypothesis is made that at the beginning of the month, all machines are empty, there is no tool on them, no raw material. This hypothesis is not restrictive. Maybe, the needed tool at the beginning of the month was already on the machine at the end of the previous month, so a setup will be avoided.

9.3.2.2 Objective Function

The objective function is used to compare solutions in the metaheurisitic. The main used criterion is the remaining quantity of all products at the end of the month, written

Algorithm 3: Principle algorithm of the list algorithm

```
Data: Initialize all variables at zero
forall the JOB in the list do
    while all quantity of JOB has not been scheduled do
        MACHINE := first machine
        while Next machines can be considered do
            DAY := first day
            while Next days can be considered do
                if MACHINE is compatible with JOB then
                    if MACHINE is available then
                        Assign JOB, the needed tool, and the needed color to MACHINE
                        during DAY
                        Update the lifetime of the tool, change the tool if needed with a
                        new tool (lifetime = 100%)
                        Update the number of setups if needed (change of tool, transition
                        from black to shiny, transition from mat to shiny)
                        Update the remaining quantity of JOB to produce
                DAY := Next day
            MACHINE := Next machine
```

C_q. This quantity would be produced next month. To provide all customer demands, our system must produce as many quantities as possible in a month. Another used criterion is the number of setup (caused by a change of tool, a change of color, or an exceeded lifetime), written C_s. A hierarchy of both criteria is used.

9.3.2.3 Results

To make our experiments, instances have been created, representing the data used by the company. Each instance is made by:

- A list of jobs to be done, with the needed quantity, the type and color, and the used tool.
- A matrix representing the effectiveness of the machines to produce the jobs, which can be null if the machine and the job are not compatible.
- The lifetime of each tool on each machine, depending on the type and color of the made product.

The experiments are made on a computer powered by an i7 CPU running at 2.6 GHz with 16 Go RAM. Table 9.3 summarizes the results of the proposed method giving the remaining quantity C_q and the number of setups C_s, with the needed computational time. The number of products is assigned to 10 machines over 28 days. Results are compared to the ones found by using the method presented by Silva and Ferreira (2006). The results show that our method gives better results than the ones from the previous method. Indeed, the previous method finds a solution given

Table 9.3 Results of our method

Instance	Silva and Ferreira (2006): (C_q; C_s)	Proposed method: (C_q; C_s; Time)
22 products	(552; 17)	(522.6; 17; 13 s)
25 products	(232; 16)	(68.0; 21; 8 s)
33 products	(387; 17)	(196.0; 19; 11 s)

by a constructive heuristic, while our method browses a set of solutions and gives the best one among all tested solutions.

9.3.3 Capacitated Lot-Sizing and Scheduling

The third solved problem with our tool is a capacitated lot-sizing and scheduling problem with setup and due dates, for the injection plastic case, presented by Silva and Ferreira (2004). Activities are jobs to schedule.

A set of N jobs has to be scheduled on shared machines with their respective mold. Each job has a given size, which determines its processing time, and an associated due date. A sequence dependent setup time is required when the production changes over from a job requiring a given mold to a job requiring a different one. A job is not allowed to be split but several jobs, requiring the same mold, may be grouped together to form one lot and, thus, saving setup costs. Due to compatibility factors, each mold can only be allocated to a subset of the available machines. Each mold is unique; thus, the same mold cannot be allocated to different machines during the same time period. The objective is to allocate jobs to each available machine and define the processing sequence in each machine in order to minimize the total tardiness.

9.3.3.1 List Algorithm

The proposed list algorithm to solve this problem is given by Algorithm 4.

9.3.3.2 Objective Function

The evaluation of a solution is made according to the value of the total tardiness. For a given solution, for each job, we compute the difference between its due date and its actual final date given by the solution. Then the sum of all tardiness for all jobs is made. The number of setups is not considered because all that matters is to minimize the tardiness while delivering the jobs.

Algorithm 4: List algorithm for an injection problem

```
Data: List of jobs (Y_i)_{i∈{1,N}}, Processing time of all jobs
forall the i do
    Order the machines according to their release date
    First machine
    while Job Y_i is not assigned do
        if Job Y_i and the machine are compatible then
            if The needed mold is available then
                if The actual used mold on the machine is the good one then
                    Actualize the release date of the machine without setup
                else
                    Actualize the release date of the machine with setup
                Assign the job
            else
                Assign the job to the machine which uses the needed mold
                Actualize the release date of that machine, taking into account the
                setup if needed
        Next machine (modulo number of machines)
```

Table 9.4 Results of the use of our tool on the injection problem

	Silva and Ferreira (2004)		Proposed method	
Instance	Aver. tardiness	Aver. setup	Aver. tardiness	Aver. setup
3 machines	68.6	21.6	53.6	35.3
5 machines	212.9	32.7	98.3	45.7
10 machines	1020	82	964	114

9.3.3.3 Results

To test the proposed tool, company historical data were collected and used to randomly generate instances of several sizes: 3 instances with 3 machines (16, 18, and 20 molds; 47, 53, and 57 jobs, respectively); 3 instances with 5 machines (25, 26, and 26 molds; 81, 80, and 79 jobs), and one instance with 10 machines (59 molds and 177 jobs). Jobs processing times follow an exponential distribution with an average of 10.75 h and due dates were generated using a uniform distribution ranging between 24 and 312 h. Setup times consider the time to dismount the current mold (ranging between 15 and 45 min) and to mount the next one (ranging between 20 and 60 min).

For small instances, 2 machines and 10 jobs and 2 machines and 15 jobs, the solutions built by our method were compared to the optimal solution obtained solving a mathematical model with CPLEX. It was found that the proposed method gives the optimal solution for the two tested instances. Next, the solutions built by our method were compared to the ones used by the injection company in real life.

Table 9.4 presents the results obtained for each instance size with the heuristic proposed in Silva and Ferreira (2004), used by the company, and with the proposed tool. Comparison considers the average tardiness and average number of setups among all instances of a given size. The results show that the developed tool is effective. An average reduction of 25% of tardiness is achieved. For some instances the reduction of tardiness is up to 50%. Nevertheless, the proposed method leads to an increase in the number of setups. It is important to note that, since the main objective of the company was to reduce the tardiness, the number of setups was not considered in the objective function of the proposed method. For other problems, an objective function considering tardiness and number of setups, with different weights, may be envisaged.

Silva and Ferreira (2004) used a two-phase algorithm: first it assigned molds to machines, and then it schedules jobs on each machine. So it was not possible to assign one mold on several machines at different times. In this way, the proposed method is less constraining which explains the better results. The proposed method is easy to develop and gives good results. For small instances, 3 and 5 machines, results are achieved in a small computational time: a few minutes. For larger instances, the computational time required to attain a good solution increases and can reach a few hours.

9.3.4 Scheduling with Precedence Constraints

The fourth application is about a scheduling problem but considering precedence constraints between activities. Activities are operations to make jobs. Several operations are needed to make a job, operations have to be done in order.

A set of metallic components are to be produced to satisfy the demand from an assembly line. Each component has to follow a processing sequence to be produced and each operation in this sequence requires a given resource (machine). The planning horizon is a week which is divided into five periods of one day. To satisfy the demand from the assembly line, a set of different lots of components is to be produced in each day of the planning horizon.

Thus, we have a set of N jobs which have to be processed on a set of M machines. Each job is defined by a sequence of operations that are associated with a particular machine. Each operation has a processing time and there is a setup time between the processing of two consecutive operations which is sequence dependent. Each job has a requested period. A penalty function is considered, with two parts:

- A storage cost (earliness) if the job is produced in a period prior to the requested one,
- A tardiness cost if the job is produced in a period after the requested one.

The objective is to define the operations sequence in each machine in order to minimize the total penalty.

This problem has been identified as a multi-period job-shop scheduling problem Silva and Klement (2017).

9.3.4.1 List Algorithm

The list algorithm developed for the case study company problem is in two steps presented in Algorithms 5 and 6. It is important to note that the list of jobs considered by the proposed algorithm is similar to the one presented in Table 9.5.

In the first step, described by Algorithm 5, the algorithm ignores the capacity constraints. All jobs are assigned to the required machine in the day they are expected

Algorithm 5: First step

```
forall the jobs in the list do
    forall the operations of the selected job j do
        Assign the selected operation i to the required machine m and production day d
        if operation i is the first one of job j then
            Start time of operation 1 of job j = release date of machine m for day d
            Finish time of operation 1 of job j = start time of operation 1 of job j +
            processing time of operation 1 of job j + setup time
            Release date of machine m = finish time of operation 1 of job j
        else
            Start time of operation i of job j = Max[release date of the machine m for day
            d; finish time of operation i-1 of job j]
            Finish time of operation i of job j = start time of operation i of job j +
            processing time of operation i of job j + setup time
            Release date of machine m = finish time of operation i of job j
```

Algorithm 6: Second step

```
forall the days do
    forall the machines do
        while capacity of machine m on day d is violated do
            Identify operation i from job j which leads to the capacity violation
            for operation 1 to i do
                for day d-1 to day 1 do
                    if capacity on selected day is available to process selected operation
                    then
                        Move operation to selected day
                        Update schedule of selected day and of day d

            for operation i to last operation of job j do
                for day d+1 to day 6 do
                    if capacity on selected day is available to process selected operation
                    then
                        Move operation to selected day
                        Update schedule of selected day
                        Update schedule of day d
```

Table 9.5 Examples of DD assignment rules

							Operation 1			Operation 2			Operation 3		
Part	Priority	Mon.	Tue.	Wed.	Thu.	Fri.	Mach.	Tool	PT	Mach.	Tool	PT	Mach.	Tool	PT
xxx	a		6				M1	T11	46						
yyy	b	2		7	5		M2	T22	3	M3	T32	2	M4	T43	43
zzz	c		18	20	16	32	M1	T11	9	M4	T42	7			
...	...	...	...	...	...	...	...	...	...	...	...	...	...	...	...

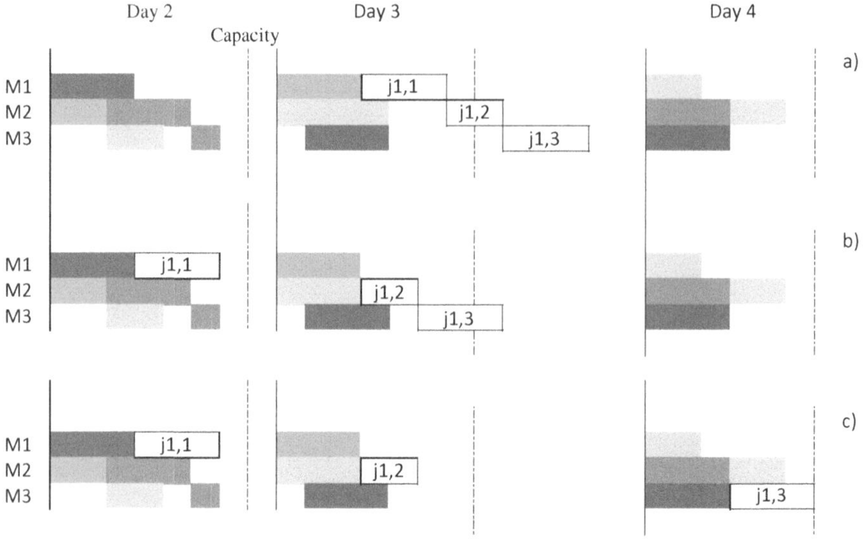

Fig. 9.3 Example of the application of step 2 from the list algorithm

to be concluded. To solve this problem and repair the schedule to respect capacity constraints, the second step described by Algorithm 6 is performed.

Note that when checking available capacity in a given machine for a given day to move a given operation, the algorithm considers not only if the required capacity is available, but also if it can be used by the moving operation while respecting precedence constraints. Figure 9.3 presents an example of the second step of the proposed list algorithm. After the first step of the algorithm, the schedule presented in Fig. 9.3a was obtained. This schedule represents several jobs (in gray scale) for which operations can be processed within the capacity limit. Operations 2 and 3 of job $j1$, represented in white, violate the capacity constraint for day 3.

Operation 1 of job $j1$, processed in machine 1, can be moved to day 2 where machine 1 has available capacity. This move leads to the schedule represented in Fig. 9.3b. After this move, the problem associated with operation 2 of job $j1$ is solved, but operation 3 continues to violate the capacity limit. Thus, operation 3 of job $j1$ is moved to day 4 where there is available capacity in machine 3 to process it. It is important to note that the algorithm considers a 6-period planning horizon, but the real planning horizon has only 5 periods. Jobs with operations assigned to period 6 will be considered as not produced and will be highly penalized.

9.3.4.2 Objective Function

The evaluation of a solution is made according to the value of the total lateness as defined by the penalty function. For each job, the difference between the period in which it is concluded and the period for which it has been required is computed. The weights of earliness and tardiness can be different.

The objective function considered for the problem consists in minimizing the total penalty which considers three parcels:

- Tardiness: priority index × number of job units × number of days delayed;
- Earliness: 0.5 × number of job units × number of days anticipated;
- Unproduced jobs: 15 × number of job units.

9.3.4.3 Results

Historical company production data show that the weekly production scheduling problem faced by the company considers, on average, 300 orders and a total of approximately 450 operations. To avoid unnecessary complexity during the development and test of the tool it was decided to generate a smaller instance.

The instance generated considers 49 jobs, each one having a number of operations ranging between 1 and 3, which can be processed in one of the 6 machines that compose the shop floor. The total number of operations to be processed is 73. Jobs and their respective operations characteristics were generated, taking into account real data from the case study company. Since the generated instance is smaller than the real company problem, the capacity limit for each period was adjusted. The capacity limit was considered constant for the five considered periods and was fixed to 5760 s. This value was chosen after some pre-test runs of the tool, and it guaranties that there is at least one solution where all the jobs may be processed within the five available periods, i.e., no operation is delayed until period 6. On the other hand, it is sufficiently tight to guaranty that most lists of jobs generated by the metaheuristic permutation process will lead to a solution where one or more operations are delayed until period 6.

Results obtained by the tool are presented in the form of a list of jobs/operations to process each day of the week, indicating: the job number, the operation number, the machine where it will be processed, the tool to be used, the start time and the finish time of each operation of the jobs. In this list, anticipated operations (produced before the job required day) or delayed (produced after the required day) are highlighted.

To solve the test instance problem, the proposed method is run for 1000 iterations. Results are obtained in less than 10 min. The test was repeated 5 times. Results obtained for each test are presented in Table 9.6. For all the solutions, all jobs are processed within the 5 days planning horizon. The total penalty obtained is on average 210 units, the number of anticipated operations (5 for the worst case) or delayed operations (3 for the worst case) is low. The solution obtained for each iteration was registered. Figure 9.4 shows the evolution of the solutions obtained for the first 500 iterations for test 1. It can be seen that more than 97% of the solutions, all the ones with a penalty higher than 250, are not feasible, i.e., they have operations delayed until period 6. This confirms that the capacity limit is tight and that the number of feasible solutions is low. Since the tool is always able to find a feasible

Table 9.6 Results obtained for the considered instance

Test	Penalty	Number of anticipated operations	Number of delayed operations
1	207	4	2
2	214	2	1
3	211	5	1
4	203.5	3	3
5	209.5	3	2

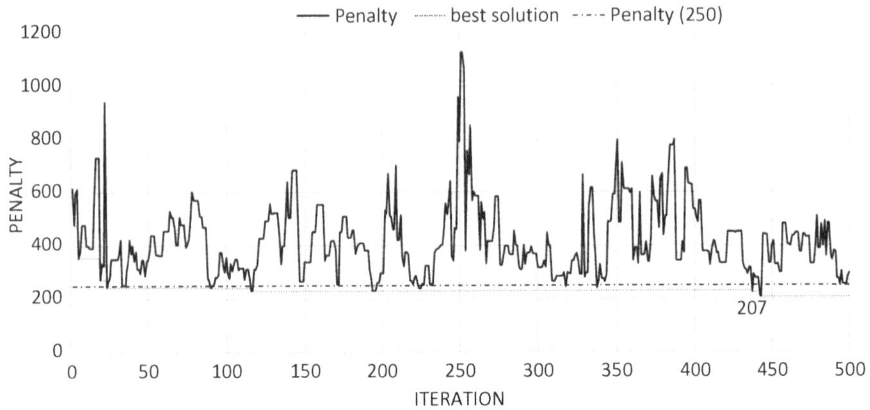

Fig. 9.4 Solutions obtained during the iterative process

solution it can be concluded that it can be effectively used to solve the problem proposed by the case study company.

9.3.5 *To Sum Up*

Table 9.7 summarizes the considered problems with their specificity. These problems have been chosen because they are from real case study and because they integrate progressive constraints.

Table 9.8 summarizes the previous results. The proposed tool has been used on real data for real use cases. It finds better result than the current solution used by the concerned company. In the hospital case, it has not been used on real data, but has been compared to literature instances, and gives the optimal solution for small instances or a solution close to the lower bound in less than half an hour.

For each of the solved problem, we got the confirmation by the stakeholder that the output given by our tool is correct and useful. The proposed tool is efficient: it finds solutions in a few seconds and minutes. It is general because it can be used to solve many problems. We still have to propose a graphical interface so it will easily be used by any managers. Our tool is efficient on such problems at tactical

Table 9.7 Sum up of the considered problems

Section	Identified problem	NP-Hard	Type	Incompatibilities	Setup	Precedence
9.3.1	Hospital (BPP)	✓	Planning	✓		
9.3.2	Textile (DLSP)	✓	Planning	✓	✓	
9.3.3	Injection (CLSP)	✓	Scheduling	✓	✓	
9.3.4	Metal (JSSP)	✓	Scheduling	✓	✓	✓

Table 9.8 Sum up of the obtained results

Section	Identified problem	Math. model	Size of instances	Resolution time	Results
9.3.1	Hospital (BPP)	Yes	$\leq$ 500 activities 10 resources 1 month	Less than 30 min	A few % than lower bound
9.3.2	Textile (DLSP)	No	$\leq$ 33 products 10 machines 1 month	A few seconds	Better than current solution
9.3.3	Injection (CLSP)	Yes	$\leq$ 200 jobs 10 machines a few days	A few minutes	Close to optimal 30% better than reality
9.3.4	Fridge (JSSP)	No	$\leq$ 50 jobs 6 machines a few days	A few minutes	A lgood feasible solution

and operational level. It is now used to deal with problems in Industry 4.0 context, for example, Reconfigurable Manufacturing Problem.

9.4 Industry 4.0

At a strategic level, the objective is to size systems for the next years, for example, to determine the needed number of resources. But because the variety and quantity of products are not stable and not easily predictable for the next years, new trends must be considered, defined by the Industry 4.0 project. Flexible, agile, reconfigurable, or changeable production systems are needed. Section 9.4.1 gives some definition of such systems, and a projection of how the proposed tool could be used to manage them. Other trends which can also be used are presented by the two next sections: Sect. 9.4.2 is about collaborative robotics and Sect. 9.4.3 about Big Data.

9.4.1 Agility and Flexibility

The current market context is characterized by global competition between industries, high product variety, and variable volumes. Those are keys that require the launch of products with a short life cycle and a high degree of customization. To do so, reconfigurable layout has been introduced. One first definition has been given by Lacksonen and Hung (1997) as "the change of an existing plant configuration to another that optimizes costs and time." Since then, many researchers are working on the subject. Benkamoun et al. (2015) summarized in Fig. 9.5 different strategies of reconfigurability.

The proposed tool could be used to solve such problems, where production system should be convertible to face the increasing product variety and extensible to face the increasing volumes. The method first starts with an analysis of the system: what are the demands or future demands and what are the constraints ruling the system. At the opposite of the previous applications, type and number of resources could be a result given by our tool, according to the demand.

Many factories or companies are actually focusing on this problematic. To help them to better understand the proposed solution, some simulation tools could also be used. The optimization part, using the proposed tool, will define one good solution to the problem. This solution is then simulated, using the data or future data of the company, with performance indicators that the company is used to understand. The hybridization between the proposed optimization and simulation can be summarized as Fig. 9.6.

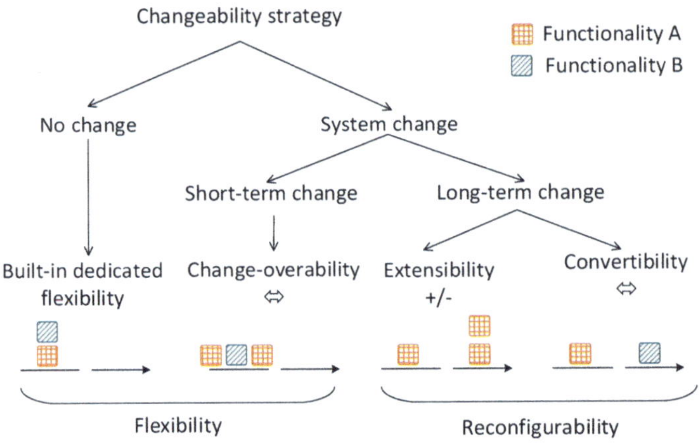

Fig. 9.5 Changeability strategies

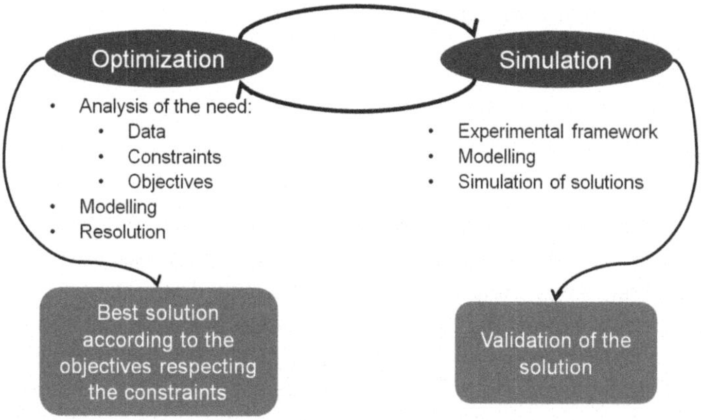

Fig. 9.6 Hybridization: optimization and simulation

9.4.2 Collaborative Robotics in Reconfigurable Manufacturing System

Thanks to collaborative robotics, robots and human can work together. Safety barriers are not needed anymore. Security has increased because these robots are more sensible. Thanks to many captors, they know when the human is close to them. So they can adapt their work and speed. If a contact is made between the robot and the human, the robot stops and needs a human action to start again its work.

These robots are less imposing than traditional robots. Because safety barriers are not needed, these robots can easily be moved from one place to another place of the layout. Mobile collaborative robotics is also existing, where the collaborative robot is put on a mobile platform. In this way, the robot can move independently.

Moreover, these robots are easily configurable. While with a traditional robot or machine, a changing of production needs a changing of tools with setup, a collaborative robot only needs to change its program. The collaborative robot can easily work on one product and then on another one.

Collaborative robotics should be used in reconfigurable manufacturing system. They will be compatible with more products, less setups and shortest setups are needed, they can move from one place to another in the layout and they are smaller than the traditional machines. All these characteristics can easily be modeled and best used with our tool.

This technology attempts to support the reconfigurability of the manufacturing systems and to contribute to adaptive operational conditions. These systems must be responsive to significant changes in demand volume and product mix. The proposed tool has been used to solve an assignment and sequencing problem of reconfigurable assembly lines, where mobile robots collaborate with human workers. The objective is to define a schedule of jobs and decide the allocation of robots to workstations and tasks to minimize the makespan. Preliminary results show that the proposed methodology can make an efficient robot allocation under high demand variety Maganha et al. (2019).

9.4.3 Big Data

Some new technologies have emerged in the last years: Internet of Things, RFID, Cloud Computing, Mobile Devices, etc. Each company can have much more information about the running of its system: it is called Big Data. Manufacturing Execution Systems exploit Big Data to collect data about the system in real time. Almada-Lobo (2016) discussed about the future of MES. It has to be decentralized, vertically integrated to consider the whole supply chain, mobile and connected, and using the cloud computing. The future evolution of our tool needs to consider Big Data:

- How to collect every available data?
- How to treat these data?
- How to exploit them?

Our tool can be used conjointly with Big Data in two ways. Big Data can feed our tool to generate more entry data to describe the system. The better the system is known, the better the proposed solution will be. Once our tool has built some solution, it can feed the MES. For example, if the proposed tool is used at an operational level, it will build the scheduling of activities day to day. This schedule will be transferred to the required resource. Another use is the response to hazards: if some resources are missing, our tool needs to be efficient enough to compute in real time a new scheduling considering the missing resource. Our tool needs to be developed in parallel to actual new technology such as Big Data.

9.5 Conclusion

Because of the quick evolution of the market, future demand in finished products is highly not predictable in quantity and variety. Production system needs to be adapted to this new fact. To do so, some new definitions of production system have been given: they need to be reconfigurable, agile, flexible, or changeable. But how to design them and how to maintain them?

A generic and modular decision support tool has been proposed in this article. It already solved many problems with a static dimension: the demand is known. After having presented the previous applications, our future work has been given: the use of our generic and modular decision support tool to consider reconfigurable manufacturing system. It can design a new layout, defining the right number of needed resources, considering the future demand. Because this demand is supposed highly variable, some new means of production need to be integrated. Collaborative robotics is one of them, the robots are flexible and may be easily adapted to the future demand. To size the system, many information is needed to describe the situation. Big Data needs to be integrated to our tool. Once the system has been designed, our tool could be daily used, to schedule the activities, facing the possible hazards. The proposed method can build a good solution in less than a few minutes.

With Industry 4.0 related technology, large amounts of data about industrial processes are available in real time. This increase in data availability and the need to process them in real time make the use of decision support tools more important than ever. DSS has been proposed and developed for more than four decades. One problem that can be pointed out to most of the proposed decision support tools is that they are highly tailored for the problem they intend to solve. This means that they are not flexible and they are hardly adaptable to different kinds of problems. Thus, a long development work is required each time a DSS has to be developed for a new problem.

The tool presented in this paper seeks to overcome these problems. It is made with two modules: a generic one that can be reused by several problems and a specific one to be developed for each specific problem. This characteristic allows for a rapid development of new decision support tools each time a new problem is proposed. In fact, in the recent past, four different problems have been solved with the help of our tool. For each problem we simply need to define and code an adequate list algorithm. Future development will pass by the ability to integrate Industry 4.0 to the proposed tool. We also intend to develop user-interfaces to facilitate the use of the proposed tool by non-programmer specialists.

References

Aarts, E. H. L., & van Laarhoven, P. J. M. (1987). *Simulated annealing: Theory and applications*. Dordrecht: Kluwer Academic Publishers.

Almada-Lobo, F. (2016). The industry 4.0 revolution and the future of Manufacturing Execution Systems (MES). *Journal of Innovation Management, 3*(4), 16–21.

Benkamoun, N., Kouiss, K., & Huyet, A.-L. (2015). An intelligent design environment for changeability management - Application to manufacturing systems. In *DS 80-3 Proceedings of the 20th International Conference on Engineering Design (ICED 15) Vol 3: Organisation and Management, Milan, 27–30.07.15* (2015)

Coello, C. A. (2000). An updated survey of GA-based multiobjective optimization techniques. *ACM Computing Surveys (CSUR), 32*(2), 109–143.

Druzdzel, M. J., & Flynn, R. R. (2010). Decision support systems. *Encyclopedia of Library and Information Sciences* (3rd ed., pp. 1458–1466).

Gourgand, M., Grangeon, N., & Klement, N. (2014). Activities planning and resource assignment on multi-place hospital system - Exact and approach methods adapted from the bin packing problem. In *HEALTHINF 2014 - Proceedings of the International Conference on Health Informatics, ESEO, Angers, Loire Valley*, 3–6 March, 2014 (pp. 117–124).

Gourgand, M., Grangeon, N., & Klement, N. (2014). An analogy between bin packing problem and permutation problem: A new encoding scheme. In *Advances in Production Management Systems. Innovative and Knowledge-Based Production Management in a Global-Local World* (Vol. 438, pp. 572–579). Berlin/Heidelberg: Springer.

Gourgand, M., Grangeon, N., & Klement, N. (2015). Activities planning and resources assignment on distinct places: A mathematical model. *RAIRO - Operations Research, 49*(1), 79–98.

Guérin, J., Gibaru, O., Nyiri, E., & Thiery, S. (2016). Learning local trajectories for high precision robotic tasks: Application to KUKA LBR iiwa Cartesian positioning. In *Industrial Electronics Society, IECON 2016-42nd Annual Conference of the IEEE* (pp. 5316–5321). New York: IEEE.

Klement, N., Gourgand, M., & Grangeon, N. (2017). Medical imaging: Exams planning and resource assignment: Hybridization of a metaheuristic and a list algorithm. In *10th International Conference on Health Informatics, Porto, Portugal* (2017).

Klement, N., Silva, C., & Gibaru, O. (2017). Solving a discrete lot sizing and scheduling problem with unrelated parallel machines and sequence dependent setup using a generic decision support tool. In *Advances in Production Management Systems. The Path to Intelligent, Collaborative and Sustainable Manufacturing - IFIP WG 5.7 International Conference, APMS 2017, Hamburg, September 3–7, 2017, Proceedings, Part I* (pp. 459–466).

Lacksonen, T. A., & Hung, C.-Y. (1997). Project scheduling algorithms for re-layout projects. *IIE Transactions, 30*(1), 91–99.

Laurent, A., & Klement, N. (2019). Bin packing problem with priorities and incompatibilities using PSO: application in a health care community. In *Manufacturing Modelling, Management and Control - 9th MIM, Berlin* (pp. 2744–2749). IFAC-online.

Lourenço, H. R., Martin, O., Stutzle, T., Glover, F., & Kochenberger, G. (2002). Iterated local search. In *Handbook of metaheuristics* (pp. 321–353).

Maganha, I., Silva, C., Klement, N., dit Eynaud, A. B., Durville, L., & Moniz, S. (2019). Hybrid optimisation approach for assignment and sequencing decision-making in reconfigurable assembly lines. In *Manufacturing Modelling, Management and Control - 9th MIM, Berlin*. IFAC-online.

March, S. T., & Smith, G. F. (1995). Design and natural science research on information technology. *Decision Support Systems, 15*(4), 251–266 (1995)

McKay, K. N., & Wiers, V. C. S. (2003). Integrated decision support for planning, scheduling, and dispatching tasks in a focused factory. *Computers in Industry, 50*(1), 5–14.

Metropolis, N., Rosenbluth, A. W., Rosenbluth, M. N., Teller, A. H., & Teller, E. (1953). Equation of state calculations by fast computing machines. *The Journal of Chemical Physics, 21*, 1087–1092.

Silva, C., & Ferreira, L. M. (2004). Microplano – a scheduling support system for the plastic injection industry. In *E-Manufacturing: Business Paradigms and Supporting Technologies* (pp. 81–89). New York: Springer.

Silva, C., & Klement, N. (2017). Solving a multi-periods job-shop scheduling problem using a generic decision support tool. *Procedia Manufacturing, 11*, 1759–1766. 27th International Conference on Flexible Automation and Intelligent Manufacturing, FAIM2017, 27–30 June 2017, Modena.

Silva, C., & Magalhaes, J. M. (2006). Heuristic lot size scheduling on unrelated parallel machines with applications in the textile industry. *Computers & Industrial Engineering, 50*(1), 76–89.

Zhu, X., & Wilhelm, W. E. (2006). Scheduling and lot sizing with sequence-dependent setup: A literature review. *IIE Transactions, 38*(11), 987–1007.

Chapter 10
The Manufacturing Planning and Control System: A Journey Towards the New Perspectives in Industry 4.0 Architectures

Andrea Grassi, Guido Guizzi, Liberatina Carmela Santillo, and Silvestro Vespoli

Abstract The work deals with the evolution of Manufacturing Planning and Control (MPC) system, from the approaches commonly adopted in the industry so far, to the new implementation schemes involved by Industry 4.0 and Cloud Manufacturing paradigms. Firstly, an introductory part presenting the different classical approaches and their scope of application is reported, together with some insights about the positioning of Industry 4.0 in future market scenarios. Then, a second section focuses the attention on the new technologies at the base of the Industry 4.0 concept, and how they can act as enablers for bridging the gap between current and future production control approaches. Finally, the last section explores the state of the art of Industry 4.0 and Cloud Manufacturing architectural implementations, also outlining further development possibilities and strategies.

10.1 Classical Approaches to Production Control and Scheduling Prior to Industry 4.0

Manufacturing systems are complex systems whose performance is significantly affected by both configuration and operational management. The goal is to achieve customer satisfaction in terms of the products delivered, expected quality, requested quantity, and desired response time. Customers, who comprise the market, and their needs are driving factors that determine, at a strategic level, the coherent configuration of the system and, at an operational level, the optimal approach for its management. Both of these aspects are strictly correlated, since the fulfilment of customers' needs can be obtained at a sustainable cost (i.e. a cost that can be

A. Grassi (✉) · G. Guizzi · L. C. Santillo · S. Vespoli
Dipartimento di Ingegneria Chimica, dei Materiali e della Produzione Industriale, Università degli Studi di Napoli Federico II, Naples, Italy
e-mail: andrea.grassi@unina.it; guido.guizzi@unina.it; liberatinacarmela.santillo@unina.it; silvestro.vespoli@unina.it

B. Sokolov et al. (eds.), *Scheduling in Industry 4.0 and Cloud Manufacturing*, International Series in Operations Research & Management Science 289,
https://doi.org/10.1007/978-3-030-43177-8_10

transferred to the customer) only by the right combination of strategic decisions and operative management.

Different theories and manufacturing system designs have been developed to face decades of new challenges imposed by market changes, as well as implement technological innovations and progress made available by the scientific and engineering fields. Hence, before exploring the new perspectives involved in the advent of Industry 4.0 and recent market trends, a general picture of which can be found in Panetto et al. (2019), it is convenient to summarise currently accepted views regarding the strategic configuration of manufacturing systems and related implications for their management. Figure 10.1 shows basic configurations that can be adopted to structure a manufacturing system, ordered along with a coherence curve that depicts the relationship between the complexity of the product to be realised and the amount of time the customer is willing to wait until the order is delivered.

These two aspects are the main driving factors that influence the configuration of the system and its management approach. The complexity of a product represents the degree of customisation requested by customers. High complexity means products are formed by many components that frequently differ. Conversely, reduced complexity indicates simple products, formed by a few standardised parts that are likely to be present in different products, typically belonging to a specific family.

The degree of complexity is inversely related to the sales volume of a single product variant (i.e. a specific final product identified by a unique part number). This aspect is the natural consequence of simple products being designed according to basic requirements that cover the needs of many people, while complex, customised products fulfil the needs of a restricted niche of customers. Thus, a standardised product experiences a total demand that is not affected by the behaviour of every customer in the market since it is derived from the sum of a large number of individual

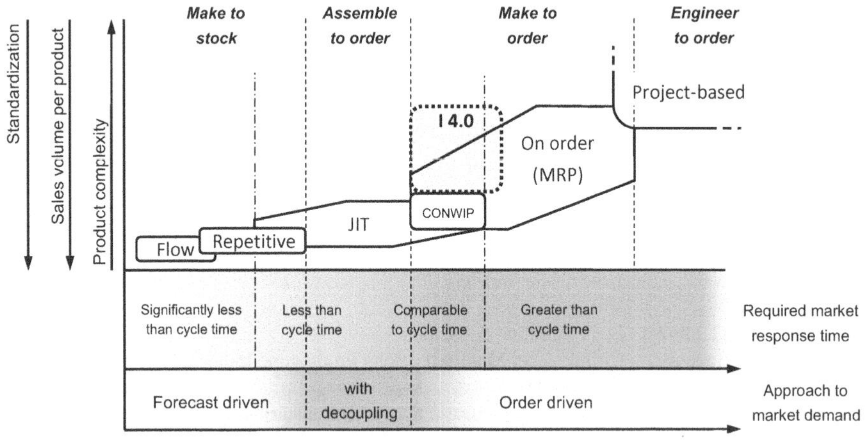

Fig. 10.1 Coherence diagram of different production system approaches

demands, resulting in a huge aggregated sales volume that also exhibits low variation and high stability over time. Conversely, a complex, customised product is affected by a low and highly variable sales volume since it is intended to fulfil the needs of a restricted and identifiable group of customers, also resulting in a demand signal that is difficult to forecast given its intrinsic variability and uncertainty.

The required market response time is the other external driving factor impacting system configuration and management. It represents the amount of time that the generic customer in the downstream (i.e. the direct client of the manufacturing system) can wait to see their order satisfied, starting from the moment they place the order with the manufacturing system. This amount of time becomes significant when compared to the (average) cycle time of manufacturing systems, which is the amount of time the generic order takes to cross the system once it has been received. The cycle time includes all the processing times, as well as every lead time and queuing time the order experiences while moving through the system. A required market response time less than the manufacturing system cycle time implies that it is not possible to wait for the customer order before starting production; stock positioning strategies have to be deployed to mask a quota (or all) of the cycle time to assure compliance with the response time required by the customer.

With respect to the two aforementioned driving factors (i.e. the product complexity/customisation and the required market response time), different approaches to the configuration and management of the manufacturing system can be adopted following a coherence criterion that is represented in Fig. 10.1 by the specific position of each different manufacturing configuration. In the left part of Fig. 10.1, the zone with the tightest request in terms of response time by the customer is represented. In this zone, the customer expects an almost immediate fulfilment of the order, making any manufacturing strategy that waits for the customer order before starting production de-facto impossible. The fulfilment of customers' orders has to be carried out by maintaining a stock of finished products since this is the only way to achieve such a rapid response. The strategy is to mask the production time to the customer, letting them experience only the time requested for the final preparation and delivery of the order.

Given the need to keep stock of finished products, a high level of standardisation is required to reduce the risk of obsolescence. Therefore, if manufacturing systems operating in these conditions are coherent (i.e. sustainable from an economic viewpoint), then they produce a restricted number of products or variants of the same product, and each product experiences a large sales volume. Given the stability of demand for such items, forecast-based production planning can be deployed efficiently, making a make-to-stock (MTS) approach possible. Hence, production plans are anticipated with respect to actual demand and computed as a trade-off between inventory costs and production costs.

The production system in this scenario is typically configured as production lines dedicated to specific products or simple variants of them. The paradigm of 'mass production' is then adopted since the lines are working at the minimum possible unit cost to produce large batches. The production system is less responsive because every batch takes a long time to be produced; however, this does not constitute an issue with respect to customer order fulfilment since the market demand is directly satisfied by the inventory of finished products, and thus the customer does not experience the cycle time of the production system. Such a manufacturing environment is referred to as 'flow' since the production line clearly defines a continuous and stable production flow. When a few different products or variants are alternated in production, realising a cyclic load of the line, the manufacturing environment is called 'repetitive'.

In the second column in Fig. 10.1, the just-in-time (JIT) manufacturing environment is found. The JIT concept was developed in the 1970s by the Toyota Motor Corporation in response to the need to reduce inventories, viewed as a source of unnecessary costs and inefficiency at the time (Ohno 1988). Toyota introduced the concept of *kanban*, developed by one of its production engineers in the mid-twentieth century, as a simple tool that enables the propagation of the demand signal along the production system by directly controlling and limiting the work-in-progress (WIP) in the system. The mechanism is simple; every time a station withdraws a predefined quantity of some item, the upstream station is allowed to produce the same amount of the withdrawn item to replenish its stock.

The original idea behind the JIT concept paved the way to define an entire class of manufacturing systems that, as can be seen in Fig. 10.1, covers the wide area of required market response times, ranging from short to comparable to the manufacturing system cycle time. In this area, different configurations of the manufacturing system are possible, as will be described below, but all of them are unified by a common denominator: creating a continuous and balanced production flow across the whole system by timing production activities to the demand pace. This is the real insight behind the JIT concept; it focuses the attention on the entire system (i.e. the flow), controlling the WIP and monitoring the resulting throughput and cycle time instead of trying to optimise each resource (i.e. station) individually. Resources are then forced to produce at the demand pace and remain synchronised with the production flow since the demand signal is propagated continuously to the upstream tanks to the adopted WIP control mechanism.

The JIT concept makes possible the flow-based management of the manufacturing system even if customisation is increased with respect to flow and repetitive production—so when some flexibility and differences in production routes are present. By maintaining flow-based control, which, in JIT manufacturing, is deployed via direct WIP measuring, it is possible to keep the system cycle time reduced and stable, resulting in increased predictability and responsiveness in the manufacturing system. Moreover, correlations among the WIP level, throughput, and cycle time can be better understood, with respect to the level of variability inflated in the system by common detractors, such as setups or failures, and by market demand, which is mainly related to mix variation rather than aggregated sales volume variation in this context.

Different kinds of configurations can be adopted to implement the basic concept of JIT manufacturing, depending on the degree of customisation, the way it is obtained, and the response time allowed by the market. The original approach developed by Toyota, which can be called a 'pure pull' system, can be placed in the bottom-left part of the JIT zone in Fig. 10.1. The pure pull system de-facto implements the original kanban mechanics, which considers the adoption of a single kanban loop for each couple of resources to limit the WIP in between them and release the production order. Hence, the presence of a stock point in any kanban loop is determined by the particular behaviour of the system. At this stock point, any item required by production must be present with an inventory level greater than zero to make production possible in the downstream processes. Items are stocked in standardised containers, typically plastic boxes, containing a specified quantity that represents the minimum possible batch size for the related item. Each box in the stock point has a kanban attached to it, and every time a box is withdrawn, the associated kanban is released and sent to the upstream process, which is then allowed to produce just the consumed quantity (i.e. the quantity needed for a new container). The kanban is then called a ‘production kanban’ since its release determines the generation of a production order. A pure pull JIT system falls in the category of MTS systems since market demand is fulfilled by the stock of finished products kept at the end of the production flow. Practically, customer orders entering the system generate the withdrawal of finished products from the final stock point, which releases production kanbans in a cascade, stage by stage, in a backward manner along the flow. The advantage of these systems is that the customisation delivered to the market can be increased while maintaining a short response time since demand is still satisfied by the inventory of finished products, and the production can be kept under strict control due to the limitation of WIP, even if a more complex production route is present. As a counterpart, given the need to keep a stock of every item used at every stage, as well as finished products, the degree of customisation cannot be too high, which prevents excessive stocks of slow-moving items and the related obsolescence risk. Moreover, demand must be stable to guarantee a steady pace in the production flow, so these systems are unable to react to significant deviation in daily demand, as well as sharp variation in the demand mix, as it will result in the emptying of some stock items at different stages of production, provoking starvation phenomena and the loss of throughput. This is also why the pure pull JIT system saw its birth in the automotive industry.

To the right of the JIT block in Fig. 10.1, the other extreme condition for the application of flow-based control is reached. Here, the level of customisation requested by customers is higher than in the pure pull case, involving a significant increase in the number of finished products to be realised and the related number of items to be managed within the system. In this situation, maintaining stock of every component and the final product is not sustainable from an economic viewpoint given the wide variety to be managed and the implicit obsolescence risk. Stocks have to be moved as upstream as possible, typically at the level of raw materials or by-products, where a certain level of standardisation is required considering their commonality, at least within the different product families. Customisation is then obtained by successive production stages, which need to be managed on an on-order basis, by adopting a make-to-order (MTO) approach. Given the increased level of customisation, the market must allow manufacturing system more time to carry out the production.

Nevertheless, at this level of customisation requested by the market, the amount of time available for production activities remains tight with small margins. In this scenario, it is strategic to adopt a flow-based configuration with production order releases based on direct WIP control to increase stability and keep the system's cycle time as brief as possible. The best configuration and control approach in this scenario is constant WIP (CONWIP; Hopp and Spearman 2011). Practically, CONWIP defines long, closed control loops involving specific parts of the manufacturing system dedicated to the production of particular families of items. Within each loop, anonymous kanbans (i.e. kanbans that are not a priori associated with specific part numbers) limit the number of production orders that can be in process simultaneously in the system. Every time an order is completed, the attached kanban is released and can move to upstream to allow a new order to enter, generally for a different type of product. Hence, stocks of customised items are not kept within the flow, while the performance of the flow itself (i.e. throughput and cycle time) can be easily controlled by directly tuning the number of kanbans. The queue of orders at the entrance of each loop is determined by the consumption of items in the downstream stages, so the pull mechanism is preserved, and prioritisation strategies can be adopted in the queue to facilitate the optimal utilisation of resources, along with the flow.

Finally, the central part of the JIT zone in Fig. 10.1, bounded by the pure pull approach to the left and CONWIP to the right, represents a transition from an MTS environment to an MTO environment. It exists on the part of the horizontal axis in which the required market response time is becoming less tight but remains at a value smaller than the cycle time of the entire manufacturing system, so a pure MTO approach cannot be adopted. Moreover, since the willingness of the customer to wait longer for the product to be delivered must be compensated in some way, the customisation level must increase. Here, a coherent configuration of the manufacturing system can be achieved by adopting a modularisation strategy in product design. If products are modularised, several different final products can be obtained by means of alternative combinations of a limited number of various modules. Customisation can then be kept downstream in the manufacturing process, mainly involving assembly activities (i.e. assembling common parts and different modules) and a few finishing production activities (e.g. painting). This approach facilitates the upstream production of modules, each of which are used in a number of different final products depending on customer orders, a characteristic similar to standardised items, making it more likely that MTS production can be adopted. At this point, the manufacturing systems is split into two parts divided by a decoupling point at which most of the stock is kept, allowing the upstream to start production in advance by exploiting an aggregated forecast for module demand estimation over the mid-term horizon, while the downstream operates on an on-order basis given that the cycle time requested by the downstream remains shorter than the required market response time. Since the downstream, which is working directly on fulfilling customer orders, mainly executes assembly activities, this kind of configuration is called assembly-to-order (ATO).

Next to the JIT block in Fig. 10.1, the 'on order' zone can be found. This zone refers to manufacturing production characterised by the highest customisation level, meaning the number of different final products delivered to the market is so large that each of them experiences a very low and unpredictable demand. The stock of final products and related components, highly customised as well, cannot be held in the manufacturing system given the high obsolescence risk. Production orders can be released only after the customer order is received, so the customer must provide the manufacturing system sufficient time to perform all the manufacturing activities. The manufacturing system is asked to maintain a cycle time as short as possible even if the production mix is very complex and formed by a large number of different small production orders moving along different routes in the system.

The high degree of fragmentation in the routes makes it impossible to identify the primary and stable production flow, so it is no longer feasible to manage the system on a flow basis, like in a JIT environment. The solution adopted to tackle the problem, due to the advent of information technology (IT), was the introduction of the Material Requirements Planning (MRP) system (Orlicky 1975). The approach adopted by the MRP is precisely the opposite of the one taken by the JIT: while the JIT considers the whole production flow, MRP focuses on resources (i.e. the stations) with the objective of maximising their utilisation as if they are operating individually. To do this, MRP discretises the planning horizon in elementary time units, also called 'time buckets' and typically sized to a 1-week duration, and by exploding the ordered products in their components, time bucket by time bucket, it computes the production orders to be released to each resource in a way that fills the available production time. The production control mechanism of MRP is also rather different from JIT manufacturing, which controls production by means of the WIP; MRP controls production by sending production orders to the resources directly based on estimations of the workload involved. Production orders are then pushed from the MRP system to the resources, defining the concept of 'push' production control, which differs from the 'pull' control of the JIT.

Since MRP relies on estimations of the production workload, its behaviour is less stable and less predictable than a JIT system, as its control mechanism is based on the direct WIP measure. Moreover, the need to create individual plans for the resources requires MRP production to create the conditions to make them operate as they were isolated from each other. This result can be obtained by increasing the WIP in the system, which means increasing and distributing inventory along the system so as to make every resource work on a dedicated queue of production orders that never empties. According to Little's law, the practical way to achieve this result is to impose fixed (significant) lead times on the production phases, resulting in the anticipation of order releases in every stage of the system. The result is a sharp increase in the entire system's cycle time because of the long periods of time production orders spend in the queue. The cycle time then becomes longer and, therefore, more variable from order to order (i.e. the system becomes less predictable), making the MRP approach sustainable only when the market is rather permissive in terms of the required response time.

Until recently, the MRP system has been universally recognised as the best system to adopt in response to a highly customised demand, making the users accept its weaknesses implicitly. However, as investigated in Sect. 10.3, recent market changes and future trends highlight critical issues with the sustainability of the MRP approach. Specifically, future market scenarios in developed countries show a continuous increase in the requested product customisation associated with a rise in order fragmentation; this is mainly due to the willingness of the actors involved in supply chains to reduce their stocks of materials, then asking for even more brief response times (Ivanov et al. 2018a). Hence, the downside compromise made by MRP to sacrifice cycle times in favour of throughput is no longer an acceptable strategy, especially considering that in modern markets, customers can agree to pay more if products are delivered more quickly.

Industry 4.0 and cloud manufacturing aim to address the need for delivering highly customised products within a short response time while maintaining high efficiency and effectiveness in production execution (Ivanov et al. 2014). Referring to Fig. 10.1, the produced effect is like crushing to the left the block of the 'on order' production, which means reducing the cycle time to the level of a system that is managed on a flow basis, even if customisation is very high. To achieve this goal, Industry 4.0 considers the massive adoption of modern technologies to empower the capability of production resources to exchange data, make decisions, improve the coordination of activities on the shop floor (Dolgui et al. 2019b), and interact with higher decisional levels. Moreover, new models and architecture for manufacturing planning and control (MPC) are needed to best exploit the new capabilities introduced by Industry 4.0, as well as enable operation at the cloud level of the manufacturing system (Ivanov et al. 2016, 2018b; Vespoli et al. 2019). In the next sections, these two aspects will be developed and discussed in detail.

10.2 The Impact of New Industry 4.0 Enabling Technologies on Production Planning, Control, and Execution

The fourth industrial revolution faces problems of an organisational and technological nature. These two aspects are intimately connected because new technologies enable the implementation of new organisational paradigms, and new organisational models in the manufacturing industry encourage the development of new technologies. Among the key elements of Industry 4.0 are interoperability, intelligence, communication, and the decentralisation of decisions. Interoperability is the ability of systems, production technologies, and operators to collaborate (Xu et al. 2018). Decentralising decisions refers to the possibility for intelligent objects (agents) to make decisions within a domain. If a conflict occurs between the decisions of agents cooperating within the same domain, this conflict must be solved via rules or a decision-maker of a higher level. Intelligence is the ability of an agent to learn and train itself using theoretical models and real information from the field. Because of this ability, the agent can make decisions or suggest them to a higher-level decision-maker. Finally, communication is the ability of machines and systems to communicate through multiple technologies. Such communications can take place locally (i.e. between devices and agents placed in a specific area) or in an extended area (i.e. between different plants placed in geographically distant locations).

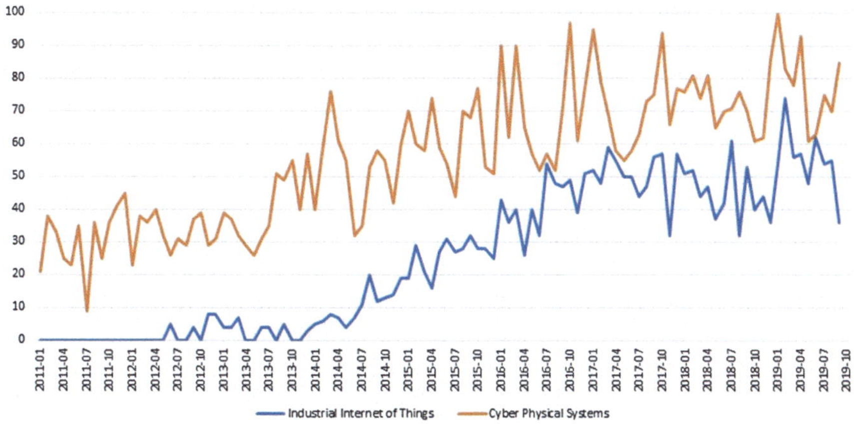

Fig. 10.2 Google trends: searches for 'industrial Internet of things' and 'cyber-physical system'. (Source: http://trends.google.it)

The aforementioned key elements explain why cyber-physical systems (CPS) and the Internet of things (IoT) are the most relevant key enabling technologies of Industry 4.0. These technologies are not the only ones introduced, but they are, indeed, the ones that characterise the Industry 4.0 paradigm. Additionally, the interest in CPS and the Industrial IoT (IIoT), which refers to migrating the IoT to the industrial context, can be assessed by consulting Google's search statistics comparing 'cyber+physical+systems' and 'industrial+Internet+of+things' keys. Figure 10.2 shows how interest in the IIoT has been growing since 2014, following the same trend as interest in CPS. The graph points out two main issues: (1) public interest in CPS preceded interest in IIoT, and (2) increasing CPS development has stimulated the tailoring of the IoT concept for use in the industrial field.

10.2.1 Cyber-Physical Systems

In the past, industrial production was mainly based on the human factor and machines. Typically, an operator was the user of a device used to manufacture a product or semi-finished product. Due to progress made in the electronics field in the 1980s, machines acquired the ability to carry out specific processes in an increasingly autonomous manner. Consequently, the operator switched from a user to a machine supervisor. The advantages of this evolution are many: higher repeatability of operations (e.g. as time passes, an operator gets tired, but a machine does not), more dangerous activities can be carried out by a machine rather than an operator, with consequent advantages in terms of safety, and the possibility of achieving massive production rates. At that time, the myth of the factories operating with the lights off grew thanks to the birth of the computer integrated manufacturing (CIM) approach, but its application failed substantially because operations technology (OT) systems, which includes hardware and software systems devoted to detecting and imposing changes in physical processes, were not fault-tolerant enough and could not be as adaptive as needed because of the technologies available. Today, production systems are becoming increasingly complex as they need to be able to quickly adapt to product changes and unpredictable demand variations, as a direct consequence of product life cycles shortening and increased product customisability requested by customers.

The emphasis on reconfigurability and adaptability in industrial processes has been the driving factor in the integration of operations technology (OT) and information technology (IT) in modern machines and industrial operations, paving the way for the birth of cyber-physical systems (CPS). Each CPS, made up of several devices or other sub-systems, has an internal logic component that utilises IT to perform evaluations and make decisions and a physical component (OT) that performs the actions. The 5C model, shown in Fig. 10.3, is a reference guideline for the development of a CPS in manufacturing processes (Lee et al. 2015). This model provides a base layer (I) where a CPS can communicate with field sensors or other CPSs.

The information exchanged between several devices is analysed in the next layer (II), which returns the semantic content of this information (e.g. the identification of information patterns that may be critical to the operation of the machine). Therefore, the system acquires the awareness of the state of its components and the devices that comprise it. The next level (III) represents a fundamental level in the model because the CPS, through the awareness acquired from the antecedent level, can compare the performance of the fleet of machines that comprise it. At this level, a digital twin of

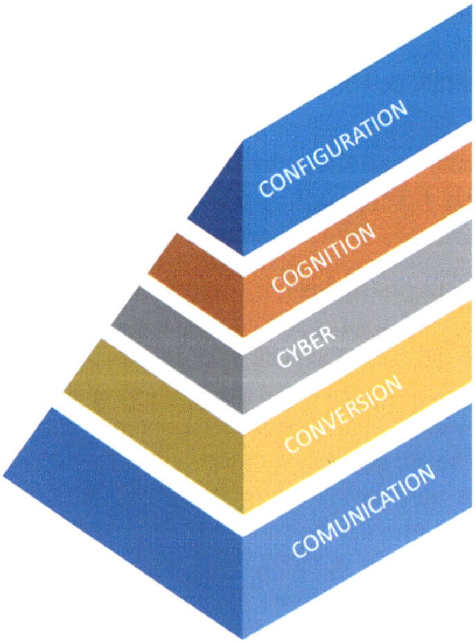

Fig. 10.3 CPS: 5C model. (Inspired from Jin et al. 2017)

the system is also available so that what-if analyses can be performed in a risk-free environment. The fourth layer of the 5C model is defined as cognition. At this level, the CPS acquires the ability to make and propose decisions. These decisions have a rank and are optimised concerning the desired objective. The top level (V) of the model is the configuration level, where decisions made in the 'cyber environment' are executed at the 'physical environment' level. The CPS, therefore, acquires the capability for flexibility and adaptability to 'real-world' conditions.

The new perspectives introduced by CPS represent a radical evolution in the concept of automation in industrial processes. CPSs can make decisions, can learn from what happened in the past (i.e. gain experience), share knowledge with other systems, and test possible decisions in virtual environments (i.e. digital twin). The latter is relevant in the context of CPS. The digital twin of the physical system can be represented by a simulation environment or a federation of simulation environments to model the CPS, integrating different techniques, such as discrete events, agents, and system dynamics. For further details, the reader can refer to Guizzi et al. (2019).

Moreover, CPSs can communicate with the company's enterprise resource planning (ERP) systems, and this is changing the role of ERP within the company's information systems. In particular, the traditional processes of factory resource planning (i.e. machines and human workers) will undergo a significant change with the introduction of CPSs. One of the characteristics of CPS systems is their ability to self-organise (Habib and Chimsom 2019). This means that having set a target, CPSs (representing factory resources in the broad sense) can self-organise and cooperate to reach the target. Thus, new architectures that bring about a change in the ERP systems we know today will be achievable. These systems need to have a high-level coordination function, while the factory processes will be managed 'autonomously' by the CPSs.

10.2.2 Industrial Internet of Things

The term Internet of things (IoT) first appeared in 1999 in a presentation by Massachusetts Institute of Technology (MIT) researcher Kevin Ashton at Procter & Gamble. Ashton was working with colleagues on RFID tags, special electronic tags that can be applied almost anywhere and read remotely with special radio equipment (Ashton 2009). Since 1999, the IoT has undergone numerous reinterpretations. In some respects, the IoT is identified as a wireless sensors network (WSN). In practice, the IoT and WSN are not synonymous. In fact, the objective of the WSN is to collect data from a network of nodes (sensors) distributed in space, while the IoT aims to perform actions, starting with information from the WSN, to achieve a goal without human intervention, interacting with other intelligent systems. Therefore, some characteristics of the IoT are: (1) the interconnection of things and their connection to the Internet, (2) uniquely identifiable things, (3) ubiquity, (4) sensing/actuation capability, (5) embedded intelligence, (6) interoperable communication capability (based on standard and interoperable communication

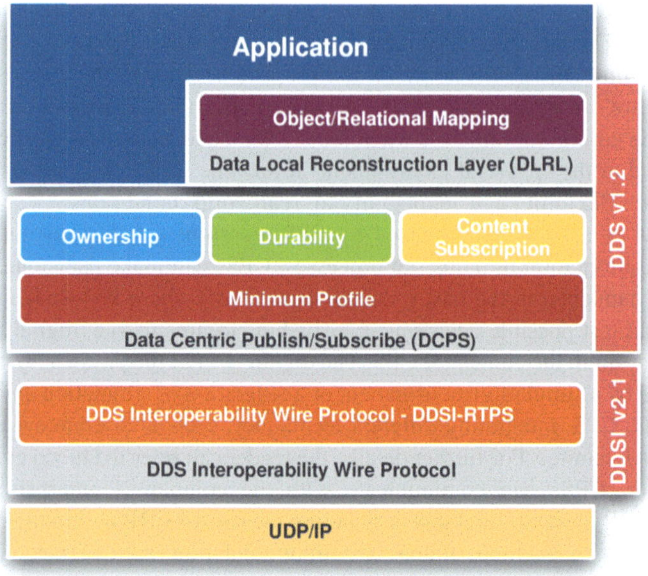

Fig. 10.4 DDS architecture. (From Corsaro and Schmidt 2012)

protocols), (7) self-configurability, and (8) programmability (Minerva et al. 2015). From another viewpoint, the IoT concept can be considered similar to CPS since the objects in an IoT system are able to implement and execute simple control actions. However, a CPS represents much more complex elements that can work as a hub for IoT systems.

In Industry 4.0 applications, the IoT is tailored to the industrial Internet of things (IIoT). The latter represents a subset of the IoT, characterised by being machine-oriented. In particular, the IIoT facilitates connections between machines, control systems, and business processes in general. Therefore, it is a critical element in the creation of a smart factory (Sisinni et al. 2018). The main communication frameworks of the IIoT are a data distribution service (DDS), an open-platform communication unified architecture (OPC-UA), and MTConnect.

The DDS is a standard of the object management group, which aims to distribute messages according to a publish/subscribe logic. It means that a set of publishers sends messages, and a set of subscribers receives them. The DDS has two levels, as shown in Fig. 10.4. The lowest level is data-centric publish/subscribe (DCPS), while the highest is the data-local reconstruction layer (DLRL). The DLRL aims to ensure communication between the application level and the DCPS. The DCPS is structured in seven entities:

1. DomainParticipantFactory—This entity represents the container of the different domains in which the information will be generated.

2. DomainParticipant—This entity represents a specific domain belonging to the DomainParticipantFactory. Within the DomainParticipant, specific topics, publishers, and subscribers are created.
3. Topics—This entity represents a part of a typed domain. It means that within a Topic, there is specific semantic content expressed through a particular data structure.
4. Publisher—This is an object that publishes data sent by different DataWriters related to it.
5. DataWriter—This is an abstract entity implemented based on the data structure of the topic to which it is uniquely associated.
6. Subscriber—This is the entity that collects the data published on topics related to it and sends them to the associated DataReaders.
7. DataReader—This is an abstract entity, like the DataWriter, which is implemented based on the topic to which it is related.

The OPC-UA, instead, is an accepted standard issued by the OPC Foundation and ratified by the International Electrotechnical Commission (IEC 62541) for the exchange of data between programmable logic controllers (PLCs), human-machine interfaces, servers, clients, and other machinery. The OPC-UA specifications are divided into two groups; the first one refers to the primary functions/services offered, and the second one refers to the particular specifications of data access. Regarding the information model, the standard provides only the infrastructure of the model information; individual vendors must model the information, generating significant fragmentation in the solutions. To avoid this problem, OPC-UA offers the ability to define extensions based on a basic information model developed by the OPC Foundation. Each manufacturer can then create solutions applicable to its devices, so clients can access this information regardless of the manufacturer because they use the same basic information model (Pfrommer et al. 2016).

The basic principles of modelling are the following:

- Using object-oriented techniques, including type and inheritance hierarchies.
- Access mode to types identical to those of instances.
- Possibility of exposing information by exploiting (fully connected) networks of nodes in many ways.
- Extensibility of type hierarchies and node links.
- No limitations on the modelling of information.
- Information modelling always placed on the server-side.

The basic concepts of information modelling in OPC-UA are nodes and references between nodes. Nodes and References are grouped in the address space, which represents the set of information exposed by the OPC-UA server and accessible by any OPC-UA client.

Comparing the two standards, we can say that the strength of DDS is the ability to share information in real-time with low sensitivity to failures and high scalability without sacrificing performance. DDS was created for the global information grid, which is the backbone of the current IoT. The OPC-UA, however, was built in the field of industrial automation to allow interoperability/integration between devices, sensors, and PLCs. Therefore, its genesis and the initial ecosystem are its strengths. Both standards, in their most recent versions, guarantee the security of information transfer. This characteristic is fundamental in the current context of Industry 4.0. To this extent, in recent years, some researchers have proposed hybrid models combining OPC-UA and DDS to identify an architecture that can overcome the limitations of individual architectures (OPC Foundation 2019).

Finally, MTConnect (MTConnect Institute 2018) is an open and royalty-free standard released by the MTConnect Institute to exchange data and information. It is based on a data dictionary defined within the standard and provides a semantic data model. The transport protocol used by the standard is HTTP, and the semantic coding of the information is done using the XML language. The three main elements of the implementation of MTConnect are equipment, an MTConnect agent, and client software. The equipment represents a data source (e.g. machine tools and workstations) that standardises the data through an adapter transferring them to one agent. The MTConnect agent is a software that collects information from data sources and replies to data requests from Clients. This software can be installed directly on the supervisory control and data acquisition (SCADA) of the machine or on a computer that acquires data from multiple systems. The client is the application that requests data from the MTConnect agent using them to support operations. MTConnect is a read-only standard. It means that the information is collected and transmitted at a level of control where it is then analysed. However, the control level does not send information to a single machine. Unlike OPC-UA, MTConnect does not comply with any IT security regulations. Currently, the OPC Foundation and the MTConnect Institute have developed MTConnect-OPC UA companion specifications (OPC Foundation 2019), which aim to provide an information management model that can be used via either the MTConnect or OPC-UA standard. In today's complex production systems involving large numbers of machines and production lines, data exchange protocols that allow easy access to information from the shop floor are preferred over legacy systems.

10.2.3 Additional Enabling Technologies

In addition to these enabling technologies, Industry 4.0 makes extensive use of a large number of technologies that did not necessarily arise in the industrial field (e.g. artificial intelligence [AI], cloud technology, blockchain technology, immersive technologies; Habib and Chimsom 2019). These technologies make it possible to implement production models that, until a few years ago, were only hypothesised in the world of research and could not find a real-world application. Artificial intelligence (AI), for example, is used in CPS for prediction and decision-making activities. Cloud technologies allow managing and coordinating CPS systems distributed across a territory; such systems can belong to the same company or different companies. Therefore, cloud technologies have a significant impact on the coordination and efficiency of supply networks. Thus, companies are increasingly integrated, which increases the need for security and privacy in the management of information. Blockchain technology satisfies this need by guaranteeing immutability, transparency, authenticity, decentralisation, distribution, the absence of intermediaries, and anonymity (Dolgui et al. 2020; Lee et al. 2019). Finally, immersive technologies are currently used in the industrial field, mainly in operator training and plant maintenance. Although they still have limitations (e.g. for health reasons, it is advisable not to use such devices for long periods of time), they can potentially make essential contributions on the shop floor (e.g. the use of collaborative robots and manual assembly activities).

10.3 New Models and Perspectives for Production Planning and Control in the Industry 4.0 Era

As discussed in the previous paragraph, it can be observed that there has never been a strong evolution of the systems of planning and control of production. In previous industrial revolutions, effort was focused on increasing productivity and the technological aspects of production. For example, except for the Taylorism or labour division and its subsequent practical application to mass production by Henry Ford (Fordism) in 1913, the first manufacturing planning and control (MPC) system was not developed until 1960. The intent, of course, was always the same: reduce production costs while maximising the use of the production plant and the exploitation of operators' working hours. After the rise and development of the information technology (IT) field, material requirements planning (MRP) and, subsequently, manufacturing resource planning (MRP-II, referred below as simply MRP) systems evolved. MRP systems were developed with the aim to tackle the increasing complexity of manufactured products and reduce the average inventory level since, through such a reduction, a considerable amount of immobilised capital could be released while simultaneously increasing plant flexibility.

Unfortunately, in the following years, such ambitions were not realised. Furthermore, many firms believe that MRP systems, despite further improvements (e.g. rough-cut capacity verification tools), have resulted in a slight increase in the average level of plant supplies (Hopp and Spearman 2011). One might ask what caused this problem. Basically, the issues are the assumptions involved in MRP: (1) the MRP assumes that the lead times of the various components are independent of the production system state (i.e. that an MRP system works at infinite capacity); (2) MRP considers a static lead time for components (i.e. it does not take into account the potential production problems or, in general, the variability of the processing cycle); and (3) a delayed job is worse than an excess in the inventory level in a real-world production environment, and thus firms' propensity to inflate components' lead times was observed. The third issue, referred to in the literature as 'lead time syndrome' (Knollmann and Windt 2013), leads to anticipating the release of production orders, with the intent to limit the delay risk involved in lead-time variability. However, the propensity to anticipate the processing of components and maximise machine utilisation, which is typical of a push control strategy, resulted in increased WIP levels, causing a vicious cycle of continuously lengthening production lead times.

As discussed in the above paragraph, in parallel to the MRP production philosophy, the Japanese 'just-in-time' approach is found. Here the firms are forced to standardise products and processes to obtain a continuous and balanced production flow across the entire system, working at the demand pace. The resulting approach focuses attention on the system as a whole rather than individual production resources, making it possible to achieve some operational advantages (e.g. shorter production cycles, reduced lead times, and lower inventories). This strategy has shown greater effectiveness in keeping the production flow organised and controlled, acting in a more targeted manner on WIP reduction. Undeniably, firms that have implemented JIT techniques have experienced significant improvements in productivity, as well as gradually reducing their WIP levels. However, this WIP reduction requires a trade-off; the standardisation of processes must be pushed continuously, which results in products having to be designed to facilitate this process and allow flow-oriented production. Additionally, this standardisation requirement, which often involves the whole production chain, is not only difficult to achieve but also pushes manufacturing companies into direct competition with industries in developing countries, where labour costs are low. This facilitates a shift in the production vision from a 'mass production' scenario, typical of the previous century, toward 'mass customisation'. Therefore, instead of achieving mere cost reduction, it is desirable to create value while meeting customers' customisation and speed of delivery requirements. Furthermore, in this dynamic context, it becomes essential to acquire and evolve the managerial capabilities of a manufacturing plant, enabling the MPC system to allocate production resources quickly and effectively (i.e. increasing their interoperability and flexibility). The impact of such a strategy is so critical that it justifies the creation of a new industrial paradigm: the 'fourth industrial revolution' and the introduction of the abovementioned Cyber-Physical System (CPS) concept. This is nowadays furthermore motivated by the need of reconfigurability and responsiveness brought to light by emerging challenges such as pandemic-related critical situations (Ivanov & Dolgui, In press).

The CPS and Internet of things (IoT) concepts are recognised as technological, infrastructural, and enabling elements of this revolution, in which it is advisable to develop new organisation models (Riedl et al. 2014). In this regard, it is possible to identify a multitude of 'reference models', 'reference architectures', and 'architectures' in the literature (Moghaddam et al. 2018). Before going on to mention some of the most important ones, we want to clarify their differences, from a taxonomic perspective, considering the various definitions available in literature. It is assumed that a 'reference model' is based on a small number of unifying concepts and can be used as a basis for the development and explanation of standards to a non-specialist. Hence, a reference model should not be linked directly to any standards, technology, or other concrete implementation details, but should try to utilise conventional semantics that can be used unambiguously through different implementations. It is then a stable model, universally recognised and recommended, based on which architectural reference models (e.g. reference architecture) can be derived for an assigned specific area. Hence, it is assumed that reference architecture is a fundamental structural model, but applicable in a particular domain, accepted as a starting point for the definition of new system-specific architectures. To this extent, it must be a sufficiently abstract framework that includes a set of basic concepts, axioms, and descriptions of the main interactions between entities in the internal and external application domain. Finally, we can refer to an 'architecture' as a well-defined system structure in greater detail regarding its elementary components, principles and relationships existing between its components (Pisching et al. 2018).

Regarding the manufacturing field, reference models available in the literature are Industry 4.0, cloud manufacturing, and the Internet of things. Liu and Xu (2017) conducted a comparative analysis of Industry 4.0 and cloud manufacturing, highlighting the similarities and differences between them. As previously stated, CPS is the enabling technology of the Industry 4.0 model, being able to adequately summarise all the technological aspects required in this production paradigm (Fig. 10.5), including the machine-to-machine (M2M) concept (e.g. inter-communication between machines), 'horizontal and vertical integration', and 'end-to-end integration'. In an Industry 4.0 environment, machines and, hence, CPSs,

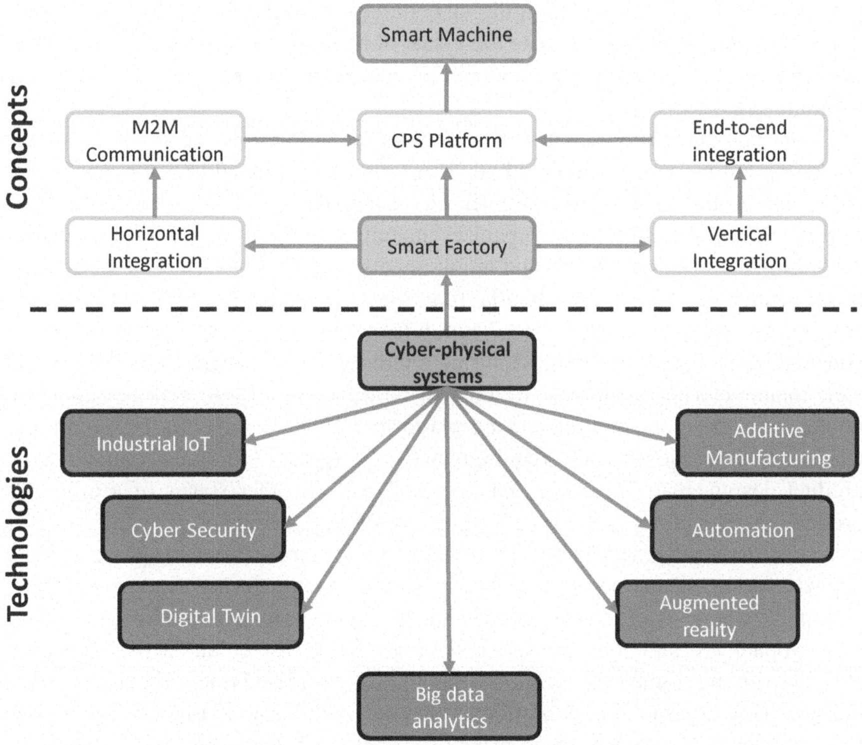

Fig. 10.5 The Industry 4.0 reference model. (Inspired from Liu and Xu 2017)

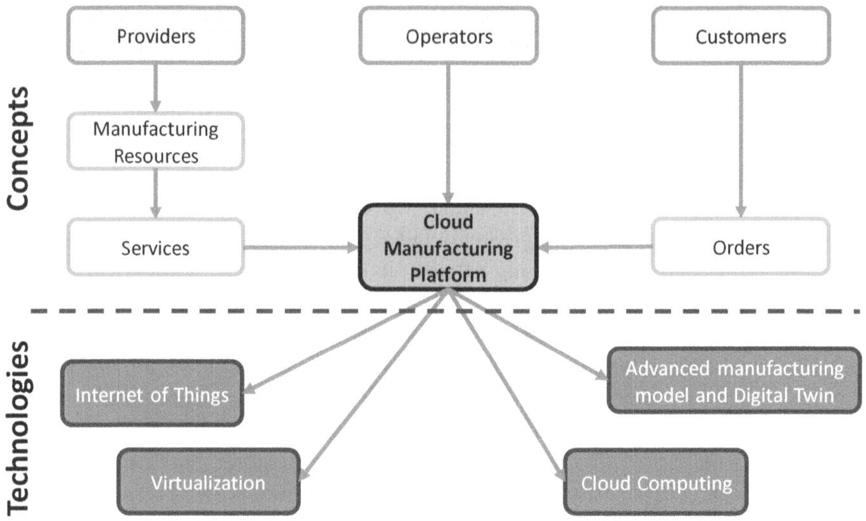

Fig. 10.6 The Cloud Manufacturing reference model. (Inspired from Liu and Xu 2017)

must be able to interact with their digital counterpart (digital twin) to optimise their operating conditions (e.g. evaluate their state of health with prognostics techniques) and with other CPSs to facilitate cooperation and achieve production objectives (i.e. schedule activities).

Cloud manufacturing is a new service-oriented business paradigm based on the cloud-computing concept and method. This term first appeared in a published work by Bo-Hu et al. (2010), which defined cloud manufacturing as 'a new networked manufacturing paradigm that organises manufacturing resources over networks according to consumers' needs and requirements to provide a variety of on-demand manufacturing services via networks (e.g. the Internet) and cloud manufacturing service platforms'. Figure 10.6 is a diagram from Liu and Xu (2017) that depicts the core concepts of the cloud manufacturing reference model, where 'providers supply their manufacturing resources, which will be transformed into services and then pooled into the cloud manufacturing platform'. Then, within this new platform, customers may submit their requirements for requested services, ranging from product design, manufacturing, management, and all other stages of a product's production. Therefore, cloud manufacturing evolves the concept of manufacturing-as-a-service (MaaS), which shifts production-oriented manufacturing processes into a service-oriented network by representing single manufacturing assets as services.

Finally, there is a plenty of reference models for the Internet of things (Bandyopadhyay and Sen 2011; Bassi et al. 2013; Xu et al. 2014). However, strictly speaking, some of these models should be considered reference architectures because their structure and definition are too detailed, meaning that it is impossible to recognise the universal, basic principles of the IoT paradigm within them. As discussed in the previous paragraph, the identified key aspects of the IoT are interconnection and interoperability. Hence, the IoT enables data acquisition from the connected 'thing' and facilitates their communication through the network and data analysis for their exploitation.

Once the reference model of a given paradigm has been defined, it is possible to derive the reference architecture. For example, as stated in the previous paragraph, considering the IoT reference model applied to the manufacturing context, we will find the industrial IoT (IIoT) reference architecture, defined specifically for the industrial context. However, since the objective here is to focus on Industry 4.0, we will analyse the reference architecture and related architectures on the latter, clarifying the various similarities among architectural proposals. Since Industry 4.0 was conceived, many working groups have been involved in its development, expressing different viewpoints. For example, the 'IBM Industry 4.0 architecture' (IBM 2018) reflects IBM's vision, introducing the division of manufacturing system architectures into three functional layers (i.e. edge, plant, and enterprise), with enhanced flexibility to deploy and move similar functionality between the three layers. The 'NIST Service-Oriented Smart Manufacturing System Architecture' (Lu et al. 2016), instead, introduced the 'Smart Manufacturing Ecosystems' concept, integrating for the first time in the same paradigm all the manufacturing competences, like production, management, design, and engineering function. Finally, 'Reference Architecture Model Industry 4.0' (RAMI 4.0 [DIN SPEC 91345:2016-04 2016]), depicted in Fig. 10.7, is derived from the CENELEC model for the Smart Grid Architecture Model (SGAM [CEN-CENELEC-ETSI 2014]), and represents the most comprehensive Industry 4.0 reference architecture, considering the many functional

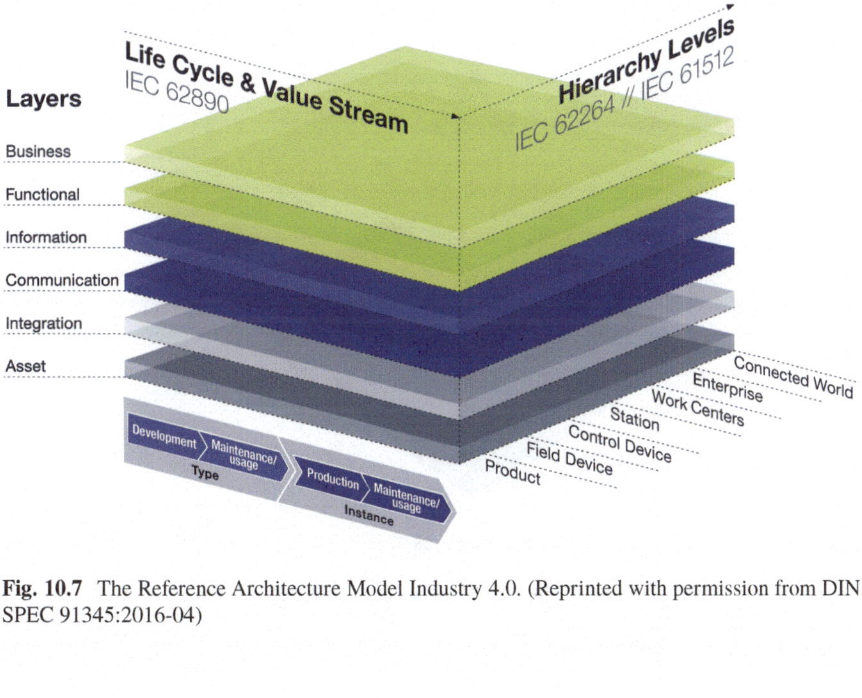

Fig. 10.7 The Reference Architecture Model Industry 4.0. (Reprinted with permission from DIN SPEC 91345:2016-04)

levels a manufacturing 'asset' may have. And it is precisely the asset concept the first definition of the architecture reference document DIN SPEC 91345:2016-04 referred to as a component or a set of components of the factory. It can be a simple sensor or a set of simple components, expressions of a processing machine. An asset may also be physical (e.g. an industrial machine or a product), as well as logical (e.g. a management system). In turn, once the asset has been positioned on the basic plane, consisting of the life cycle and value stream axis, defined with respect to the IEC 62890, and a second axis characteristic of the managerial hierarchical level, expanded from IEC 61512, it is possible to characterise its 'functionalities' among the different provided layers. This model considers seven levels of functional interaction and can describe all the structural properties of the asset: business, functional, information, communication, and integration, as well as the asset itself.

After explaining the reference architecture concept, the discussion now focuses on the analysis and definition of the architecture of industrial production planning and control systems (i.e. MPC systems). The problem is that, to the authors' best knowledge, there are not many examples of Industry 4.0 architecture, and, more importantly, none are derived from a specific reference architecture. Thus, to derive an MPC system architecture model from the proposed RAMI 4.0, it becomes clear that an asset or, more precisely, a set of assets may represent an MPC system. Interestingly, from the structural viewpoint, the aforementioned reference architectures betray their assumptions. These reference architecture models are the result of the generalisation effort from existing architectural models. In particular, the ANSI/ISA95 standard shown in Fig. 10.8, falling into the category of RAMI 4.0 specifications, represents a specific application, also if it was developed before the reference architecture. The problem is that ANSI/ISA95 was defined with the

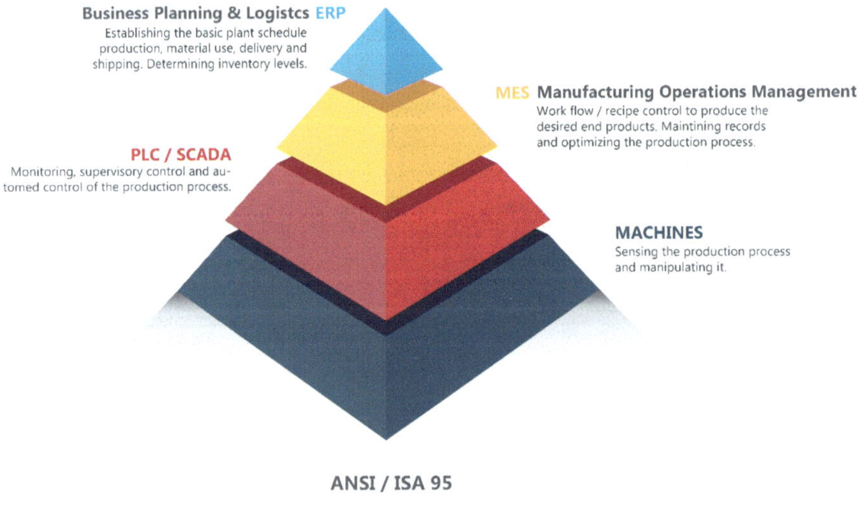

Fig. 10.8 The ANSI/ISA95 hierarchical architecture of an MPC system

aim to rationalise the different competencies of the different management levels of a production plant, clarifying the interactions between the different levels and identifying the communication standards among them (e.g. OPC-UA). In this attempt, the ANSI/ISA95 standard provides a hierarchical view of a production planning and control system.

Moreover, it is from this standard that MRP systems started their development process, with all the limitations that we have previously exposed. Their ability to react to a problem is limited to only the same predefined and basic rules that an advanced manufacturing executing system (MES) may have. With the intention to generalise, we may say that the limits of hierarchically structured architectures are that (1) most real-world manufacturing systems are too complex to be modelled in exact terms and solved optimally (e.g. optimising operations is an NP-hard problem in the industrial field), and (2) a breakdown between the enterprise resource planning (ERP) system and the MES creates a situation in which, if the best mathematical optimisation is found at the ERP level, it might not be performed optimally for the specific state of the shop floor (e.g. an unexpected delay or a machine failure; Guizzi et al. 2017).

RAMI 4.0, like all reference architectures based on the Industry 4.0 reference model, is structured to include the hierarchical organisational system as a particular case, (e.g. the ERP level described from the ANSI/ISA95 may be represented from a logical asset located within the enterprise base plane of the RAMI 4.0, with high-level functionality), while being open to all possible system configurations. Hence, the intention is to overcome the rigid structure of previous MPC systems when possible, proposing a general framework in which all the managerial structures may be accepted. Duffie and Prabhu (1996) propose the first complete taxonomy of management architectures that an MPC system may have, identifying four possible alternatives (Fig. 10.9): hierarchical, oligarchical, semi-heterarchical, and heterarchical.

All these alternatives are allowed in RAMI 4.0 and, by analysing the scientific literature, it is possible to find studies that, even prior to RAMI 4.0, propose MPC system transitions from a centralised hierarchical approach (i.e. MRP-based) to a decentralised and heterarchical one (Scholz-Reiter 2004).

However, due to the greater complexity and dynamism required by continuously increasing product customisation requirements, shortening the product life cycle, and reducing the delivery time, the transition from a pure hierarchical structure to a more heterarchical one is not a trivial or quick task (Bozarth et al. 2009). The

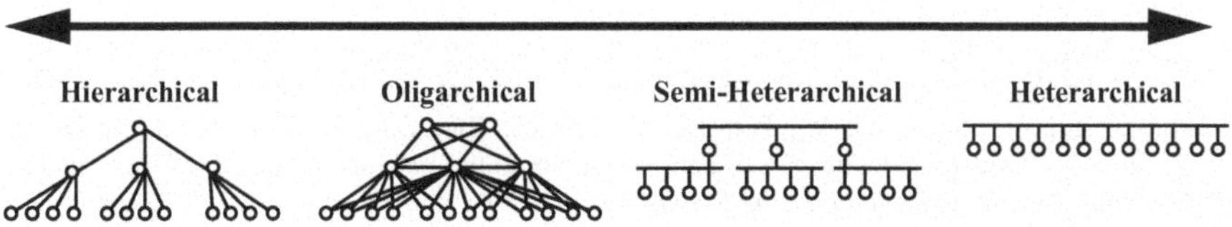

Fig. 10.9 Spectrum of distributed manufacturing control systems. (From Duffie and Prabhu 1996)

underlying problem of such a transition is that a considerable amount of decision-making has to be transferred to entities localised at the bottom levels of the system. When switching such an increased level of decision-making autonomy to low-level entities (i.e. the assets), productivity can be increased only up to a certain threshold, after which it is agreeable to assume that the behaviour of the individual entities will tend to drive the objectives toward achieving local goals, short-sighting the system's objectives (Philipp et al. 2006). This happens because the behaviour of a completely heterarchical system strongly depends on local decisions and, therefore, the logics that each system component makes (Jeken et al. 2012). In a completely decentralised system, decisions are made with only partial information regarding the system since each entity having global knowledge is an unrealisable objective because it would cause the deployment of the entire system's complexity on every component. For the above-stated reasons, more in-depth scientific studies are necessary in this regard, mainly because most of the proposed Industry 4.0 architectures consider purely decentralised (i.e. heterarchical) management architectures.

In conclusion, even in an Industry 4.0 empowered environment, it is still necessary to gain a clear understanding of the dynamics involved in the complex interactions taking place in a manufacturing system. It is also advisable to learn as much as possible from the expertise and advances achieved by academics and practitioners regarding manufacturing-related reference architectures. Therefore, considering RAMI 4.0 as a commonly accepted, established reference architecture for Industry 4.0, future research effort could be focused on formulating architectures that go beyond the limits and problems associated with a strict hierarchical scheme while attempting to avoid the complete delegation of autonomy to entities within the plant.

It is, thus, advisable that future studies may also consider the possibility of developing new MPC configurations based on intermediate approaches, such as oligarchic or semi-heterarchical ones (Grassi et al. 2020). These latter approaches can better support the merging of the predictive and managerial advantages of a centralised structure with the reactivity of a fully decentralised one, given their intrinsic capability to allow functional and physical hybridisation deployable in the design of an Industry 4.0-oriented MPC system.

References

Ashton, K. (2009). That 'internet of things' thing. *RFID Journal, 22*, 97–114.

Bandyopadhyay, D., & Sen, J. (2011). Internet of things: Applications and challenges in technology and standardization. *Wireless Personal Communications, 58*(1), 49–69.

Bassi, A., Bauer, M., Fiedler, M., Kramp, T., van Kranenburg, R., Lange, S., & Meissner, S. (Eds.). (2013). *Enabling things to talk*. Berlin, Heidelberg: Springer Berlin Heidelberg.

Bo-Hu, L., Lin, Z., Shi-Long, W., Fei, T., Jun-Wei, C., Xiao-Dan, J., Xiao, S., & Xu-Dong, C. (2010). Cloud manufacturing: A new service-oriented networked manufacturing model. *Computer Integrated Manufacturing Systems, 16*, 1–7.

Bozarth, C. C., Warsing, D. P., Flynn, B. B., & Flynn, E. J. (2009). The impact of supply chain complexity on manufacturing plant performance. *Journal of Operations Management, 27*, 78–93. https://doi.org/10.1016/j.jom.2008.07.003.

CEN-CENELEC-ETSI. (2014). Smart Grid Coordination Group—Smart Grid Information Security, December 2014.
Corsaro, A., & Schmidt, D. C. (2012). The data distribution service the communication middleware fabric for scalable and extensible systems-of-systems, Chapter 2. In A. V. Gheorghe (Ed.), *System of systems*. Rijeka: IntechOpen.
DIN SPEC 91345:2016-04. (2016). Reference Architecture Model Industrie 4.0 (RAMI4.0).
Dolgui, A., Ivanov, D., Potryasaev, S., Sokolov, B., Ivanova, M., & Werner, F. (2020). Blockchain-oriented dynamic modelling of smart contract design and execution in the supply chain. *International Journal of Production Research, 58*(7), 2184–2199.
Dolgui, A., Ivanov, D., Sethi, S. P., & Sokolov, B. (2019b). Scheduling in production, supply chain and Industry 4.0 systems by optimal control. *International Journal of Production Research, 57*(2), 411–432.
Duffie, N., & Prabhu, V. V. (1996). Heterarchical control of highly distributed manufacturing systems. *International Journal of Computer Integrated Manufacturing, 9*, 270–281. https://doi.org/10.1080/095119296131562.
Grassi, A., Guizzi, G., Santillo, L. C., & Vespoli, S. (2020). A semi-heterarchical production control architecture for industry 4.0-based manufacturing systems. *Manufacturing Letters, 24*, 43–46. https://doi.org/10.1016/j.mfglet.2020.03.007.
Guizzi, G., Vespoli, S., & Santini, S. (2017). On the architecture scheduling problem of Industry 4.0. In *CEUR Workshop Proceedings*.
Guizzi, G., Falcone, D., & De Felice, F. (2019). An integrated and parametric simulation model to improve production and maintenance processes: Towards a digital factory performance. *Computers and Industrial Engineering, 137*, 106052.
Habib, M., & Chimsom, C. (2019). Industry 4.0: Sustainability and design principles. In *Proceedings of the 2019 20th International Conference on Research and Education in Mechatronics, REM 2019*.
Hopp, W. J., & Spearman, M. L. (2011). *Factory physics* (3rd ed.). Long Grove, IL: Waveland Press.
IBM. (2018). IBM Industrie 4.0 Architecture—SmartFactoryXL.
Ivanov, D., Sokolov, B., & Dilou Raguinia, E. A. (2014). Integrated dynamic scheduling of material flows and distributed information services in collaborative cyber-physical supply networks. *International Journal of Systems Science: Operations & Logistics, 1*(1), 18–26.
Ivanov, D., Sokolov, B., Dolgui, A., Werner, F., & Ivanova, M. (2016). A dynamic model and an algorithm for short-term supply chain scheduling in the smart factory Industry 4.0. *International Journal of Production Research, 54*(2), 386–402.
Ivanov, D., Das, A., & Choi, T.-M. (2018a). New flexibility drivers for manufacturing, supply chain and service operations. *International Journal of Production Research, 56*(10), 3359–3368.
Ivanov, D., Sethi, S., Dolgui, A., & Sokolov, B. (2018b). A survey on the control theory applications to operational systems, supply chain management and Industry 4.0. *Annual Reviews in Control, 46*, 134–147.
Ivanov, D., & Dolgui, A. (In press). Viability of intertwined supply networks: Extending the supply chain resilience angles towards survivability. A position paper motivated by COVID-19 outbreak. *International Journal of Production Research*.
Jeken, O., Duffie, N., Windt, K., Blunck, H., Chehade, A., & Rekersbrink, H. (2012). Dynamics of autonomously acting products and work systems in production and assembly. *CIRP Journal of Manufacturing Science and Technology, 5*, 267–275. https://doi.org/10.1016/j.cirpj.2012.09.012.
Jin, W., Liu, Z., Shi, Z., Jin, C., & Lee, J. (2017). CPS-enabled worry-free industrial applications. In *2017 Prognostics and System Health Management Conference, PHM—Harbin 2017—Proceedings*.
Knollmann, M., & Windt, K. (2013). Control-theoretic analysis of the lead time syndrome and its impact on the logistic target achievement. *Procedia CIRP, 7*, 97–102. https://doi.org/10.1016/j.procir.2013.05.017.

Lee, J., Bagheri, B., & Kao, H.-A. (2015). A cyber-physical systems architecture for industry 4.0-based manufacturing systems. *Manufacturing Letters, 3*, 18–23.
Lee, J., Azamfar, M., & Singh, J. (2019). A blockchain enabled cyber-physical system architecture for industry 4.0 manufacturing systems. *Manufacturing Letters, 20*, 34–39.
Liu, Y., & Xu, X. (2017). Industry 4.0 and cloud manufacturing: A comparative analysis. *Journal of Manufacturing Science and Engineering, 139*(3), 034701. https://doi.org/10.1115/1.4034667.
Lu, Y., Morris, K., & Frechette, S. (2016) Current Standards Landscape for Smart Manufacturing Systems. https://doi.org/10.6028/NIST.IR.8107.
Minerva, R., Biru, A., & Rotondi, D. (2015). Towards a definition of the internet of things (IoT). IEEE Internet Initiative.
Moghaddam, M., Cadavid, M. N., Kenley, C. R., & Deshmukh, A. V. (2018). Reference architectures for smart manufacturing: A critical review. *Journal of Manufacturing Systems, 49*, 215–225. https://doi.org/10.1016/j.jmsy.2018.10.006.
MTConnect Institute (2018). MTConnect Standard, Version 1.4.0 (ANSI/MTC1.4-2018).
Ohno, T. (1988). *Toyota production system: Beyond large-scale production* (1st ed.). Boca Raton, FL: CRC Press.
OPC Foundation (2019). OPC 30070-1 OPC UA Companion Specification release 2.00.00.
Orlicky, J. (1975). *Material requirements planning: The new way of life in production and inventory management*. New York: McGraw-Hill.
Panetto, H., Iung, B., Ivanov, D., Weichhart, G., & Wang, X. (2019). Challenges for the cyber-physical manufacturing enterprises of the future. *Annual Reviews in Control, 47*, 200–213.
Pfrommer, J., Gruner, S., & Palm, F. (2016). Hybrid OPC UA and DDS: Combining architectural styles for the industrial internet. In *IEEE International Workshop on Factory Communication Systems—Proceedings, WFCS, June* (Vol. 2016).
Philipp, T., Böse, F., & Windt, K. (2006). Evaluation of autonomously controlled logistic processes. In *Proceedings of the Fifth CIRP International Seminar on Intelligent Computation in Manufacturing Engineering* (pp. 347–352).
Pisching, M. A., Pessoa, M. A. O., Junqueira, F., dos Santos Filho, D. J., & Miyagi, P. E. (2018). An architecture based on RAMI 4.0 to discover equipment to process operations required by products. *Computers and Industrial Engineering, 125*, 574–591. https://doi.org/10.1016/j.cie.2017.12.029.
Riedl, M., Zipper, H., Meier, M., & Diedrich, C. (2014). Cyber-physical systems alter automation architectures. *Annual Reviews in Control, 38*, 123–133. https://doi.org/10.1016/j.arcontrol.2014.03.012.
Scholz-Reiter, B. (2004). Autonomous logistic processes: New demands and first approaches. In: *Proceedings of the 37th CIRP International Seminar on Manufacturing Systems* (pp. 357–362).
Sisinni, E., Saifullah, A., Han, S., Jennehag, U., & Gidlund, M. (2018). Industrial internet of things: Challenges, opportunities, and directions. *IEEE Transactions on Industrial Informatics, 14*(11), 4724–4734.
Vespoli, S., Grassi, A., Guizzi, G., & Santillo, L. C. (2019). Evaluating the advantages of a novel decentralised scheduling approach in the industry 4.0 and cloud manufacturing era. *IFAC-PapersOnLine, 52*(13), 2170–2176. https://doi.org/10.1016/j.ifacol.2019.11.527.
Xu, L. D., He, W., & Li, S. (2014). Internet of things in industries: A survey. *IEEE Transactions on Industrial Informatics, 10*, 2233–2243. https://doi.org/10.1109/TII.2014.2300753.
Xu, L., Xu, E., & Li, L. (2018). Industry 4.0: State of the art and future trends. *International Journal of Production Research, 56*(8), 2941–2962.

Chapter 11
Multi-Criteria Single Batch Machine Scheduling Under Time-of-Use Tariffs

Junheng Cheng, Feng Chu, and Peng Wu

Abstract Most of the industrialized countries are moving towards the fourth stage of industrialization. This development has provided immense opportunities for industrial sustainability. As the largest energy consumer in the world, most of the industrial sector's consumption is in the form of electricity. In recent years, to strengthen the peak load regulation capability, time-of-use (ToU) pricing has been implemented in many countries to encourage consumers to shift their use from peak to mid- and off-peak periods such that their energy bills can be reduced. In this chapter, we first introduce a basic single machine batch scheduling problem under ToU electricity tariffs. Then it is extended by further considering machine on/off switching. Finally, a single machine batch scheduling problem under ToU tariffs in a continuous processing environment is investigated. For the three considered problems, appropriate mathematical models are established, and their problem properties and complexities are demonstrated.

11.1 Introduction

The growing population brings great opportunities for economic and social development, but also presents enormous challenges to limited resources. Globally, facing the contradiction between the continuously increasing worldwide demand for consumer goods and sustainable evolvement of human existing environment,

J. Cheng
School of Economics, Fujian Normal University, Fuzhou, Fujian, China
e-mail: junheng.cheng@fjnu.edu.cn

F. Chu (✉)
University of Evry-Val d'Essonne, University of Paris-Saclay, Evry, France
e-mail: feng.chu@ibisc.fr

P. Wu
School of Economics and Management, Fuzhou University, Fuzhou, Fujian, China
e-mail: wupeng88857@fzu.edu.cn

B. Sokolov et al. (eds.), *Scheduling in Industry 4.0 and Cloud Manufacturing*,
International Series in Operations Research & Management Science 289,
https://doi.org/10.1007/978-3-030-43177-8_11

industrial value creation must turn towards sustainability (Stock and Seliger 2016). Recent technological leaps have paved way for highly mechanization and automation in industry, most of the early industrialized countries are moving towards the era of Industry 4.0. This development provides tremendous opportunities for the realization of sustainable manufacturing, where energy saving takes a very important position. Industrial sector is the largest energy consumer in the world. Improving their energy usage efficiency is highly important to enhance the competitiveness of manufacturing enterprises in the new era and promote sustainable development of the economy.

Electricity is one of the most widely used energies, which accounts for about 30% of total energy consumption in the area of Asia Pacific Economic Cooperation (Fang et al. 2011). As a renewable resource, it is cleaner, safer, and more convenient than many nonrenewable resources, such as coal and oil. With the acceleration of electrification process, electricity will play a more important role in modern manufacturing industry. However, electricity cannot be efficiently stored. Thus its production, transmission, and consumption has to be conducted simultaneously. In addition, electricity demand distribution is uneven over time. To improve peak load regulation ability, demand response (DR) strategy has been widely implemented in the world. It encourages customers to change their normal consumption patterns to respond to the varying electricity prices and save total power cost. ToU electricity tariffs are one of the most important DR strategies. It provides varying electricity prices, for example, low price in off-peak periods and high price in on-peak periods.

Under ToU tariffs, manufacturing enterprises can simultaneously achieve the optimization on production efficiency and energy cost via more reasonable production scheduling, and this topic has attracted increasing interest from researchers recently. For single-machine environment, Shrouf et al. (2014) proposed a genetic algorithm for jobs with a given processing order to optimize total electricity cost with turn-on/off strategy under ToU electricity prices. Fang et al. (2016) considered energy cost saving for uniform-speed and speed-scalable machines and proposed several heuristics to obtain near-optimal solution. Che et al. (2016) developed a mixed-integer linear programming (MILP) model and a greedy insertion heuristic to minimize total electricity cost. For parallel machines, Moon et al. (2013) investigated an unrelated parallel machine scheduling problem under varying electricity prices. Ding et al. (2016) presented a MILP model and a column generation based heuristic to minimize total electricity cost respecting a given makespan. Later, Cheng et al. (2018) improved Ding's model. For flow shops, Luo et al. (2013) developed a novel ant colony optimization based meta-heuristic to minimize both electricity cost and makespan. Zhang et al. (2014) formulated a time-indexed MILP model to minimize total electricity cost and carbon emissions while ensuring the production throughput at the same time. All of the above studies focused on classical machine environment.

Batch processing manufacturing system representing a typical production environment has been widely encountered in modern manufacturing industries, such as steel manufacturing (Tang et al. 2001; Wang et al. 2019), semiconductor manufacturing (Jia et al. 2015; Uzsoy et al. 1994), and aircraft industry (Xu and Bean 2016), and most of them are energy-intensive ones. A specific feature of batch processing is that a processing machine can process multiple jobs at a time.

As a result, batch scheduling is usually more complex than traditional production scheduling, because it needs to optimally group the jobs into batches and schedule the formed batches. A majority of batch scheduling problems have been proved to be NP-hard, even under single-machine environment. Considering ToU tariffs, a traditional batch processing machine scheduling problem will be more complicated; hence, it is significant to investigate such problems in theoretical perspective to guide the practice in manufacturing industries. This chapter discusses some most recent advances and issues in this field including single batch machine scheduling under ToU tariffs (SBMS-ToU) (Cheng et al. 2016), SBMS-ToU considering machine on/off switching (SBMS-ToU-on/off) (Cheng et al. 2017), and SBMS-ToU in continuous processing (SBMS-ToU-CP) (Cheng et al. 2016). Their problem descriptions and formulations will be introduced in the following sections.

11.2 Single Batch Machine Scheduling Under ToU Tariffs

Single batch machine is a basis of more complicated batch processing systems and has been widely encountered in real production environments, such as semiconductor manufacturing industry (Lee et al. 1992; Wang and Uzsoy 2002) and shoe manufacturing industry (Fanti et al. 1996). A variety of single batch machine scheduling (SBMS) problems have been proved to be NP-hard, where most of them focus on optimizing production efficiency, e.g., makespan, total completion time, maximum tardiness. Taking ToU tariffs into consideration, a traditional SBMS problem will be further complicated, since it has to not only group jobs into batches and determine batch processing sequence, but also position batch processing time period. In order to simultaneously optimize environmental benefits as well as production efficiency, total electricity cost and makespan are considered as the two optimization objectives.

11.2.1 Problem Description

A bi-objective single batch machine scheduling problem under ToU tariffs (SBMS-ToU) can be described as follows:

A given set of $J = \{1, 2, \ldots, |J|\}$ jobs is to be processed on a single batch processing machine within a horizon $I = \{1, 2, \ldots, |I|\}$. The duration of period $i \in I$ is denoted as S_i. Job $j \in J$ is nonpreemptive and has a processing time p_j. Any p_j is less than the duration of any period i; i.e., $S_i \gg p_j, \forall j \in J, \forall i \in I$. Without loss of generality, we assume that the jobs are numbered in nonincreasing order of the processing times; i.e.,

$$p_1 \geq p_2 \geq \cdots \geq p_{|J|}.$$

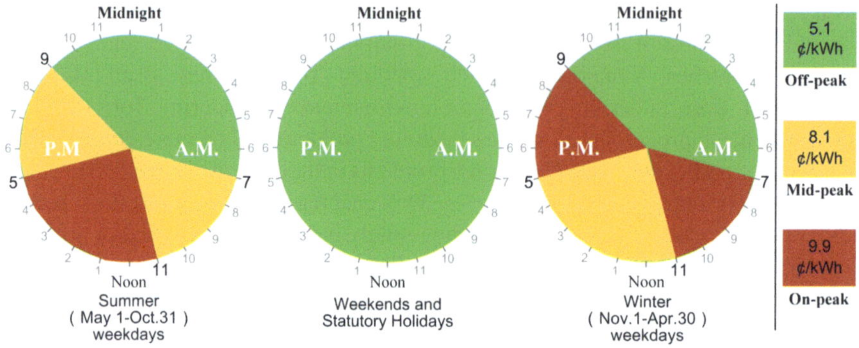

Fig. 11.1 An example of time-of-use tariffs (Source: Ontario Energy Board)

The jobs can be regrouped to $|B|$ (to be optimized) batches and each batch can contain at most C jobs. Therefore, we must have $\lceil |J|/C \rceil \leq |B| \leq |J|$ (Ikura and Gimple 1986). The processing time of a batch is determined by the longest processing time job in the batch.

The processing of a batch should be completed before the end of a period or it must wait for the beginning of another period. Each period can be regarded as a workshift, whose unit electricity cost can be calculated according to the tariff information. Take the TOU tariffs of Ontario (see Fig. 11.1) for example, for a three-shift in a work day: 8–16 h, 16–0 h, 0–8 h, the corresponding unit electricity costs are as follows: $e_{8\,\mathrm{h}-16\,\mathrm{h}} = (11.4 * 3 + 14.0 * 5)/8$ ¢/kWh $* 1$ kWh/h $= 13.0250$ ¢/h, $e_{16\,\mathrm{h}-0\,\mathrm{h}} = 9.4125$ ¢/h, and $e_{0\,\mathrm{h}-8\,\mathrm{h}} = 8.1625$ ¢/h. Generally, a work day is often composed of two or three work shifts according to the types of tariffs.

11.2.2 Mathematical Model

The parameters and decision variables for model formulation can be summarized as follows:

Indices

j: index of jobs, $j \in J = \{1, 2, \dots, |J|\}$;
b: index of batches, $b \in B = \{1, 2, \dots, |B|\}$;
i: index of periods, $i \in I = \{1, 2, \dots, |I|\}$;

Parameters

J: set of all jobs, $J = \{1, 2, \dots, |J|\}$;
B: set of batches, $B = \{1, 2, \dots, |B|\}$;
I: set of time periods on the planning horizon, $I = \{1, 2, \dots, |I|\}$;
C: capacity of a batch;

p_j: processing time of job j, $\forall j \in J$;
s_i: starting time of period i, $\forall i \in I$;
S_i: duration of period i, $\forall i \in I$, in which $S_i = s_{i+1} - s_i$;
e_i: unit electricity cost of period i, $\forall i \in I$;

Decision Variables

$|B|$: number of batches;
$x_{j,b,i}$: $= 1$, if job j is assigned to batch b and processed in period i; otherwise 0; $\forall j \in J$, $\forall b \in B$, $\forall i \in I$;
$y_{b,i}$: $= 1$, if batch b is assigned to period i; otherwise 0; $\forall b \in B$, $\forall i \in I$;
z_i: $= 1$, if at least one batch is assigned to period i; otherwise 0; $\forall i \in I$;
$P_{b,i}$: $= P_b = \max\{p_j \mid j \in b\}$, if batch b is processed in period i, where P_b is the processing time of batch b; otherwise 0; $\forall b \in B$, $\forall i \in I$;
E: total electricity cost for completing all jobs;
C_{max}: completion time of the last job.

In the work, the number of batches $|B|$ is initially set as $|J|$. A batch is opened if there is at least one job allocated to the batch. On the contrary, a batch is closed without any job and its corresponding processing time equals to 0, i.e., $P_b = 0$. The considered problem can be formulated as the following bi-objective MILP model $\mathcal{P}_1'$:

$$\mathcal{P}_1' : \min E \tag{11.1}$$

$$\min C_{max} \tag{11.2}$$

$$s.t. \sum_{i \in I} \sum_{b \in B} x_{j,b,i} = 1, \forall j \in J \tag{11.3}$$

$$\sum_{i \in I} y_{b,i} = 1, \forall b \in B \tag{11.4}$$

$$\sum_{j \in J} x_{j,b,i} \leq C y_{b,i}, \forall b \in B, \forall i \in I \tag{11.5}$$

$$x_{j,b,i} p_j \leq P_{b,i}, \forall j \in J, \forall b \in B, \forall i \in I \tag{11.6}$$

$$\sum_{b \in B} P_{b,i} \leq S_i z_i, \forall i \in I \tag{11.7}$$

$$\sum_{i \in I} e_i \sum_{b \in B} P_{bi} \leq E \tag{11.8}$$

$$s_i z_i + \sum_{b \in B} P_{b,i} \leq C_{max}, \forall i \in I \tag{11.9}$$

$$x_{j,b,i}, y_{b,i}, z_i \in \{0, 1\}, \forall j \in J, \forall b \in B, \forall i \in I \tag{11.10}$$

$$P_{b,i} \geq 0, E \geq 0, C_{max} \geq 0, \forall b \in B, \forall i \in I. \tag{11.11}$$

Objective (11.1) is to minimize the total electricity cost E on the horizon I. Objective (11.2) is to minimize the makespan C_{max}, which is the completion time of the last batch. Constraints (11.3) ensure that job j, $\forall j \in J$, is assigned to only one batch. Constraints (11.4) guarantee that each batch $b, \forall b \in B$, is processed in only one period. Constraints (11.5) assume that the number of jobs assigned to any batch should not exceed the batch capacity C, and any job $j, \forall j \in J$, cannot be assigned to period i if its corresponding batch is not processed in this period. Constraints (11.6) limit processing time of each batch. Constraints (11.7) ensure that the total processing time of batches in period $i, \forall i \in I$, should not exceed its duration, and $z_i = 1$ if there is at least one batch assigned to period i. Constraint (11.8) calculates the total electricity cost. Constraint (11.9) defines the makespan C_{max}. Constraints (11.10)–(11.11) enforce the restrictions on decision variables.

11.2.3 Property Analysis and Improved Model

As mentioned above, the number of batch $|B|$ equals to the number of jobs $|J|$ in model $\mathcal{P}'_1$. According to the preliminary results, it is very time-consuming to directly solve model $\mathcal{P}'_1$. Now we try to analyze properties of the problem to reduce solution space. We show that the formation of batches can be solved independent of the scheduling of batches, with two objectives considered in the problem.

A solution of the problem is uniquely defined by $(|B|, \{J_b, 1 \le b \le |B|\}, \{\tau_b, 1 \le b \le |B|\})$, where $|B|$ is the number of batches, J_b and τ_b are the set of jobs involved in the batch b and the period the batch is processed in, respectively.

We consider in particular those solutions where the batches are formed with a so-called LPT-based rule. In this rule, any job j with $(b-1)C < j \le bC$ and $1 \le b \le \lceil |J|/C \rceil - 1$ is put into batch b and the remaining jobs to batch $\lceil |J|/C \rceil$, where $\lceil x \rceil$ denotes the smallest integer greater than or equal to x. Thus, the processing time of batch b equals to that of job $(b-1)C+1$. Figure 11.2 gives a simple example to illustrate the rule.

The following theorem shows that we only need to consider such solutions in order to find the Pareto front.

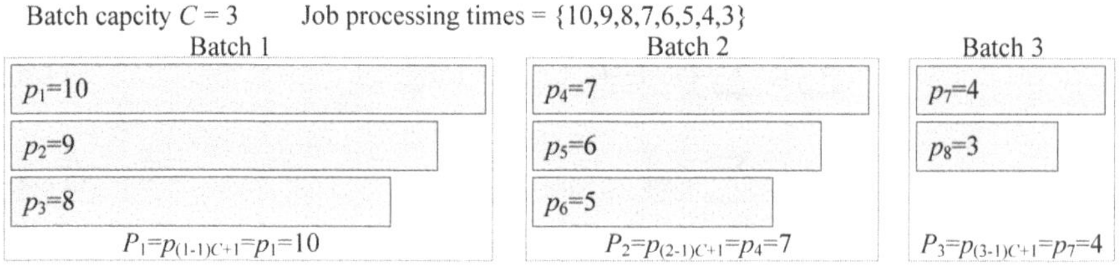

Fig. 11.2 An example of batches formed with the LPT-based rule

Theorem 11.1 *Any solution in which the batches are different from those formed with the LPT-based rule is (at least weakly) dominated.*

Proof. To facilitate the proof, the following notations are used:

J_b: the set of jobs contained in batch b, $J_b \subseteq J$;
$n(J_b)$: the serial number of the least indexed job (thus with the largest processing time) in set J_b, i.e., $n(J_b) = \min\{j | j \in J_b\}$;
$P(J_b)$: the processing time of batch b (the processing time of the least indexed job in J_b), i.e., $P(J_b) = \max_{j \in J_b} p_j = p_{n(J_b)}$.

Let $|B^*|$ and J_b^* represent the number of batches formed with LPT-based rule and the set of jobs involved in batch b $(1 \le b \le |B^*|)$, respectively. We have

$$|B^*| = \lceil |J|/C \rceil,$$
$$J_b^* = \{(b-1)C+1, (b-1)C+2, \ldots, bC\}, \quad b = 1, 2, \ldots, |B^*|-1,$$
$$J_{|B^*|}^* = \{(|B^*|-1)C+1, (|B^*|-1)C+2, \ldots, |J|\}.$$

With the above construction, we have the following equations for batch b, $1 \le b \le |B^*|$:

$$n(J_b^*) = (b-1)C+1, \tag{11.12}$$
$$P(J_b^*) = p_{(b-1)C+1}. \tag{11.13}$$

Consider a feasible solution $\hat{S}$ with $(|\hat{B}|, \{\hat{J}_b, 1 \le b \le |\hat{B}|\}, \{\hat{\tau}_b, 1 \le b \le |\hat{B}|\})$ in which the batches are different from those formed with the LPT-based method. Obviously, we must have

$$|\hat{B}| \ge \lceil |J|/C \rceil = |B^*|. \tag{11.14}$$

In other words, there are at least as many batches as those formed with the LPT-based method. Without loss of generality, we renumber the batches in an increasing order of $n(\hat{J}_b)$'s; i.e.,

$$n(\hat{J}_1) < n(\hat{J}_2) < \cdots < n(\hat{J}_{|\hat{B}|}). \tag{11.15}$$

Then for any batch b such that $1 \le b \le |\hat{B}|$, there exists

$$n(\hat{J}_b) = n(\widetilde{J}), \text{ where } \widetilde{J} = \hat{J}_b \cup \hat{J}_{b+1} \cup \cdots \cup \hat{J}_{|\hat{B}|}. \tag{11.16}$$

Owing to the fact that for any subset $\widetilde{J}$ belonging to set $J = \{1, 2, \ldots, |J|\}$, we have

$$n(\widetilde{J}) \le |J| - |\widetilde{J}| + 1, \quad \forall \widetilde{J} \subseteq J, \tag{11.17}$$

where $|J|$ denotes the number of jobs contained in set J. As for schedule $\hat{S}$, in which the job set has $|\hat{B}|$ disjoint subsets that are sorted as (11.15). According to (11.16) and (11.17), we must have

$$n(\hat{J}_b) = n(\widetilde{J}) \leq |J| - \sum_{\beta=b}^{|\hat{B}|} |\hat{J}_\beta| + 1, \quad \widetilde{J} \subseteq J. \tag{11.18}$$

Due to $J/\widetilde{J} = J/\{\hat{J}_b \cup \hat{J}_{b+1} \cup \cdots \cup \hat{J}_{|\hat{B}|}\} = \hat{J}_1 \cup \hat{J}_2 \cup \cdots \cup \hat{J}_{b-1}$, Equation (11.18) can be further written as:

$$n(\hat{J}_b) \leq \sum_{\beta=1}^{b-1} |\hat{J}_\beta| + 1 \leq \sum_{\beta=1}^{b-1} C + 1 = (b-1)C + 1. \tag{11.19}$$

Since the jobs are indexed in nonincreasing order of their processing times, thus Eq. (11.19) implies that

$$P(\hat{J}_b) = p_{n(\hat{J}_b)} \geq p_{(b-1)C+1} = P(J_b^*), 1 \leq b \leq |B^*|. \tag{11.20}$$

In other words, the processing time of the batches is at least as long as those formed with the LPT-based method.

Construct a new solution by removing batches $|B^*| + 1, \ldots, |\hat{B}|$, if any, and replacing each batch $\hat{J}_b$ $(1 \leq b \leq |B^*|)$ by the corresponding one formed with the LPT-based method (i.e., batch J_b^*), without changing the starting time. In other words, consider solution $(|B^*|, \{J_b^*, 1 \leq b \leq |B^*|\}, \{\hat{\tau}_b, 1 \leq b \leq |B^*|\})$. Relation (11.20) implies that this new solution is also feasible. Furthermore, due to the fact that some batches are removed and the processing times of the remaining batches are reduced, neither the electrical consumption cost nor the makespan is increased, which means that the initial solution is (at least weakly) dominated by the new one. □

As a consequence, by considering batches formed with the LPT-rule as new jobs, the problem is transformed into a classical production scheduling problem without a batching machine. Due to the fact that each batch (new job) should be entirely executed in one period, these new jobs are nonpreemptive. There is an (infinitely short) unavailability period between two successive periods. This latter problem has been proved to be NP-hard in the strong sense, even when the single objective is to minimize the makespan. Hence, we have the following property.

Property 11.1 The SBMS under ToU is strongly NP-hard.

According to Theorem 11.1, optimal solutions will not lose by separately solving batch formation and batch scheduling. That is, jobs can be first batched with LPT-based rule, i.e., $P_b = p_{(b-1)C+1}, b \in B^*$, and $B^* = \{1, 2, \ldots, \lceil |J|/C \rceil\}$. Then decision variables can be restricted to $y_{b,i}$'s and z_i's. The initial model can be equivalent to the following one $\mathcal{P}_1$:

$$\mathcal{P}_1: \quad \min E$$

$$\min C_{max}$$

$$s.t. \sum_{i \in I} y_{b,i} = 1, \forall b \in B^* \tag{11.21}$$

$$\sum_{b \in B^*} P_b y_{b,i} \leq S_i z_i, \forall i \in I \tag{11.22}$$

$$\sum_{i \in I} \sum_{b \in B^*} e_i P_b y_{b,i} \leq E \tag{11.23}$$

$$s_i z_i + \sum_{b \in B^*} P_b y_{b,i} \leq C_{max}, \forall i \in I \tag{11.24}$$

$$y_{b,i}, z_i \in \{0, 1\}, \forall b \in B, \forall i \in I, E \geq 0, C_{max} \geq 0. \tag{11.25}$$

Constraints (11.21) ensure that a formed batch b, $\forall b \in B^*$ is allocated into exactly one period. Constraint (11.22) guarantees that the total processing time of the batches in period i does not exceed its duration and $z_i = 1$ if there is at least one batch allocated to period i, $\forall i \in I$. Constraints (11.23) and (11.24) restrict the total electricity cost and makespan, respectively. Constraint (11.25) specifies the restrictions on the variables.

Remark 11.1 Compared with the initial model $\mathcal{P}'_1$, the improved model $\mathcal{P}_1$ reduces $|I| \cdot |J|^2 + (|J| - \lceil |J|/C \rceil) \cdot |I|$ binary variables and $|I| \cdot |J|$ real variables as well as $(2 + |I| \cdot |J| + |I|) \cdot |J| - (|J| \cdot |I| + |I| + 1) \cdot \lceil |J|/C \rceil$ constraints.

Taking an instance with $|J| = 100$, $C = 10$, and $|I| = 10$ as an example, $\mathcal{P}_1$ can reduce 100900 binary variables, 1000 real variables, and 91090 constraints compared with $\mathcal{P}'_1$. Owing to such reduction of variables and constraints, the search space for Pareto optimal solutions via the improved model $\mathcal{P}_1$ is significantly reduced.

11.3 SBMS-ToU Considering Machine On/Off Switching

By observing the scheduling results of problem SBMS-ToU, we can find that an idle duration may exist in some periods. In some circumstances, machine turn-on and -off may not consume energy and it is always turned off when finishing processing in each period and is restarted when needed. This kind of problem can be directly solved by the model $\mathcal{P}'_1$. However, turning on machines can consume a great amount of energy in some manufacturing environments, e.g., steel manufacturing. As indicated by Mouzon et al. (2007), the resulted non-processing energy (NPE) consumption related to machine turn-on, turn-off, and idling constitutes a significant part (over 30%) of the total energy consumption for certain scheduling environment. Remarkably, it has been shown in Mouzon and Yildirim (2008), Mouzon et al. (2007), and Yildirim and Mouzon (2012), the NPE consumption can be significantly reduced by rationalized machine turn-on/off. Thus, to optimize the whole electricity consumption cost including processing cost as well as machine on/off switching cost is highly important in practice. In other words, it is worthful to extend the problem SBMS by considering machine on/off switching strategy.

This section introduces a bi-criteria single machine batch scheduling problem with machine on/off switching under TOU pricing, SBMS-ToU-on/off in short. It can be described as follows. A batch processing machine with C jobs' capacity is associated with processing energy consumption rate P^{proc} and idling energy consumption rate P^{idle}. It is assumed that $P^{proc} > P^{idle}$. Turning on the machine needs a relatively short time and consumes P^{on} units electricity, while turning off the machine is assumed to require no energy. A given set $J = \{1, 2, \ldots, |J|\}$ independent nonpreemptive jobs has to be processed on the single batch processing machine within a scheduling horizon with $|I|$ periods. Each job j is available at time 0 and has a processing time p_j. Without loss of generality, we assume that all the jobs are numbered in nonincreasing order of the processing times, i.e., $p_1 \geq p_2 \geq \cdots \geq p_{|J|}$. All jobs can be regrouped into $|B|$ (to be determined) batches, where $\lceil |J|/C \rceil \leq |B| \leq |J|$. The processing time of a batch is determined by the largest job processing time in the batch.

Similarly to the problem SBMS in the previous section, a period $i \in I$ is considered as a work shift with duration S_i. Let s_i denote the starting time of period i, respectively. The length of the scheduling horizon is $s_{|I|+1}$. Job processing time p_j is less than the period duration $S_i, \forall i \in I$. The unit electricity price of period i, denoted by Pr_i, is calculated as the average price of period i based on the tariffs information as problem SBMS. Thus, the unit electricity cost for processing jobs in period i, denoted by e_i^p, is calculated as $e_i^p = P^{proc} \times Pr_i$. Similarly, the unit electricity cost of machine staying idle is $e_i^s = P^{idle} \times Pr_i$ and cost for machine turning on is $e_i^o = P^{on} \times Pr_i$, respectively. To determine the NPE cost, we need to analyze the turn-on/off strategies between any two adjacent periods. All cases are analyzed as follows:

Case 1: there exists processing in period $i \in I \backslash \{|I|\}$ while period $i+1$ has not, then the machine will be turned off when finishing processing in period i and thus the machine idling cost in period i and the NPE cost in period $i+1$ are 0;

Case 2: no processing exists in period $i \in I \backslash \{|I|\}$ while period $i+1$ has, then the machine will be turned on in period $i+1$ as it is in a shut-down state in period i and thus the NPE cost in period i is 0 and there exists turn-on energy cost e_{i+1}^o in period $i+1$;

Case 3: there exists processing in both periods $i \in I \backslash \{|I|\}$ and $i+1$, then if $e_i^s d_i^{idle} > e_{i+1}^o$, where d_i^{idle} denotes the idling duration in period i, then the machine will be turned off in period i when finishing processing and thus the machine idling cost in period i is 0 and there exists turn-on energy cost e_{i+1}^o in period $i+1$; otherwise, the machine will be kept idling in period i and thus the machine idling cost in period i is $e_i^s d_i^{idle}$ and there is no turn-on energy cost in period $i+1$.

The objective of the problem is to find an optimal schedule that consists of batching the jobs, allocating the batches to periods, and deciding whether the machine should be turned off or kept running idle for an idle duration in order to optimize the total electricity cost (E) and the makespan (C_{max}) simultaneously.

11.3.1 Mathematical Formulation

The problem SBMS-ToU-on/off considered in this section is a natural extension of problem SBMS and continues to use part of notations in the previous section, which include indices $j, b\ i$, parameters C, p_j, $s_i\ S_i$, and decision variables $x_{j,b,i}$, $y_{b,i}$, z_i, $|B|$, $P_{b,i}$. New parameters and decision variables for SBMS-ToU-on/off are listed below.

Parameters

e_i^p: electricity cost for processing jobs per unit time in period i , $\forall i \in I$;
e_i^s: electricity cost for machine idling per unit time in period i, $\forall i \in I$;
e_i^o: electricity cost produced by turning on the machine in period i, $\forall i \in I$;

Decision Variables

v_i: 0 if the machine is turned off or no job is processed in period i, or no job is to be processed in period $i+1$; 1 otherwise, $\forall i \in I$, $v_0 = 0$;
u_i: > 0 if the machine is not turned off in period i; 0 otherwise, $\forall i \in I \backslash \{|I|\}$.

With the notations and variables defined above, the investigated problem can be formulated as the following model $\mathcal{P}_2'$:

$$\mathcal{P}_2': \ \min\ E \tag{11.26}$$

$$\min\ C_{max} \tag{11.27}$$

s.t. Constraints (11.3)–(11.7),

$$z_i \le \sum_{b\in B} y_{b,i}, \forall i \in I \tag{11.28}$$

$$v_i \le z_{i+1}, \forall i \in I\backslash\{|I|\} \tag{11.29}$$

$$v_i \le z_i, \forall i \in I \tag{11.30}$$

$$u_i \ge e_i^s (S_i v_i - \sum_{b\in B} P_{b,i}), \forall i \in I \tag{11.31}$$

$$\sum_{i\in I}\sum_{b\in B} e_i^p P_{b,i} + \sum_{i\in I} e_i^o (z_i - v_{i-1}) + \sum_{i\in I/|I|} u_i \le E \tag{11.32}$$

$$s_i z_i + \sum_{b \in B} P_{b,i} \leq C_{max}, \forall i \in I \tag{11.33}$$

$$x_{j,b,i}, y_{b,i}, z_i, v_i \in \{0, 1\}, \forall j \in J, \forall b \in B, \forall i \in I \tag{11.34}$$

$$P_{b,i} \geq 0, u_i \geq 0, E \geq 0, C_{max} \geq 0, \forall b \in B, \forall i \in I. \tag{11.35}$$

Objectives (11.26) and (11.27) are to minimize total electricity cost E and makespan C_{max}, respectively. Constraints (11.3)–(11.7) state the limitations on allocating jobs to batches and periods. The difficulty of model formulation is controlling the machines' on/off status, which can be achieved by constraints (11.28)–(11.35). Equation (11.28) exactly determines if there is any batch processed in each period. Specifically, $z_i = 1$ if any batch is processed in period i, otherwise 0. Binary variable v_i, $i \in I$ equals to 1 if the machines stay running idle in period i according to its definition. Thus we use constraints (11.29) and (11.30) to, respectively, guarantee Cases 1 and 2 of turn-on/off strategies. That is, $v_i = 0$ if no batch is processed in period i or $i + 1$. Constraints (11.31) calculate the idling electricity cost in period $i \in I$. It only works when the machines stay running idle, i.e., $v_i = 1$. Constraint (11.32) calculates total electricity cost, which includes processing, machine turn-on, and idling cost. Constraint (11.33) defines the makespan. Constraints (11.34) and (11.35) are the restrictions on decision variables. Note that the number of batches $|B|$ is initially considered as its upper bound $|J|$ to derive a linear model.

11.3.2 Optimal Batch Rule Analysis

In this section, we devote our attention to reducing the search space for optimal solutions by analyzing the properties of the problem. In what follows, we demonstrate that batch formation can still be solved independent of batch allocation with LPT-based batch rule.

A solution of the problem $TOU, 1|on/off, B|E, C_{max}$ can be uniquely defined by $(|B|, \{J_b, 1 \leq b \leq |B|\}, \{\tau_b, 1 \leq b \leq |B|\}, \{v_i, i \in I\})$, where $|B|$, J_b, and τ_b are the number of batches, the set of jobs allocated into batch b ($J_1 \cup J_2 \ldots \cup J_{|B|} = J$), and the period in which batch b is processed, respectively. v_i denotes the machine status in the idle duration of period i, i.e., the machine is kept idling or turned off. Theorem 11.2 shows that we only need to consider the solutions with LPT-based batch formation to derive the Pareto front of the considered problem.

Theorem 11.2 *Any solution of SBMS-ToU-on/off in which the batches differ from those formed with the LPT-based rule is (at least weakly) dominated.*

Proof. To facilitate the proof, we first recall the following notations:

J_b: the set of jobs contained in batch b, $J_b \subseteq J$;
$n(J_b)$: the serial number of the least indexed job (thus with the largest processing time) in set J_b, i.e., $n(J_b) = \min\{j | j \in J_b\}$;

$P(J_b)$: the processing time of batch b (the processing time of the least indexed job in J_b), i.e., $P(J_b) = \max_{j \in J_b} p_j = p_{n(J_b)}$.

Let $|B^*|$ and J_b^* represent the number of batches formed with LPT-based method and the set of jobs involved in the batch b $(1 \leq b \leq |B^*|)$, respectively. We have

$$|B^*| = \lceil |J|/C \rceil,$$
$$J_b^* = \{(b-1)C+1, (b-1)C+2, \ldots, bC\}, \quad b = 1, 2, \ldots, |B^*|-1,$$
$$J_{|B^*|}^* = \{(|B^*|-1)C+1, (|B^*|-1)C+2, \ldots, n\}.$$

With the above construction, we have

$$n(J_b^*) = (b-1)C+1, \tag{11.36}$$

$$P(J_b^*) = p_{(b-1)C+1}. \tag{11.37}$$

Suppose there is a feasible schedule $\hat{S}$ with solution $(|\hat{B}|, \{\hat{J}_b, 1 \leq b \leq |\hat{B}|\}, \{\hat{\tau}_b, 1 \leq b \leq |\hat{B}|\}, \{\hat{v}_i, i \in I\})$, in which the batches differ from those formed with LPT-based rule. With similar proof of Theorem 11.1 for problem SBMS from formulas (11.14) to (11.19), we can conclude that the processing time of the batches are at least as long as those formed with the LPT-based rule, i.e.,

$$P(\hat{J}_b) = p_{n(\hat{J}_b)} \geq p_{(b-1)C+1} = P(J_b^*), 1 \leq b \leq |B^*|. \tag{11.38}$$

Renew schedule $\hat{S}$ to S^* with the batches formed with LPT-based rule, the new solution $(|B^*|, \{J_b^*, 1 \leq b \leq |B^*|\}, \{\hat{\tau}_b, 1 \leq b \leq |\hat{B}|\}, \{\hat{v}_i, i \in I\})$ can be achieved by removing batches $|B^*|+1, \ldots, |\hat{B}|$, if any, and replacing each batch $\hat{J}_b (1 \leq b \leq |B^*|)$ by the corresponding one formed with the LPT-based method (i.e., batch J_b^*) without changing the starting time. Relation (11.38) indicates that the new schedule S^* is also feasible. Because some batches are removed and the processing time of the rest batches is reduced, the makespan is not increased. Next, we prove that the total electricity cost is also not increased.

For any period $i \in I$ that involves job-processing, i.e., $z_i = 1$, according to (11.38), the total processing time in period i of schedule $\hat{S}$, calculated by $\sum_{b=1}^{|\hat{B}|} \hat{P}_{b,i}$, and that of schedule S^*, calculated by $\sum_{b=1}^{|B^*|} P_{b,i}^*$, must have the following relation:

$$\sum_{b=1}^{|\hat{B}|} \hat{P}_{b,i} \geq \sum_{b=1}^{|B^*|} P_{b,i}^*. \tag{11.39}$$

Let $\hat{E}_i$ (resp. E_i^*) denote the total processing and idling cost of period i and the turn-on cost of period $i+1$ of schedule $\hat{S}$ (resp. S^*). Since $z_i = 1$, the magnitude relationship of $\hat{E}_i$ and E_i^* can be analyzed through the following three cases:

Case 1: $z_{i+1} = 0$, then, $\hat{E}_i - E_i^* = e_i^p \sum_{b=1}^{|\hat{B}|} \hat{P}_{b,i} - e_i^p \sum_{b=1}^{|B^*|} P_{b,i}^* \geq 0$.

Case 2: $z_{i+1} = 1$ and $e_i^s(S_i - \sum_{b=1}^{|\hat{B}|} \hat{P}_{b,i}) > e_{i+1}^o$, then, we have

$$\hat{E}_i = e_i^p \sum_{b=1}^{|\hat{B}|} \hat{P}_{b,i} + e_{i+1}^o. \tag{11.40}$$

According to (11.39), we have

$$e_i^s(S_i - \sum_{b=1}^{|B^*|} P_{b,i}^*) \geq e_i^s(S_i - \sum_{b=1}^{|\hat{B}|} \hat{P}_{b,i}) > e_{i+1}^o,$$

thus, for the solution of E_i^*, we have

$$E_i^* = e_i^p \sum_{b=1}^{|B^*|} P_{b,i}^* + e_{i+1}^o. \tag{11.41}$$

Comparing (11.40) with (11.41), it is obvious that $\hat{E}_i - E_i^* \geq 0$.

Case 3: $z_{i+1} = 1$ and $e_i^s(S_i - \sum_{b=1}^{|\hat{B}|} \hat{P}_{b,i}) \leq e_{i+1}^o$, then we have

$$\begin{aligned}
&\hat{E}_i - E_i^* \\
&= e_i^p \sum_{b=1}^{|\hat{B}|} \hat{P}_{b,i} + e_i^s(S_i - \sum_{b=1}^{|\hat{B}|} \hat{P}_{b,i}) - (e_i^p \sum_{b=1}^{|B^*|} P_{b,i}^* + \min\{e_i^s(S_i - \sum_{b=1}^{|B^*|} P_{b,i}^*), e_{i+1}^o\}) \\
&\geq e_i^p \sum_{b=1}^{|\hat{B}|} \hat{P}_{b,i} + e_i^s(S_i - \sum_{b=1}^{|\hat{B}|} \hat{P}_{b,i}) - (e_i^p \sum_{b=1}^{|B^*|} P_{b,i}^* + e_i^s(S_i - \sum_{b=1}^{|B^*|} P_{b,i}^*)) \\
&= (e_i^p - e_i^s)(\sum_{b=1}^{|\hat{B}|} \hat{P}_{b,i} - \sum_{b=1}^{|B^*|} P_{b,i}^*).
\end{aligned}$$

Since $e_i^p > e_i^s$ and $\sum_{b=1}^{|\hat{B}|} \hat{P}_{b,i} \geq \sum_{b=1}^{|B^*|} P_{b,i}^*$, thus we have $\hat{E}_i - E_i^* \geq 0$.

The above results of the three cases indicate that the total electricity cost of the new schedule S^* is not greater than that of schedule $\hat{S}$. Consequently, neither electricity cost nor makespan is increased in the schedule with batches formed with LPT-based rule, which means that the initial schedule $\hat{S}$ is (at least weakly) dominated by the new one. □

Besides, the following property also holds.

Property 11.2 The batch scheduling problem SBMS-ToU-on/off is strongly NP-hard.

Proof. Consider a special case that $e_i^s = 0, \forall i \in I$ and $e_i^o = 0, \forall i \in I$, problem SBMS-ToU-on/off reduces to problem SBMS, which has been proved to be NP-hard in the strong sense in the previous section. Therefore, the problem SBMS-ToU-on/off is also strongly NP-hard. □

11.3.3 An Improved MILP Model

By pre-processing the batches of SBMS-ToU-on/off with the LPT-based rule according to Theorem 11.2, we have $P_b = p_{(b-1)C+1}$ and $|B^*| = \lceil |J|/C \rceil$, a new MILP model, denoted by $\mathcal{P}_2$, can be derived as follows:

$$\mathcal{P}_2 : \min E$$

$$\min C_{max}$$

$$\text{s.t.} \sum_{i \in I} y_{b,i} = 1, \forall b \in B^* \tag{11.42}$$

$$\sum_{b \in B^*} P_b y_{b,i} \leq S_i z_i, \forall i \in I \tag{11.43}$$

$$z_i \leq \sum_{b \in B^*} y_{b,i}, \forall i \in I \tag{11.44}$$

$$u_i \geq e_i^s (S_i v_i - \sum_{b \in B^*} P_b y_{b,i}), \forall i \in I \tag{11.45}$$

$$\sum_{i \in I} e_i^p \sum_{b \in B^*} P_b y_{b,i} + \sum_{i \in I} e_i^o (z_i - v_{i-1}) + \sum_{i \in I/|I|} u_i \leq E \tag{11.46}$$

$$s_i z_i + \sum_{b \in B^*} P_b y_{b,i} \leq C_{max}, \forall i \in I \tag{11.47}$$

$$y_{b,i}, z_i, v_i \in \{0, 1\}, \forall b \in B^*, \forall i \in I \tag{11.48}$$

$$u_i \geq 0, C_{max} \geq 0, E \geq 0, \forall i \in I \tag{11.49}$$

and constraints (11.29) and (11.30),

where $B^* = \{1, 2, \ldots, |B^*|\}$ is the set of batches formed with the LPT-based method, $P_b = p_{(b-1)C+1}$. Constraints (11.42) state that a formed batch $b, \forall b \in B^*$ should be entirely processed in one period. Constraints (11.43) ensure that total processing time in period $i \in I$ cannot exceed its duration. Constraints (11.44) ensure that variable z_i takes the value of 1 only if there are batches to be processed in period

$i, \forall i \in I$. Constraints (11.45) denote the total electricity cost when the machine is left running idle. Constraints (11.46) and (11.47) calculate total electricity cost E and makespan C_{max}, respectively. Constraints (11.48) and (11.49) enforce the restrictions on decision variables. Since part of variables and constraints are removed, the search space for Pareto optimal solutions of the initial problem is significantly reduced. To be more specific, model $\mathcal{P}_2$ reduces $|I| \cdot |J|^2 + (|J| - \lceil |J|/C \rceil) \cdot |I|$ binary variables, $|I| \cdot |J|$ real variables, and $(2 + |I| \cdot |J| + |I|) \cdot |J| - (|J| \cdot |I| + |I| + 1) \cdot \lceil |J|/C \rceil$ constraints comparing to model $\mathcal{P}'_2$.

11.4 SBMS Under ToU in Continuous Processing

The two problems introduced in the previous sections both assumed each job has to be completed in one work shift. However, continuously processing manufacturing systems are more widespread and practical in real production environments. In this section, we consider a batch scheduling problem under ToU tariffs where a nonpreemptive job is allowed to be processed in multiple periods. For the convenience of expression, we denote the problem as SBMS-ToU-CP.

11.4.1 Problem Description

Problem SBMS-ToU-CP can be described as follows. There are J jobs to be processed on a single BPM within a given planning horizon H. The machine and all the jobs are available from time 0 to H. The processing time of job j, $1 \leq j \leq J$, is denoted by p_j. Without loss of generality, we label the jobs in nonincreasing order of their processing times, i.e., $p_1 \geq p_2 \geq \ldots \geq p_J$.

The machine is able to process up to C jobs simultaneously. Thus, all jobs can be grouped into B ($1 \leq B \leq \lceil J/C \rceil$) batches (to be determined). The processing time of batch b, $1 \leq b \leq B$, is given by the longest processing time job in this batch. Once a batch is being processed, it cannot be interrupted until its processing is completed.

In the planning horizon, unit electricity price varies over the time according to ToU tariffs or real-time electricity pricing. In other words, electricity cost incurred by the processing in unit time is calculated based on the present electricity price and power rate of the given machine. It is high in peak periods and low in off-peak periods.

The scheduling is to determine the batch formation and the processing position of each batch in the horizon, such that total electricity cost E and makespan C_{max} of the jobs are minimized.

It is obvious that batch formation is one of the key decisions for solving SMBSC-ToU. According to the preliminary analysis, we find that LPT-based rule can be applicable to batch formation for the considered problem, and we have the following lemma.

Lemma 11.1 *Any solution of SBMS-ToU-CP in which the batches differ from those formed with LPT-based rule is (at least weakly) dominated.*

Proof. The theorem can be proved in a similar way as Theorem 11.1. To be specific, it has been proved by Eqs. (11.14)–(11.19) that the processing time of randomly formed batches are at least as long as those formed with LPT-based rule. Consequently, for any schedule with the batches that do not satisfy LPT-based rule, adjusting the jobs with LPT-based method will not deteriorate the schedule. In other words, Lemma 11.1 holds. □

11.4.2 Mathematical Models

With the above theorem, optimal solutions will not lose by preforming the batches with LPT-based rule. By considering the formed batches as new jobs, SMBSC-ToU is reduced to a single machine scheduling problem under ToU tariffs, and the number of new jobs equals to the number of batches B^*.

To calculate total electricity cost, the planning horizon has to be divided into several segments to position the start processing times of the batches. Based on two different division ways, two models, respectively, named as time-index-based model and time-interval-based model, are developed. The common parameters for the two models are as follows:

H: the duration of a given planning horizon;
b, c: index of batches;
i, k: index of time periods or intervals;
B^*: total number of batches;
P_b: processing time of batch b, $1 \leq b \leq B^*$.

11.4.2.1 Time-Index-Based MILP Model

This is an intuitive modeling way. Specifically, we first discrete the scheduling horizon H into $|H|$ unit time periods, then exactly determine the processing position of each batch on the horizon while making sure that the processing is not interrupted and adjacent jobs are not overlapped. For each time period i, $1 \leq i \leq H$, the duration and the electricity cost are 1 and e^i, respectively. Note that e_i may equal to e_{i+1}. Then the schedule can be achieved by determining the following decision variables:

$x_{b,i}$: equal to 1 if batch b is in processing at time period i, 0 otherwise; $1 \leq b \leq B^*$; $1 \leq i \leq H$;

y_{bc}: equal to 1 if batch b is processed (maybe not immediately) before batch c, 0 otherwise; $1 \le b, c \le B^*$;
t_b: the start time of batch b; $1 \le b \le B^*$;
t'_b: the completion time of batch b; $1 \le b \le B^*$.

With the above notations, time-index-based model is presented as follows:

$$\mathcal{P}_3: \quad \min E \tag{11.50}$$

$$\min C_{max} \tag{11.51}$$

$$\text{s.t.} \quad t_b + P_b = t'_b, \forall b \in B^* \tag{11.52}$$

$$i \ge t_b - (1 - x_{b,i})|H|, \forall i \in H; \forall b \in B^* \tag{11.53}$$

$$i \le t'_b - 1 + (1 - x_{b,i})|H|, \forall i \in H; \forall b \in B^* \tag{11.54}$$

$$\sum_{i \in I} x_{b,i} = P_b, \forall b \in B^* \tag{11.55}$$

$$y_{bc} + y_{cb} = 1, \forall b, c \in B^*, \forall b \neq c \tag{11.56}$$

$$t'_b \le t_c + (1 - y_{bc})|H|, \forall b, c \in B^*, \forall b \neq c \tag{11.57}$$

$$E = \sum_{i \in H} \sum_{b \in B^*} e^i x_{b,i} \tag{11.58}$$

$$C_{max} \ge t'_b, \forall b \in B^* \tag{11.59}$$

$$x_{b,i}, y_{bc} \in \{0, 1\}, \forall i \in H; \forall b, c \in B^*, b \neq c \tag{11.60}$$

$$t_b, t'_b \in \mathbb{Z}^+, \forall b \in B^*. \tag{11.61}$$

The objective functions (11.50) and (11.51) are to minimize total electricity cost E and makespan C_{max}. Constraints (11.52) imply that the processing of a batch is not interrupted, i.e., its completion time is the sum of its starting time and processing time. Constraints (11.53) and (11.54) determine the processing time periods of each batch by limiting the value of binary variable $x_{b,i}$. That is, the two constraints imply that if processing of batch b is not started or has been completed, $x_{b,i}$ must be 0. Constraints (11.55) guarantee that $x_{b,i}$ takes value 1 if time period i is between the start time and completion time of batch b. To sum up, constraints (11.52)–(11.55) ensure that the processing of each batch is not interrupted. Constraints (11.56) and (11.57) make sure that the processing of adjacent batches do not overlap. Specifically, constraints (11.56) express that batch b either precedes or follows batch c. Constraints (11.57) imply that if batch b precedes batch c, the start processing time of batch c must be equal or larger than the completion time of b. Constraints (11.60) and (11.61) denote the restrictions on variables. In model $\mathcal{P}_3$, there are $[|B^*|^2 + (|H| + 2)|B^*|]$ variables and $[2|B^*|^2 + (2|H| + 3)|B^*| + 1]$ constraints to be determined.

11.4.2.2 Time-Interval-Based MILP Model

Another dividing way for the planning horizon is based on time-of-use pricing information. That is, divide the horizon into $|I|$ pricing intervals, where interval i, $1 \le i \le |I|$, is associated with a starting time s_i, a duration $S_i = s_{i+1} - s_i$, and a unit electricity cost e_i, note that $e_i \neq e_{i+1}$. Obviously, $s^{|I|+1} = |H|$. Then SMBS-ToU can be solved by determining how long each batch is processed in each interval; meanwhile, the uninterrupted processing and machine availability are guaranteed. For the modeling, the following decision variables are defined:

$x_{b,i}$: equals to 1 if batch b is processed in interval i, 0 otherwise; $\forall i \in I; \forall b \in B^*$;
$w_{b,i}$: equals to 1 if batch b is simultaneously processed in interval i and $i+1$, 0 otherwise; $\forall i \in I; \forall b \in B^*$;
$t_{b,i}$: processing duration of batch b in interval i; $\forall i \in I; \forall b \in B^*$.

Now the time-interval-based model, called Model 4 hereafter, can be formulated as follows:

$$\mathcal{P}_4: \min E$$

$$\min C_{max}$$

$$\text{s.t. } t_{b,i} \le x_{b,i} P_b, \forall i \in I; \forall b \in B^* \tag{11.62}$$

$$\sum_{i \in I} t_{b,i} = P_b, \forall b \in B^* \tag{11.63}$$

$$\sum_{b \in B^*} t_{b,i} \le s_{i+1} - s_i, \forall i \in I \tag{11.64}$$

$$w_{b,i} \ge x_{b,i} + x_{b,i+1} - 1, \forall i \in I/\{|I|\}; \forall b \in B^* \tag{11.65}$$

$$\sum_{b \in B^*} w_{b,i} \le 1, \forall i \in I/\{|I|\} \tag{11.66}$$

$$t_{b,i} + s_i x_{b,k} - s_{i+1} x_{b,i} + t_{b,k} \le P_b + (2 - x_{b,i} - x_{b,k})H,$$
$$\forall i \in I/\{|I|\}; \forall k \in I; \forall b \in B^* \tag{11.67}$$

$$E = \sum_{i \in I} \sum_{b \in B^*} e_i t_{b,i} \tag{11.68}$$

$$x_{b,i}, w_{b,i} \in \{0, 1\}, \forall i \in I; \forall b \in B^* \tag{11.69}$$

$$t_{b,i} \ge 0, \forall i \in I; \forall b \in B^*. \tag{11.70}$$

The objectives are to optimize total electricity cost and makespan. Constraints (11.62) restrict that processing duration of batch b in interval i does not exceed batch processing time, and processing duration $t_{b,i}$ takes value 0 if batch b is not processed in interval i. Constraints (11.63) guarantee that all the batches are completed within the planning horizon. Constraints (11.64) limit that the total processing time in a given interval does not exceed the interval duration. Constraints (11.65) and (11.66) state that any two consecutive intervals can be crossed by only one batch. Constraint (11.67) ensures that the processing of any job is not interrupted. It states that once a batch b simultaneously processed in interval i and k ($i < k$), the total processing time of the job in the two intervals plus the distance between the ending time of interval i and starting time of interval k should be less than the processing time of batch b. Constraints (11.69) and (11.70) enforce the restrictions on decision variables.

11.5 Conclusion

The goal of this chapter is to provide an insight into the domain of batch scheduling under ToU tariffs. Three bi-objective batch scheduling problems under ToU electricity tariffs are introduced, which aims to design production plans for batch processing machines under fluctuating electricity prices, with the objectives of simultaneously optimizing total electricity cost and production efficiency. For each of the considered problem, appropriate mathematical model is formulated and the problem property is analyzed.

In the future, more complicated problems under ToU tariffs are worth of further investigation, such as the problems involving other machine environments (e.g., unrelated parallel machines, flow shop, job shop), job characteristics (e.g., dynamic release times, non-identical due dates), more regular objective functions (e.g., total completion time, maximum lateness), and production features (e.g., serial batching, maintenance activity).

Acknowledgments This work was supported by the National Natural Science Foundation of China, under Grants 71901069, 71571061, 71701049.

References

Che, A., Zeng, Y., & Lyu, K. (2016). An efficient greedy insertion heuristic for energy-conscious single machine scheduling problem under time-of-use electricity tariffs. *Journal of Cleaner Production, 129*, 565–577.

Cheng, J., Chu, F., Chu, C., & Xia, W. (2016). Bi-objective optimization of single-machine batch scheduling under time-of-use electricity prices. *RAIRO – Operations Research, 50*(4–5), 715–732.

Cheng, J., Chu, F., Liu, M., Wu, P., & Xia, W. (2017). Bi-criteria single-machine batch scheduling with machine on/off switching under time-of-use tariffs. *Computers & Industrial Engineering, 112*, 721–734.

Cheng, J., Chu, F., Liu, M., & Xia, W. (2016). Single-machine batch scheduling under time-of-use tariffs: New mixed-integer programming approaches. In *2016 IEEE International Conference on Systems, Man, and Cybernetics (SMC)* (pp. 3498–3503). New York: IEEE.

Cheng, J., Chu, F., & Zhou, M. (2018). An improved model for parallel machine scheduling under time-of-use electricity price. *IEEE Transactions on Automation Science and Engineering, 15*(2), 896–899.

Ding, J. Y., Song, S., Zhang, R., Chiong, R., & Wu, C. (2016). Parallel machine scheduling under time-of-use electricity prices: New models and optimization approaches. *IEEE Transactions on Automation Science and Engineering, 13*, 1138 – 1154.

Fang, K., Uhan, N., Zhao, F., & Sutherland, J. W. (2011). A new approach to scheduling in manufacturing for power consumption and carbon footprint reduction. *Journal of Manufacturing Systems, 30*(4), 234–240.

Fang, K., Uhan, N. A., Zhao, F., & Sutherland, J. W. (2016). Scheduling on a single machine under time-of-use electricity tariffs. *Annals of Operations Research, 238*(1–2), 199–227.

Fanti, M. P., Maione, B., Piscitelli, G., & Turchiano, B. (1996). Heuristic scheduling of jobs on a multi-product batch processing machine. *International Journal of Production Research, 34*(8), 2163–2186.

Ikura, Y., & Gimple, M. (1986). Efficient scheduling algorithms for a single batch processing machine. *Operations Research Letters, 5*(2), 61–65.

Jia, W., Jiang, Z., & Li, Y. (2015). Combined scheduling algorithm for re-entrant batch-processing machines in semiconductor wafer manufacturing. *International Journal of Production Research, 53*(6), 1866–1879.

Lee, C.-Y., Uzsoy, R., & Martin-Vega, L. A. (1992). Efficient algorithms for scheduling semiconductor burn-in operations. *Operations Research, 40*(4), 764–775 (1992)

Luo, H., Du, B., Huang, G. Q., Chen, H., & Li, X. (2013). Hybrid flow shop scheduling considering machine electricity consumption cost. *International Journal of Production Economics, 146*(2), 423–439.

Moon, J.-Y., Shin, K., & Park, J. (2013). Optimization of production scheduling with time-dependent and machine-dependent electricity cost for industrial energy efficiency. *The International Journal of Advanced Manufacturing Technology, 68*(1–4), 523–535.

Mouzon, G., & Yildirim, M. B. (2008). A framework to minimise total energy consumption and total tardiness on a single machine. *International Journal of Sustainable Engineering, 1*(2), 105–116.

Mouzon, G., Yildirim, M. B., & Twomey, J. (2007). Operational methods for minimization of energy consumption of manufacturing equipment. *International Journal of Production Research, 45*(18–19), 4247–4271.

Shrouf, F., Ordieres-Meré, J., García-Sánchez, A., & Ortega-Mier, M. (2014). Optimizing the production scheduling of a single machine to minimize total energy consumption costs. *Journal of Cleaner Production, 67*, 197–207.

Stock, T., & Seliger, G. (2016). Opportunities of sustainable manufacturing in industry 4.0. *Procedia CIRP, 40*, 536–541.

Tang, L., Liu, J., Rong, A., & Yang, Z. (2001). A review of planning and scheduling systems and methods for integrated steel production. *European Journal of Operational Research, 133*(1), 1–20.

Uzsoy, R., Lee, C.-Y., & Martin-Vega, L. A. (1994). A review of production planning and scheduling models in the semiconductor industry part II: Shop-floor control. *IIE Transactions, 26*(5), 44–55.

Wang, C. S., & Uzsoy, R. (2002). A genetic algorithm to minimize maximum lateness on a batch processing machine. *Computers & Operations Research, 29*(12), 1621–1640.

Wang, Y., Li, T.T., & Wang, B. (2019). Heuristic for hot-rolled batch scheduling of seamless steel tubes with machine maintenance and tardiness. *International Journal of Performability Engineering, 15*(7), 1849–1859.

Xu, S., & Bean, J. C. (2016). Scheduling parallel-machine batch operations to maximize on-time delivery performance. *Journal of Scheduling, 19*(5), 583–600.

Yildirim, M. B., & Mouzon, G. (2012). Single-machine sustainable production planning to minimize total energy consumption and total completion time using a multiple objective genetic algorithm. *IEEE Transactions on Engineering Management, 59*(4), 585–597.

Zhang, H., Zhao, F., Fang, K., & Sutherland, J. W. (2014). Energy-conscious flow shop scheduling under time-of-use electricity tariffs. *CIRP Annals–Manufacturing Technology, 63*(1), 37–40.

Chapter 12
Service Composition in Cloud Manufacturing: A DQN-Based Approach

Haifeng Zhang, Yongkui Liu, Huagang Liang, Lihui Wang, and Lin Zhang

Abstract Cloud manufacturing is a new service-oriented manufacturing model that integrates distributed manufacturing resources to provide on-demand manufacturing services over the Internet. Service composition that builds larger-granularity, value-added services by combining a number of smaller-granularity services to satisfy consumers' complex requirements is an important issue in cloud manufacturing. Meta-heuristic algorithms such as genetic algorithm, particle swarm optimization, and ant colony algorithm are frequently employed for addressing service composition issues in cloud manufacturing. However, these algorithms require complex design flows and lack adaptability to dynamic environment. Deep reinforcement learning provides an alternative approach for solving cloud manufacturing service composition issues. This chapter proposes a deep Q-network (DQN) based approach for service composition in cloud manufacturing, which is able to find optimal service composition solutions through repeated training and learning. Results of experiments that take into account changes of service scales and service unavailability reveal the scalability and robustness of the DQN algorithm-based service composition approach.

Keywords Cloud manufacturing · Service composition · Deep reinforcement learning · Deep Q-network

---

H. Zhang · H. Liang
School of Electronics and Control, Chang'an University, Xi'an, China

Y. Liu
School of Mechano-Electronic Engineering, Xidian University, Xi'an, China

L. Wang (✉)
Department of Production Engineering, KTH Royal Institute of Technology, Stockholm, Sweden
e-mail: lihuiw@kth.se

L. Zhang
School of Automation Science and Electrical Engineering, Beihang University, Beijing, China

B. Sokolov et al. (eds.), *Scheduling in Industry 4.0 and Cloud Manufacturing*, International Series in Operations Research & Management Science 289, https://doi.org/10.1007/978-3-030-43177-8_12

12.1 Introduction

Cloud manufacturing is a new model for enabling aggregation of distributed manufacturing resources (e.g., manufacturing software tools, manufacturing equipment, and manufacturing capabilities) and ubiquitous, convenient, on-demand network access to a shared pool of configurable manufacturing services that can be rapidly provisioned and released with minimal management effort or service operator and provider interaction (Liu et al. 2019a). In cloud manufacturing, distributed manufacturing resources are virtualized and encapsulated into manufacturing cloud services and aggregated in a cloud manufacturing platform (Lin et al. 2014; Chen et al. 2019; Dolgui et al. 2019; Liu et al. 2017). Service composition is an important issue in cloud manufacturing as it supports building of larger-granularity, value-added services from smaller-granularity services and thus enables cloud manufacturing to provide consumers with services in an on-demand manner (Liu et al. 2018; Pisching et al. 2015; Panetto et al. 2019).

Cloud manufacturing service composition (CMfg-SC) plays a key role in providing services to consumers in an on-demand way. It is a typical NP-hard problem and also a supporting technique for scheduling in cloud manufacturing (Liu et al. 2019b; Ivanov et al. 2016). To date, much research work on this issue has been done, and most of existing research adopts meta-heuristic algorithms such as genetic algorithms, particle swarm optimization, artificial bee colony optimization, and ant colony optimization algorithms. Typically, Tao et al. (2011) reviewed service composition methods in cloud manufacturing and discussed the problems with service composition in cloud manufacturing. Liu et al. (2013) addressed hierarchical manufacturing cloud service composition based on the hierarchical manufacturing implementation processes, and developed a simulated annealing algorithm for solving that issue. Xiang et al. (2016) analyzed the difficulties and solutions of service composition and optimal selection of big data and proposed a two-phase service composition and optimal selection method based on case libraries to solve the large-scale optimal selection problem. A fuzzy soft set-based decision-making method enhanced with volatility analysis was proposed by Li et al. (2018), which is the method of autonomy-oriented service composition and optimal selection in cloud manufacturing to break limitations of existing centralized mode. Ding et al. (2020) described manufacturing services in multiple levels, modeled from the level of resource services, functional services, and process services. A service-composition preferred quality-assessment function is constructed to improve accuracy of service modeling and combinatorial optimization in cloud manufacturing environment. Methods adopted for CMfg-SC in current research are mainly meta-heuristic algorithms, such as ant colony algorithm, particle swarm algorithm, and genetic algorithms (Yang et al. 2019; Xu et al. 2017; Li et al. 2019; Ghomi et al. 2019). However, these algorithms have a number of limitations: (1) their design process is complex and they lack adaptability, (2) they are easily trapped into local optimization, and (3) their parameter adjustment process is complex and depends on experience of algorithm designers.

During the past years, artificial intelligence (AI) technologies underwent rapid development (Jackson 2019), and deep reinforcement learning (DRL) as an emerging AI technology, in particular, has been one of the research hotspots in the area of AI since the advent of AlphaGo who defeated a professional human Go player (Mnih et al. 2015; Silver et al. 2016). DRL is an approach that combines deep learning and reinforcement learning. The basic idea of the former is to combine low-level features through multilayer network structure and nonlinear transformations to form an abstract and easily distinguishable high-level representation so as to discover the distributed features, and the basic idea of the latter is to let the agent learn to make wise decisions by trial-and-error interactions with the environment (Li 2017). DRL combines reinforcement learning and neural networks and is therefore able to admit high-dimensional task and resource data and extract their high-level features. Furthermore, DRL has high adaptability to dynamic environments and high efficiency in the presence of large-scale services. DRL provides a novel approach for solving the CMfg-SC issue. A typical DRL algorithm is deep Q-network (DQN), which utilizes a convolutional neural network to learn successful control policies in complex reinforcement learning environments. The network is trained with a variant of the Q-learning (Watkins and Dayan 1992) algorithm, with stochastic gradient descent to update weights. Moreover, to alleviate the problems of correlated data and nonstationary distributions, an experience replay mechanism with uniform sampling is introduced, which thereby smoothens the training distribution over many past behaviors. Therefore, this chapter applies DQN to CMfg-SC with the aim to explore a new DQN-based CMfg-SC approach.

The rest of this chapter is organized as follows. Mathematical descriptions of our model are presented in Sect. 12.2. Section 12.3 presents a DQN-based CMfg-SC model. In Sect. 12.4, a series of experiments are conduct to validate the feasibility and effectiveness of the model proposed. Section 12.5 concludes this chapter.

12.2 Basic CMfg-SC Model

CMfg-SC is a process of combining multiple basic smaller-granularity manufacturing services to build value-added larger-granularity services (i.e., service composition solutions) and selecting the one with the highest quality of service (QoS) to best satisfy complex manufacturing requirements with or without constraints. As a result, CMfg-SC is essentially a two-stage process that consists of composition and optimal solution selection. CMfg-SC is task-oriented, and from the task perspective, CMfg-SC overall undergoes three phases from task submission to the completion of the service composition process: (1) task decomposition (i.e., a task is decomposed into a number of subtasks, atomic or not, each of which can be fulfilled by a service), (2) service search and matching (i.e., a candidate service set for each subtask that matches its functional requirement is established), and (3) service composition and optimal selection (i.e., building service composition solutions by selecting a service from each candidate service set and then selecting the optimal one as the composition solution) (Tao et al. 2015).

12.2.1 Task

This chapter considers CMfg-SC for a single composite task T, which can be described as follows:

$$T = \{st_1, \cdots, st_k, \cdots st_n\} \tag{12.1}$$

where st_k represents the kth subtask of T, and the required type of service is r_k. There are four basic execution flows of subtasks, namely, sequential, parallel, selective, and circular. A task's execution flow could be any one of them or their combination.

Each subtask has a certain workload (Wang et al. 2014). The workload of st_k can be expressed as

$$wl_k = a_k \times \alpha_0 \times t_k \tag{12.2}$$

where a_k represents unit quantity of the required service, t_k is the required service time, and α_0 is the benchmark efficiency coefficient. The total workload of T is therefore $\sum_{k=1}^{n} wl_k$.

12.2.2 Enterprise and Service

Consider a cloud manufacturing system with I enterprises $\{E_1, \ldots, E_i, \ldots, E_I\}$. E_i $(1 \le i \le I)$ provides l_i $(1 \le l_i \le j)$ types of manufacturing services with the sth $(1 \le s \le l_i)$ type of service being $s_{i,s}$. Attributes of service $s_{i,s}$ considered include type $R_{i,s}$, quantity $A_{i,s}$, cost for unit amount of service $c_{i,s}$, efficiency coefficient α_i, and reliability $Rel_{i,s}$ (Duflou et al. 2012; Vidayev et al. 2014). For simplicity but without loss of generality, here we assume that all services of E_i have the same efficiency coefficient. To facilitate the calculation of the time required for $s_{i,s}$ to undertake a subtask, the concept of the enterprise capability $Cap_{i,s}$ is introduced, which is defined as

$$Cap_{i,s} = A_{i,s} \times \alpha_i \tag{12.3}$$

12.2.3 Time, Cost, and Reliability

QoS metrics of services considered include time T, cost C, and reliability Rel. The total time for performing T includes machining time and logistics time. The time required for $s_{i,\,s}$ to complete st_k can be calculated as follows:

$$ST_{i,s}^{k} = wl_k / Cap_{i,s} \tag{12.4}$$

Logistics time between two consecutive subtasks st_k and $st_{k\,+\,1}$ depends on three factors: geographical distance $d_{i,i'}$ between the two companies E_i and E_i' that undertake the two adjacent subtasks, logistics time t_{ud} per unit distance. The logistics time between st_k and $st_{k\,+\,1}$ can thus be expressed as follows:

$$LT_{k,k+1}^{i,i'} = \delta_{k,k+1}^{i,i'} \times t_{ud} \times d_{i,i'} \tag{12.5}$$

where $\delta_{k,k+1}^{i,i'}$ is a Boolean variable introduced to characterize whether logistics exists between st_k and $st_{k\,+\,1}$.

The cost of performing T includes machining cost and logistics cost. The cost for $s_{i,\,s}$ to complete st_k can be calculated as follows:

$$SC_{i,s}^{k} = A_{i,s} \times c_{i,s} \tag{12.6}$$

The logistics cost between st_k and its next adjacent $st_{k\,+\,1}$ can be expressed as:

$$LC_{k,k+1}^{i,i'} = \delta_{k,k+1}^{i,i'} \times c_l \times w_{k,k+1}^{i,i'} \times d_{ii'} \tag{12.7}$$

where $\delta_{k,k+1}^{i,i'}$ indicates whether logistics is required, c_l represents the logistics cost per unit weight and unit distance, $w_{k,k+1}^{i,i'}$ is the weight of the workpiece/product that needs to be transported between st_k and $st_{k\,+\,1}$, and $d_{ii'}$ stands for the physical distance between the selected enterprises E_i and E_i'.

Reliability indicates the probability that a manufacturing service can work normally. When $s_{i,\,s}$ is selected for st_k, the reliability of performing st_k is equal to the reliability of $s_{i,\,s}$, i.e.,

$$\mathrm{Rel}_k = Rel_{i,s} \tag{12.8}$$

Since attribute values of the above-mentioned performance indicators are of different ranges, normalization is required for calculating QoS values of a service composition solution. QoS metrics fall into two categories: positive (the larger the better, such as reliability) and negative (the smaller the better, such as time and cost). For a positive metric q_i, the normalization method is as follows:

$$Nq_i^+ = \frac{q_i - q_i^{min}}{q_i^{max} - q_i^{min}} \tag{12.9}$$

For a negative metric q_i, the normalization method is as follows:

$$Nq_i^- = \frac{q_i^{max} - q_i}{q_i^{max} - q_i^{min}} \tag{12.10}$$

12.3 DQN-Based CMfg-SC Model

Consider the general setting shown in Fig. 12.1, where an agent interacts with an environment. At each time step t, the agent takes an action a_t according to the state s_t of the current environment. Once the action is performed, the environment will transition to a new state s_{t+1}. At the same time, the agent will receive a reward that reflects the value of the state transition. The goal is to maximize the cumulative discount reward: $\mathbb{E}\left[\sum_{t=0}^{\infty} \gamma^t r_t\right]$, where $\gamma \in (0, 1]$ is the reward discount factor. The state transitions and rewards are stochastic and are assumed to have the Markov property (i.e., the state transition probabilities and rewards depend only on the state of the environment s_t and the action taken by the agent a_t).

In order to obtain the maximum cumulative reward (Mnih et al. 2013), a deep Q-network model-based CMfg-SC model is proposed in Sect. 12.3.1.

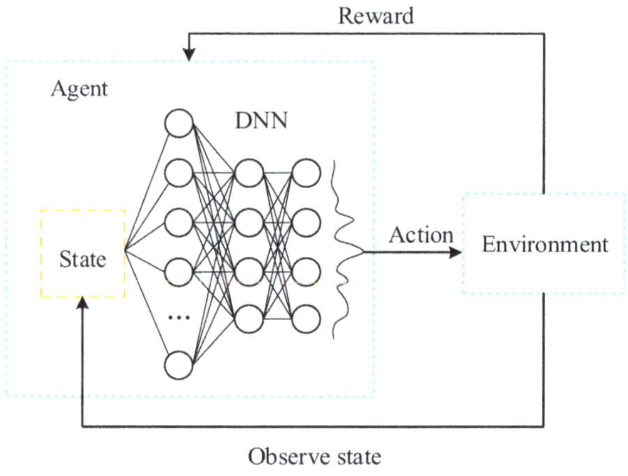

Fig. 12.1 Reinforcement learning and deep neural networks

12.3.1 DQN-Based Service Composition Model

For the storage of high-dimensional Q tables in Q-learning (Watkins and Dayan 1992), Mnih et al. (2015) proposed the DQN model creatively by combining deep neural network and Q-learning, which turned the Q-Table update problem in Q-Learning into a function fitting problem, and used a deep neural network to fit the Q-value function. DQN adopts the experience replay mechanism during the training process (de Bruin et al. 2015). At each time step, the transition samples obtained by interactions of the agent with environment are stored in the experience pool (i.e., replay buffer). When training a deep neural network, mini-batch samples are used for training the neural network model. In addition to using a deep neural network to approximate the current value function, DQN also uses another network separately to generate target Q-value. As shown in Fig. 12.2, $Q(s, a; \theta)$ represents the output of the current value network, and $\max_{a'} Q\left(s_{j+1}, a'; \theta'\right)$ represents the output of the target value network, then $Y_j = r_{j+1} + \gamma \max_{a'} Q\left(s_{j+1}, a'; \theta'\right)$ approximates the optimization goal of the value function, that is, the target Q-value. Parameters of the current value network are updated in real time, but the parameters of the current value network are copied to the target value network every C time steps. Within the C time steps, the target value network parameters do not change, which reduces correlations of transition samples and improves the stability of the algorithm. The overall network structure is shown in Fig. 12.2.

The DQN-based single task-oriented service composition algorithm:

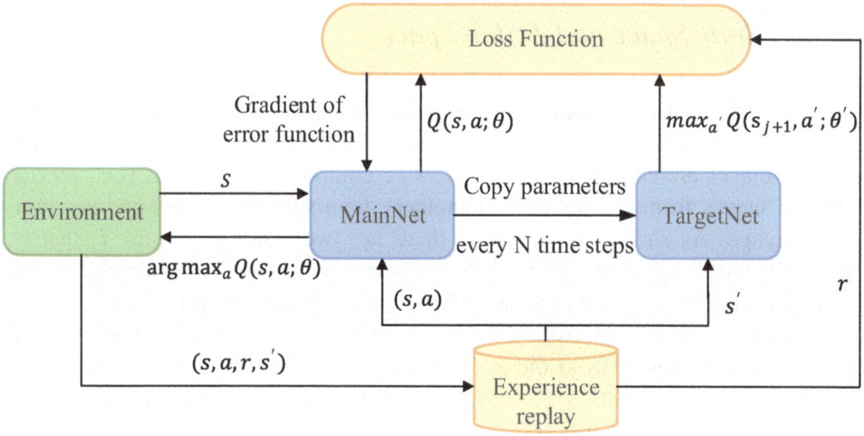

Fig. 12.2 DQN network structure

1. Initialize replay memory D to capacity
2. Initialize action-value function Q with random weights θ.
3. Initialize target action-value function with weights $\theta' = \theta$
4. For episode=1, M do
5. Initialize subtasks and service sequences
6. For st_k (k=1 … n), do
7. With probability p, select a random service $s_{i,s}$
8. Otherwise select $s_{i,s} = \arg\max_{s_{k,u}} Q\left(st_k, s_{k,u}; \theta\right)$
9. Observe reward r_k and the system moves to the next subtask
10. Store transition (st_k, $s_{i,s}$, r_k st_{k+1}) in D
11. Sample with random experience replay of transitions (st_j, $s_{l,s}$, r_j st_{j+1}) from D
12. Calculate the action-value function in the current state $Target\ Q = r + \gamma \max_{s_{i,s}'} Q\left(st_k', s_{i,s}'; \theta'\right)$
13. Calculate the loss function $L(\theta) = E[(T\arg etQ - Q(st_j, s_{l,s}; \theta))^2]$ of neural network and the parameters are updated by back propagation
14. Each C step updates the target network according to the parameters of the evaluation network
15. End for
16. End for

12.3.2 State Space and Action Space

CMfg-SC for a single composition task can be modeled as a Markov decision process in which a service is selected for each subtask until the selection of services for all subtasks are completed. As the service composition proceeds (i.e., service selection moves from one subtask to another), both subtasks' states and services' states change. As shown in Fig. 12.3, there are two states for st_k of T, that is, $\left\{s_k^{se}, s_k^{un}\right\}$, where s_k^{se} means that service selection for st_k has been completed, and s_k^{un} means that no service has been selected for st_k. When service selection for st_k is completed, service composition for T will move to the next subtask st_{k+1}, and the system will transfer from the state $S_k = \left\{s_1^{se}, \cdots, s_k^{se}, s_{k+1}^{un}, \cdots, s_n^{un}\right\}$ to the state $S_{k+1} = \left\{s_1^{se}, \cdots, s_k^{se}, s_{k+1}^{se}, s_{k+2}^{un}, \cdots, s_n^{un}\right\}$. As a result, in terms of subtasks, there are $n + 1$ states, that is, $\{S_0, S_1, \cdots, S_k, \cdots, S_n\}$, which is the state space for subtasks. Similarly, service $s_{k,j}$ from the service set ss_k for st_k also has two states, that is, $\left\{sr_{k,j}^{se}, sr_{k,j}^{un}\right\}$, where $sr_{k,j}^{se}$ represents the state that $s_{k,j}$ has been selected for st_k, whereas $sr_{k,j}^{se}$ stands for the state that $s_{k,j}$ has yet been selected for st_k. As the service composition process proceeds, services' states alsochange. All services'

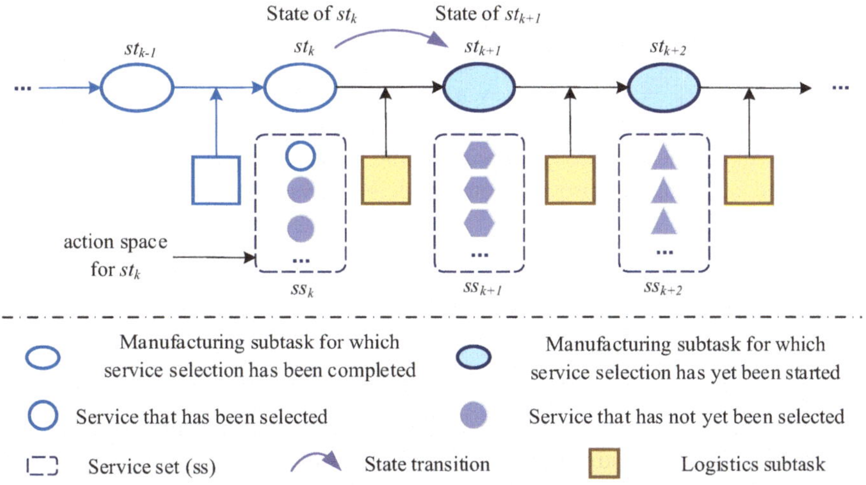

Fig. 12.3 Schematic diagram of state, state space, action, and action space

states constitute the entire states of services. The state of the whole system consists of all subtasks' states and all services' states.

There is an action space for each subtask. An action corresponds the selection of a service for a subtask, and the result of an action is that a service is selected. Hence, the size of the action space for st_k is equal to the number of services in the corresponding service set. The size of the action space for the whole system is equal to the number of all candidate services from service set for all subtasks.

12.3.3 Reward Function

In the current DQN-based CMfg-SC model, the reward function is used to guide the agent to make appropriate actions (Christiano et al. 2017). The reward is calculated based on the normalized QoS values. Because of the involvement of logistics, the calculation of the reward should consider logistics. As the aim of CMfg-SC is to find the service composition solution with the highest QoS, rewards are closely related to the QoS of services (including manufacturing services and logistics services). In order to take into account the effect of logistics, logistics prior to each subtask is also considered. Therefore, the performance of a single candidate manufacturing service can be expressed as:

$$q_k = \left(w_t NST_{i,s}^k + w_c NSC_{i,s}^k + w_{rel} NRel_k\right) + \left(w_t NLT_{k,k+1}^{i,i'} + w_c NLC_{k,k+1}^{i,i'}\right) \tag{12.11}$$

where w_t, w_c, and w_{rel} are weights of normalized time$NST^k_{i,s}$, $NLT^{i,i'}_{k,k+1}$, normalized cost $NSC^k_{i,s}$, $NLC^{i,i'}_{k,k+1}$, and normalized reliability $NRel_k$, respectively, and $w_t + w_c + w_{rel} = 1.0$. In this way, a service with a higher QoS value and a short logistics distance will be preferentially selected. In this way, the optimized CMfg-SC solution will be selected through repeated training (Table 12.1).

Therefore, after each episode is completed, the QoS of the composite task can be expressed as:

$$q = \sum_{k=1}^{n} q_k \tag{12.12}$$

Due to the fact that the total time, cost, and reliability are all calculated through summation, it is easy to prove that by calculating the reward according to Eq. (12.12) one is able to find the optimal service composition solution.

12.3.4 Action Selection and State Transition

The training process iterates episode after episode. An episode starts from the selection of a service for the first subtask, proceeds with service selection for the remaining subtasks sequentially, and finally ends with the completion of the service selection for the last subtask. A DRL agent needs to estimate the Q-value based on the current state and the selected service according to the deep neural network. An epsilon-greedy method is used to select actions, that is, with probability ϵ services are randomly selected for subtasks, and with probability $1 - \epsilon$, a service with the highest Q-value is selected. The training starts with a large ϵ so that the DRL agent randomly selects a service for each subtask, irrespective of Q-values of services. When enough state-action sequence pairs are stored in the experience replay buffer, ϵ decreases gradually following $\epsilon_{i+1} = \epsilon_i - \sigma$. This method enables the agent to find the optimal service composition solution through repeated training. During the training process, the system keeps shifting from one state to another. State shifting occurs every time service selection for a subtask is completed.

12.3.5 Model Training

The model training is based on the DQN algorithm, which has a neural network with two fully connected layers, and each layer of the network has 128 neurons. The size of the experience replay buffer is 10,000. A uniform random sample policy is adopted, that is, a mini-batch of 32 samples is uniformly sampled from the replay buffer, regardless of the significance of transitions. During the training process, the current value network parameters are copied to the target value network every 5000

Table 12.1 QoS aggregation formulas for different task flows with the consideration of logistics

Structure	Time	Cost	Reliability
Sequence	$\sum_{k=1}^{n}\left(NST_{k,u}+NLT_{k,k+1}^{i,i'}\right)$	$\sum_{k=1}^{n}\left(NSC_{k,i,s}+NLC_{k,k+1}^{i,i'}\right)$	$\sum_{k=1}^{n}NRel_{k,u}$
Parallel	$\max\left\{NST_{k,u}+NLT_{k,k+1}^{i,i'}\right\}$	$\max\left\{NSC_{k,i,s}+NLC_{k,k+1}^{i,i'}\right\}$	$\sum_{k=1}^{n}NRel_{k,u}$
Selective	$\sum_{k=1}^{n}\left[\left(NST_{k,u}+NLT_{k,k+1}^{i,i'}\right)\times\lambda_j\right]$	$\sum_{k=1}^{n}\left[\left(NSC_{k,i,s}+NLC_{k,k+1}^{i,i'}\right)\times\lambda_j\right]$	$\sum_{k=1}^{n}\left(NRel_{k,u}\times\lambda_j\right)$
Circular	$\beta\times\sum_{k=1}^{n}\left(NST_{k,u}+NLT_{k,k+1}^{i,i'}\right)$	$\beta\times\sum_{k=1}^{n}\left(NSC_{k,i,s}+NLC_{k,k+1}^{i,i'}\right)$	$\beta\times\sum_{k=1}^{n}NRel_{k,u}$

Note: λ_j is the probability of selecting st_k with $\sum_{j=1}^{n}\lambda_j=1$, and β is the number of cycles steps. The initial learning rate is 0.001, the reward value decaying factor is set to 0.9, and the RMSProp optimizer is used to update the network parameters. The training starts with $\epsilon = 0.9$. After reaching the capacity of the experience replay buffer, the value of ϵ decreases by 0.0005 at each step until $\epsilon = 0.01$.

12.4 Experimental Results

12.4.1 Experiment 1

The objective of this experiment is to demonstrate that the DQN method has better adaptability than the ant colony algorithm in solving the CMfg-SC issue. In this experiment, the number of services for each subtask is varied from 20 to 40 to investigate the performance of the proposed DQN-based algorithm for different problem sizes while keeping the number of subtasks in the composite task and its logical structure, and the parameters of the algorithm unchanged. For the sake of comparison, experimental results obtained with the ant colony algorithm are also shown in Fig. 12.4.

Figures 12.4 and 12.5, respectively, show the adjustment capabilities of the DQN-based CMfg-SC algorithm and the ant colony algorithm-based CMfg-SC approach. The task considered has ten subtasks, among which the fifth and the sixth subtasks have a parallel structure (i.e., they will be executed in parallel when a service composition solution is put into execution). Experiments with 20, 30, and 40 services for each subtask are conducted. It can be observed that, irrespective of the number of services for each subtask, the curves for DQN-based CMfg-SC algorithm are able to reach a stable value adaptively after undergoing an adjusting process. But for the ant colony algorithm, with the increase in the number of iterations of episodes, the

Fig. 12.4 Performance curves for the DQN-based CMfg-SC algorithm with varying numbers of services for each subtask

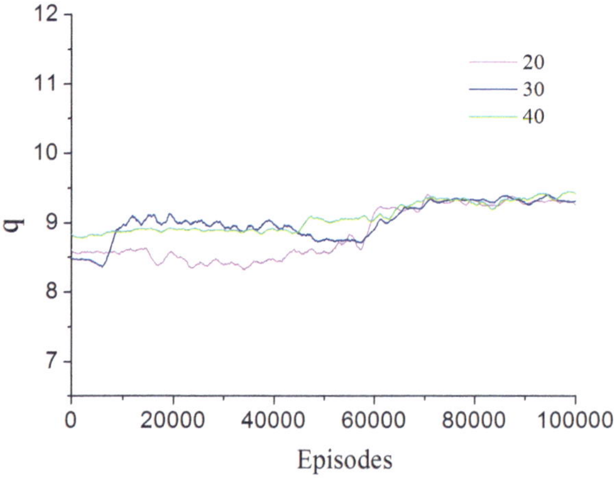

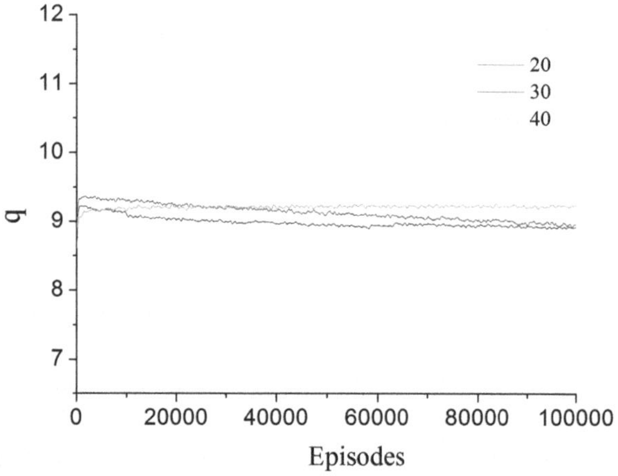

Fig. 12.5 Performance curves for the traditional ant colony algorithm with varying numbers of services for each subtask

QoS of the generated service composition solution decreases gradually, indicating that parameter adjustments of the ant colony algorithm are more complicated than DQN, and, moreover, the suitable iteration termination condition for the ant colony algorithm is more difficult to determine. All these factors seriously hinder applications of traditional heuristic algorithms (including ant colony algorithms) to CMfg-SC.

12.4.2 Experiment 2

A cloud manufacturing system is highly dynamic in which many unexpected events such as machine breakdown occur frequently. These dynamic events cause unavailability of manufacturing resources in cloud manufacturing. As a result, there is a need to study the robustness of the DQN algorithm in solving the CMfg-SC issue. The result is shown in Fig. 12.6, for which the number of subtasks of the task is 40, and the number of services for each subtask is 50. The robustness of the proposed DQN-based CMfg-SC algorithm is investigated by randomly disabling a percentage of services (4% here) at 60,000th time step during the training process, and then observing the dynamic adjusting process. As shown in Fig. 12.6, after a dynamic adjustment process, the curve recovers to the initial level automatically, indicating that the system is able to learn a new CMfg-SC solution based on the current available services without introducing any special mechanism. This reflects robustness and adaptability of the proposed algorithm.

The above experiments indicate that although the model training process of the DQN-based CMfg-SC algorithm is more time-consuming in comparison with the DQN algorithm, the DQN-based model and approach is more universal in the sense that fewer parameters need to be adjusted and it is more adaptive to dynamic environment.

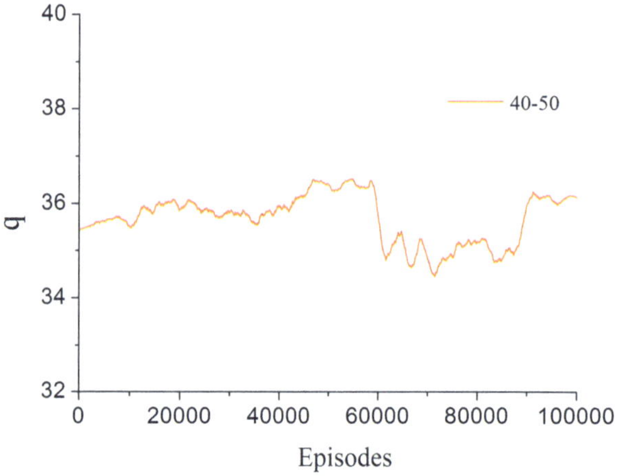

Fig. 12.6 Adjusting curve of the proposed DQN-based CMfg-SC algorithm

12.5 Conclusions and Discussion

In this chapter, we have proposed a DQN-based approach for CMfg-SC with the aim to overcome issues with traditional meta-heuristic algorithms in solving CMfg-SC. First, a DQN-based CMfg-SC model was built, including the overall architecture of the DQN algorithm, the state space and action space, the reward function, action selection and state transitions, and modeling training. Then, two experiments were conducted, which focused on the effects of problem sizes and dynamic service unavailability, respectively. The results indicated the effectiveness, adaptability, and robustness of the proposed algorithm.

Deep reinforcement learning such as DQN provides an effective and powerful method for solving service management issues in cloud manufacturing, including service composition and scheduling, etc. Currently, research on CMfg-SC with DRL is just beginning, and much work needs to be done in the future. Our future work will extend this chapter in the following three aspects. First, this chapter just presents some preliminary results, and in the future more detailed results and analysis will be given. Second, extended DQN algorithms such as DQN with prioritized replay and dueling DQN will be adopted for achieving better system performance. Third, other service management issues, scheduling in particular, will also be addressed using DRL.

Acknowledgments This work was supported by the National Natural Science Foundation of China (NSFC) under Grant Nos. 61973243 and 61873014.

References

Chen, J., Huang, G. Q., Wang, J., & Yang, C. (2019). A cooperative approach to service booking and scheduling in cloud manufacturing. *European Journal of Operational Research, 273*(3), 861–873.

Christiano, P. F., Leike, J., Brown, T. B., Martic, M., Legg, S., & Amodei, D. (2017). Deep reinforcement learning from human preferences. arXiv preprint arXiv:1706.03741v3.

de Bruin, T., Kober, J., Tuyls, K., & Babuška, R. (2015). The importance of experience replay database composition in deep reinforcement learning. In *Deep reinforcement learning workshop, NIPS*.

Ding, T., Yan, G., Lei, Y., & Xiangyu, X. (2020). A niching behaviour-based algorithm for multi-level manufacturing service composition optimal-selection. *Journal of Ambient Intelligence and Humanized Computing, 11*, 1177–1189.

Dolgui, A., Ivanov, D., Sethi, S. P., & Sokolov, B. (2019). Scheduling in production, supply chain and Industry 4.0 systems by optimal control: Fundamentals, state-of-the-art and applications. *International Journal of Production Research, 57*(2), 411–432.

Duflou, J. R., Sutherland, J. W., Dornfeld, D., Herrmann, C., Jeswiet, J., Kara, S., Hauschild, M., & Kellens, K. (2012). Towards energy and resource efficient manufacturing: A processes and systems approach. *CIRP Annals, 61*(2), 587–609.

Ghomi, E. J., Rahmani, A. M., & Qader, N. N. (2019). Service load balancing, scheduling, and logistics optimization in cloud manufacturing by using genetic algorithm. *Concurrency and Computation: Practice and Experience, 31*(20), e5329.

Ivanov, D., Dolgui, A., Sokolov, B., Werner, F., & Ivanova, M. (2016). A dynamic model and an algorithm for short-term supply chain scheduling in the smart factory industry 4.0. *International Journal of Production Research, 54*(2), 386–402.

Jackson, P. C. (2019). *Introduction to artificial intelligence*. New York: Courier Dover Publications.

Li, Y. (2017). Deep reinforcement learning: An overview. arXiv preprint arXiv:1701.07274.

Li, C., Guan, J., Liu, T., Ma, N., & Zhang, J. (2018). An autonomy-oriented method for service composition and optimal selection in cloud manufacturing. *The International Journal of Advanced Manufacturing Technology, 96*(5–8), 2583–2604.

Li, Y., Yao, X., & Liu, M. (2019). Cloud manufacturing service composition optimization with improved genetic algorithm. *Mathematical Problems in Engineering, 2019*, 7194258.

Lin, Z., Luo, Y., Tao, F., Li, B. H., Ren, L., Zhang, X., Guo, H., Cheng, Y., Hu, A., & Liu, Y. (2014). Cloud manufacturing: A new manufacturing paradigm. *Enterprise Information Systems, 8*(2), 167–187.

Liu, W., Ma, G., & Liu, B. (2013). Study on hierarchical service composition in cloud manufacturing. *China Mechanical Engineering, 24*(10), 1349–1356.

Liu, Y., Xu, X., Srinivasan, A., & Zhang, L. (2017). Enterprises in cloud manufacturing: A preliminary exploration. In *ASME 2017 12th International Manufacturing Science and Engineering Conference collocated with the JSME/ASME 2017 6th International Conference on Materials and Processing, June 4–8, 2017, Los Angeles, CA, USA*.

Liu, Y., Wang, L., & Wang, X. V. (2018). Cloud manufacturing: Latest advancements and future trends. *Procedia Manufacturing, 25*, 62–73.

Liu, Y., Wang, L., Wang, X. V., Xu, X., & Jiang, P. (2019a). Cloud manufacturing: Key issues and future perspectives. *International Journal of Computer Integrated Manufacturing, 32*(9), 858–874.

Liu, Y., Wang, L., Wang, X. V., Xu, X., & Lin, Z. (2019b). Scheduling in cloud manufacturing: State-of-the-art and research challenges. *International Journal of Production Research, 57*(15–16), 4854–4879.

Mnih, V., Kavukcuoglu, K., Silver, D., Graves, A., Antonoglou, I., Wierstra, D., Riedmiller, M., et al. (2013). Playing atari with deep reinforcement learning. arXiv preprint arXiv:1312.5602.

Mnih, V., Kavukcuoglu, K., Silver, D., et al. (2015). Human-level control through deep reinforcement learning. *Nature, 518*(7540), 529–533.

Panetto, H., Iung, B., Ivanov, D., Weichhart, G., & Wang, X. (2019). Challenges for the cyber-physical manufacturing enterprises of the future. *Annual Reviews in Control, 47*, 200–213.

Pisching, M. A., Junqueira, F., Santos Filho, D. J., & Miyagi, P. E. (2015). Service composition in the cloud-based manufacturing focused on the industry 4.0. In *Doctoral Conference on Computing, Electrical and Industrial Systems* (pp. 65–72). Cham: Springer.

Silver, D., Huang, A., Maddison, C. J., et al. (2016). Mastering the game of Go with deep neural networks and tree search. *Nature, 529*(7587), 484–489.

Tao, F., Lin, Z., Guo, H., Luo, Y., & Ren, L. (2011). Typical characteristics of cloud manufacturing and several key issues of cloud service composition. *Computer Integrated Manufacturing Systems, 17*(3), 477–486.

Tao, F., Lin, Z., Liu, Y., Cheng, Y., Wang, L., & Xun, X. (2015). Manufacturing service management in cloud manufacturing: Overview and future research directions. *Journal of Manufacturing Science and Engineering, 137*(4), 040912.

Vidayev, I. G., Martyushev, N. V., Ivashutenko, A. S., & Bogdan, A. M. (2014). The resource efficiency assessment technique for the foundry production. *Advanced Materials Research, 880*, 141–145.

Wang, T., Guo, S., & Lee, C.-G. (2014). Manufacturing task semantic modeling and description in cloud manufacturing system. *The International Journal of Advanced Manufacturing Technology, 71*(9–12), 2017–2031.

Watkins, C. J. C. H., & Dayan, P. (1992). Q-learning. *Machine Learning, 8*(3–4), 279–292.

Xiang, F., Jiang, G., Xu, L., & Wang, N. (2016). The case-library method for service composition and optimal selection of big manufacturing data in cloud manufacturing system. *The International Journal of Advanced Manufacturing Technology, 84*(1–4), 59–70.

Xu, L., Jiawei, D., & Ming, H. (2017). Research on hybrid cloud particle swarm optimization for multi-objective flexible job shop scheduling problem. In *2017 IEEE 6th International Conference on Computer Science and Network Technology (ICCSNT), October 21–22, 2017, Dalian, China.*

Yang, Y., Yang, B., Wang, S., Liu, F., Wang, Y., & Shu, X. (2019). A dynamic ant-colony genetic algorithm for cloud service composition optimization. *The International Journal of Advanced Manufacturing Technology, 102*(1–4), 355–368.

Chapter 13
The Tolerance Scheduling Problem in a Single Machine Case

Daniel Alejandro Rossit, Fernando Tohmé, and Gonzalo Mejía Delgadillo

Abstract This chapter introduces the *Tolerance Scheduling* problem, which involves the decision-making issues in rescheduling processes. The solutions to this problem can be incorporated in the design of Decision Support Systems (DSS) in Industry 4.0 environments. We present here the mathematical foundations for the solutions of the Tolerance Scheduling problem as well as the technical requirements of their embodiment in Industry 4.0's DSS. We illustrate these ideas in a case study with a single machine, in which we analyze the performance of the model at different tolerances.

13.1 Introduction

The last years have witnessed an explosive increase in the penetration of digital technologies like the Internet of Things (IoT), facilitating significant advances in production systems by augmenting their flexibility and the autonomy of their control systems (Lee et al. 2015). This, in turn, promotes new business models on the basis of the intensive and massive personalization of production processes (Monostori et al. 2016; Bortolini et al. 2017). The autonomy of control systems allows the autonomous

D. A. Rossit (✉)
Departamento de Ingeniería, Universidad Nacional del Sur, Bahía Blanca, Argentina

INMABB UNS CONICET, Instituto de Matemática de Bahía Blanca, Bahía Blanca, Argentina
e-mail: daniel.rossit@uns.edu.ar

F. Tohmé
INMABB UNS CONICET, Instituto de Matemática de Bahía Blanca, Bahía Blanca, Argentina

Departamento de Economía, Universidad Nacional del Sur, Bahía Blanca, Argentina
e-mail: ftohme@criba.edu.ar

G. M. Delgadillo
Universidad de la Sabana, Faculty of Engineering, Campus Universitario del Puente del Común, Chía, Cundinamarca, Colombia
e-mail: gonzalo.mejia@unisabana.edu.co

B. Sokolov et al. (eds.), *Scheduling in Industry 4.0 and Cloud Manufacturing*, International Series in Operations Research & Management Science 289,
https://doi.org/10.1007/978-3-030-43177-8_13

solution of problems that previously required human intervention, enriching the capacities of decision support systems (Ivanov et al. 2016; Rossit and Tohmé 2018).

Cyber-Physical Systems (CPS) are the main drivers of this accelerated automatization process (Monostori 2014). A CPS is a single system that integrates the physical part of the device (which carries out the actual production) with a digital component (Lee 2008). These components communicate through IoT technologies, increasing the autonomy of the system, thanks to the incorporation of large information-processing capacities into the production process (Lee et al. 2015). Many of the classical operations in manufacturing planning will become delegated to CPS to either fully automatized them or to provide a powerful support to decision makers (Almada-Lobo 2016).

These considerations apply naturally to scheduling processes, which will also be affected by these technologies (Monostori 2014; Ivanov et al. 2016; Rossit et al. 2019b). In this sense, numerous advances have been already made in this field (Ivanov et al. 2018; Zhang et al. 2019; Rossit et al. 2019d). In this chapter we present some tools that will profit from their implementation in Industry 4.0 technologies. They are intended to solve scheduling problems, in particular *Rescheduling* ones. The problem of rescheduling may arise once the production process is getting executed according a predefined schedule and an unexpected (not anticipated by the plan) event happens, which affects the performance of the production (Vieira et al. 2003). To solve this problem, corrective actions have to be exerted, according to the magnitude of the impact of the event. The possibilities are either to initiate a rescheduling process or not (Pinedo 2012). Our proposal involves analyzing it in the light of the Tolerance Scheduling Problem (Rossit et al. 2019b). Basically, the tolerance scheduling problem captures explicitly how a human modeler faces a rescheduling situation. Its formal structure allows incorporating its solution into a Decision Support System (DSS) applied to production process (Rossit et al. 2019b). In the following text, we will analyze in detail the fundamentals of this problem and of its incorporation in DSS, evaluating the approach in the context of a study case.

13.2 Scheduling and Rescheduling

Scheduling is the process of assigning resources to jobs for specified time periods in order to optimize a single or multiobjective function. The resources and jobs may vary from organization to organization. The former may include the machines in a workshop, squads of workers dedicated to maintain or build things, IT people, and airfields. In turn, jobs may include the operations in a production process, stages in a construction project, runs of computer programs, takeoff and landings in airports.

The objectives may also differ between organizations. A usual one is the minimization of the finishing time of the last job. With such flexible and all-embracing view, it can be said that scheduling problems are pervasive in all fabrication and production systems as well as in information-processing activities.

Among the main features of scheduling processes, we can highlight the following ones: (1) they are complex decision-making processes requiring the detailed planning of production orders (assigning jobs or operations to resources and machines). The trend toward deep customization of production processes only increases their complexity, (2) the lifetimes of schedules is short, requiring their repetition over short temporal horizons, (3) a critical feature is the delivery dates of the products, affecting directly the production costs, and (4) schedules obtained from the most structured decision-making processes in an organization, requiring well-defined data, constraints, and objects (Framinan et al. 2014).

Among the decision processes in scheduling we can distinguish two levels, associated with different time horizons: one involves the classical problem of designing a production schedule *before* production starts. The other level involves the case in which a schedule has already been executed, but new issues arise in real time (e.g., a machine breaks down or the delivery of input materials is delayed). The latter is known as the Rescheduling problem (Vieira et al. 2003; Peng et al. 2018). This problem has been studied widely in the literature since it is a feature of the real praxis of production processes (Vieira et al. 2003). In real-world production contexts, rescheduling is almost mandatory for the minimization of the impact of perturbations on the plan. These perturbations induce delays in processing times, problems in the quality of products and the lack of necessary material. In order to face these events, there exist different possibilities. One approach is to try to foresee these events and endow the schedule with the ability to "absorb" them. Another approach involves facing each of the perturbations once they occur, handling them in a singular and reactive way (Ouelhadj and Petrovic 2009).

A distinctive predictive approach involves using robust scheduling processes, which compute the optimal schedule taking into account the probabilities of the possible perturbations. The resulting schedules are able to absorb or at least minimize the impact of future events. The solutions are usually obtained by using either stochastic optimization (Birge et al. 1990; Arnaout 2014) or robust optimization (Al-Hinai and ElMekkawy 2011; Rahmani and Heydari 2014). Despite its obvious benefits, this approach requires detailed statistical information that is not always fully available in industrial contexts.

The reactive approach addresses issues only when they appear. A usual strategy involves using dispatch rules (Ouelhadj and Petrovic 2009). These generate schedules by comparing and ranking the jobs, for example, according to which one requires less time or has closer due dates. Solutions are accordingly always found in a fast way (Pinedo 2012). A shortcoming is that these dispatch rules are very sensitive to the kind of problem at hand. For different objective functions the performance of the same dispatch rule may vary considerably (Vieira et al. 2003). Even so, dispatch rules are common in industrial contexts in which it is necessary to rearrange quickly and consistently the production schedule (Framinan et al. 2014).

Most rescheduling strategies are event-driven. That is, disruptive events trigger the application of rescheduling mechanisms (Dong and Jang 2012). This event-driven logic is embodied by the Decision Support System that helps managing schedules in a shop floor (Rossit and Tohmé 2018). The human scheduler, for instance, when

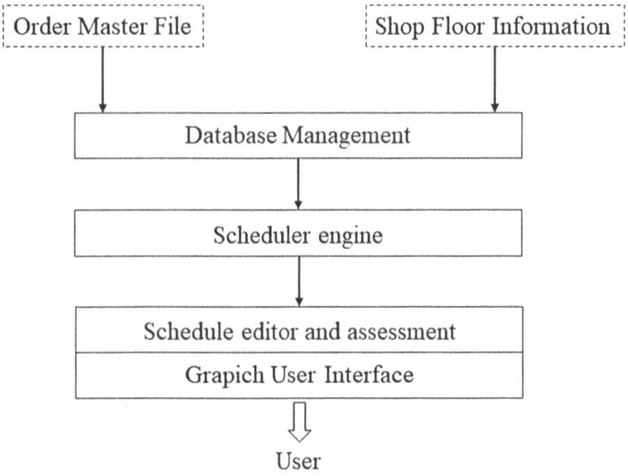

Fig. 13.1 Scheduling system (Pinedo 2012)

noticing a difference between the due date and the actual date of delivery of the production, uses the DSS to simulate potential scenarios and the impact on them of the delay. Upon this assessment the scheduler decides whether or not to initiate the rescheduling process, either following a predefined strategy or, if possible, choosing the best possible strategy (Katragjini et al. 2013).

In the decision-making process of the scheduler, we can identify two initial stages. In the first one she analyzes the disruptive event. In the second stage she checks whether the event requires triggering a rescheduling process (Rossit et al. 2019a). If not, the alarm is dismissed. Otherwise, a next stage of analysis evaluates the impact of the event on the current schedule. Then, again, the scheduler assesses the costs and benefits of rescheduling and according to that decides whether to trigger or not the rescheduling process (Rossit et al. 2019b). Figure 13.1 depicts the rescheduling process according to the system described in the study by Pinedo (2012). It shows that the system feeds in data from the production objectives (Order Master file) and from the shop floor.

13.3 The Impact of Industry 4.0

The concept of Industry 4.0 tries to capture all the relevant features of the transformations that are changing the classically accepted manufacturing practices (Lee et al. 2015). These changes are mostly due to the advent of cyber-physical systems (CPSs) and Internet of Things (IoT), which allow redesigning manufacturing processes oriented towards highly personalized products (Ivanov et al. 2016; Rossit et al. 2019a). CPSs integrate the physical world with virtual environments in which the information drawn from the former can be analyzed using Big Data tools (Lee 2008). This makes possible the real-time analysis of the physical system. This, in turn, endows the production system with the possibility of adapting flexibly itself to new circumstances. IoT, in turn, provides connections among distributed CPSs allowing the transmission of data collected by a CPS to other CPSs or to a DSS (Dolgui et al. 2019; Ivanov et al. 2019). All these possibilities give rise to the concept of Cyber-Physical Production Systems (CPPSs) (Monostori 2014). An Industry 4.0 environment running under a CPPS will be managed by CPSs, given that they can incorporate AI programs able to solve highly complex problems (Monostori et al. 2016).

This implies that CPSs constitute a key technology for Industry 4.0. It becomes highly relevant, thus, to understand how they work. Lee et al. (2015) present the basic architecture of a CPS based on layers establishing a parallel between the physical and virtual spaces working in consonance inside the CPS. Table 13.1 shows a representation of this. In the first level we find the Connection Level that connects the virtual to the physical world in real time, in such a way that all the information processed by the CPS is obtained at this level. The data are collected either through wireless sensors or by plugged-in ones. The CPS processes the data at the Conversion Level. In this layer, the CPS transforms the data in information giving them a meaning in terms of the environment in which it operates. A multidimensional analysis fits the information to a model of how the CPS works. This, in turn, yields the inputs for the Cyber Level of the architecture. At this layer the CPS evaluates whether the process runs as planned or not. At this level it becomes also possible to compare the internal states of the CPS working together on a single production process. The next layer up is the Cognition Level, in which the models of the previous stage are used to simulate potential cases and choose the best courses of action for the system. Finally, the results are conveyed to the Configuration Level for the autonomous implementation of the actions decided on at the Cognition Level.

Table 13.1 Cyber-physical 5C's architecture

Level	Attribute
I. Connection level	• Plug-in • Tether-free communication • Sensor network
II. Conversion level	• Data-to-information • Multidimensional data correlation • Smart analytics
III. Cyber level	• Virtual modeling • Clustering information • Controllability
IV. Cognition level	• Integrated simulation and synthesis • Collaborative diagnostics and decision making • Early awareness
V. Configuration level	• Self-configuration • Self-optimization

13.3.1 Cyber-Physical Production Systems

The possibility of integrating the functionalities of industrial information systems with physical production assets in the CPS allows the production system to be conceived globally as a Cyber-Physical Production System (CPPS) (Monostori 2014; Monostori et al. 2016; Rossit and Tohmé 2018). The CPPS incorporates the different business functions of an industrial organization into a single CPS-based system (Monostori 2014). Among these built-in business functions are some functions with direct interference in production such as Manufacturing Execution Systems (MES), which are responsible for programming, executing, and programming production processes. To show how a CPPS can program production processes (almost autonomously), we will use the well-known ANSI/ISA-95 control standard as a reference architecture.

The ANSI/ISA 95 standard allows to link control systems and production facilities through an automated interface. It can provide a common environment for the communication of all participants in a production process and offers a representation to model and use the information. It organizes the different levels of decision making hierarchically. This standard is based on the "Purdue Enterprise Reference Architecture" (PERA), which establishes five different hierarchies, as shown in Fig. 13.2. Level 0 is associated with the physical manufacturing process. Level 1 involves the smart devices that measure and manipulate the physical process. Typical instruments at this level are sensors, analyzers, effectors, and related instruments. Level 2 represents the control and supervision of the underlying activities, a typical case of ISA-95. Level 2 systems are Supervisory Control and Data Acquisition (SCADA) or Programmable Logic Controllers (PLC). Level 3

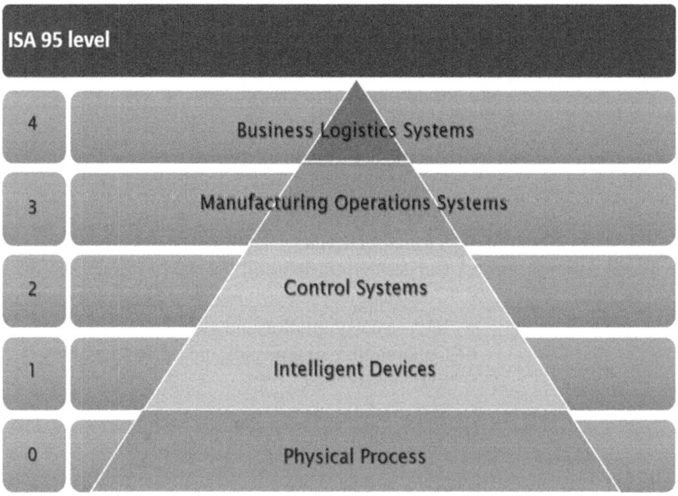

Fig. 13.2 Control structure ANSI/ISA 95 (Rossit and Tohmé 2018)

involves the management of operations and the workflow with respect to production. A clear case of systems belonging to level 3 are manufacturing execution/operations management systems (MES/MOMS). Therefore, this level is of special importance for our work, since this is where the scheduling process is developed. Finally, level 4 is associated with the commercial activities of the entire company. This architecture represents the different activities and functions of a production system hierarchically. In addition, determine the way in which the different levels communicate. In traditional production environments, in particular, each level interacts only with its adjacent levels (Rossit and Tohmé 2018).

13.3.2 Decision Making in CPPS

CPPSs, due to their functional characteristics, will directly impact the decision-making processes of industrial planning and control. To present our perspective on this, we show in Fig. 13.3 the levels of ISA 95 that would be managed by a CPPS in an integrated way. This integration is derived from the capabilities of CPPSs, which can implement physical processes (level 0), measure and manage the instruments that read physical processes (level 1), and implement control actions on their operations (level 2). In addition, given the computational capability of CPPSs, they can also plan, evaluate, and manage the whole production process flow (level 3).

Being able to integrate these functionalities within the CPPS will allow increasing the capacity to respond to unexpected events that occur in production, endowing it with high flexibility. In addition to other advantages, one that should be highlighted is the improvement of the transmission of information, since the same system that

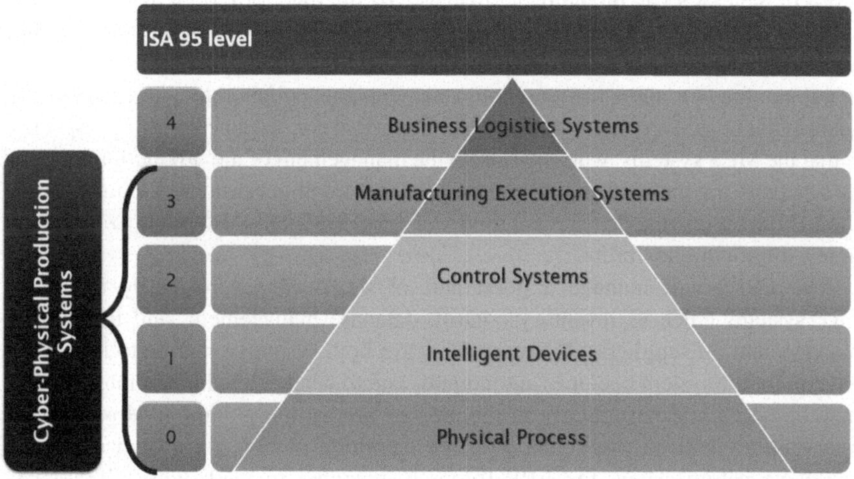

Fig. 13.3 Control structure of a CPPS

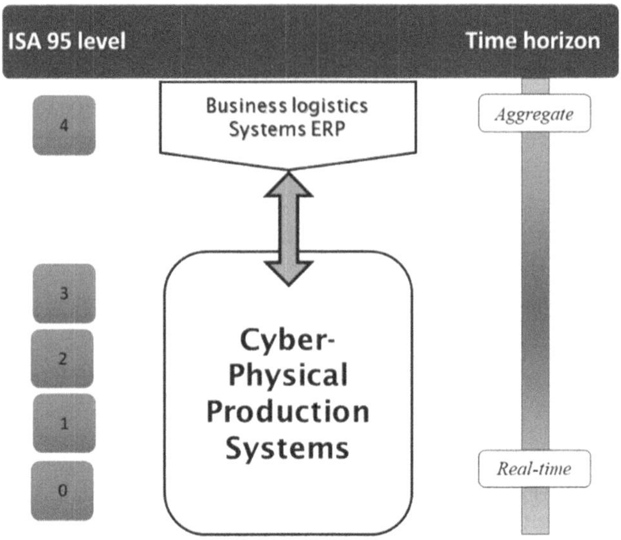

Fig. 13.4 Distribution of ISA 95 levels between ERP and CPPS. The representation of time is drawn from the model of the Manufacturing Enterprise Solutions Association (MESA) International

collects the data is the one that processes and analyzes them. This reduces the impacts of the restrictions imposed by adjacencies inherited from the PERA architecture.

In addition, the integration of these functionalities will have a direct impact on decision making related to production planning. This will lead to the ISA 95 architecture levels being managed by two large systems: the ERP (Enterprise Resource Planning) and the CPPS, as shown in Fig. 13.4. It can be seen there that the levels of the ISA 95 architecture are managed by the ERP for level 4, where the most strategic and aggregate-level decisions are made, while the rest of the levels will be administered by the CPPS. In this perspective, a CPPS should be considered as a set of autonomous (or almost autonomous) elements that collaborate with each other to achieve the objectives set by the ERP. An important consequence of this perspective is that the MES systems, whose tasks are the management of the production flow and its schedule, are embedded within the CPPS. This will generate better information, useful both for making decisions at this level (level 3, MES) and to minimize response times, improving flexibility.

The CPPS will manage a good part of the decisions made by traditional ERP systems (such as inventory control, database management, and information management on suppliers). However, we leave both systems separate to indicate at what point the system becomes autonomous and to what extent human interventions may be necessary, particularly in the area of production planning. Most of these interventions will be associated with the definition of objectives or decisions of aggregate nature. Then, the ERP (properly prepared to work with CPPSs) will translate these decisions to the CPPSs that handle the production flow and the schedule. Consequently, the CPPSs will not be completely autonomous, since they run on an open loop with the ERP, at least on production planning decisions (Rossit et al. 2019d). In this sense, Almada-Lobo (2016) states that although Industry 4.0 integrates production execution systems, this integration will not include the highest levels of management that define more general aspects of manufacturing.

Within the levels that will be integrated by the CPPSs, it is important to point out that the ISA-95 level 3 will be integrated, even horizontally, allowing access to all the systems at that level: inventory management systems, quality control, historical records, and maintenance planning. This allows CPPSs to work on both scheduling and material inventory problems, which can affect the performance of the production process (Kuo and Kusiak 2019). The systems associated with maintenance work better when monitored in real time, since overrequirements may result in temporary maintenance stops, affecting their availability (Lee et al. 2017; Rossit et al. 2019c). While many of these aspects of maintenance or the lack of materials can have an indirect effect on the schedule, access to that information improves the results of computing the optimal schedule. The proposed design seeks to incorporate these functions into CPPSs, in order to improve the quality of scheduling processes.

In addition to integrating other functions, CPPSs will have access to the data and systems used as inputs for traditional scheduling, as detailed in Fig. 13.1, although with a larger advantage over traditional systems. Essentially, CPPSs will improve the management of the physical processes, integrating execution, control, and analysis in the same system. Moreover, if the decision-making process is delegated to the CPPSs, they will collect the data, make decisions, and execute them. This will result in a complete vertical integration, improving the quality of the results and the responsiveness of the system. The DSS proposed in this work aims to benefit from this capacity of CPPSs and is intended as a step toward implementing the aforementioned vertical integration in Smart Manufacturing environments.

This integration allows to manage the different industries increasingly in a virtual way. This lends the production system resilience in the face of extreme disruptions such as the one generated by the COVID-19 pandemic of 2020 (Ivanov and Das 2020). The possibility of managing the system virtually/remotely allows to reduce the number of people in the shop floor, as well as to run a large number of critical operations online (Ivanov and Dolgui 2020). This virtualization alone, does not by itself guarantee the protection of all personnel, but certainly provides a stronger shield against a pandemic scenario.

13.4 Tolerance Scheduling Problem

In this section we present a summary of how Smart Scheduling Scheme works. Basically, we illustrate the concepts of classical and inverse scheduling. Then, we show how inverse scheduling allows gaining new insights on the scheduling problem. From these insights, we derive the notion of tolerance. Finally, we present the flow of decisions to be made to address the tolerance scheduling problem.

13.4.1 Inverse and Reverse Scheduling

The problem of inverse scheduling is based on the concepts developed for inverse optimization in the study by Ahuja and Orlin (2001). The essential idea of inverse optimization is to start from a given solution (i.e., defined values of the decision variables), to adjust the value of the parameters as to make optimal the given solution. That is, in inverse optimization, the roles are reversed, the variables become fixed, and the parameters become "adjustable." These parameters can be the coefficients of the objective function or the coefficients of the inequalities of the constraints of the classic optimization problem. The problems of inverse scheduling tend to be similar to the latter, modifying internal parameters of the restrictions, being typical cases where the adjustable parameters are due dates (Brucker and Shakhlevich 2009) or processing times (Koulamas 2005).

Unlike the classic scheduling problems in which the parameters are known in advance, in inverse scheduling problems they are unknown and must be determined to make a given schedule optimal (Brucker and Shakhlevich 2011). Generally, the values that these parameters can take are restricted to certain intervals. For example, Brucker and Shakhlevich (2009) seek to analyze schedules in a single machine case with the goal of minimizing the maximum tardiness. The tardiness of a job j (T_j) is defined as the excess of the actual processing time with respect to its due date. It is calculated by

$$T_j(\pi, d) = \max\left\{C_j(\pi) - d_j, 0\right\} \tag{13.1}$$

Here π is a given schedule, d_j is the due date of job j, and C_j the completion time of job j under schedule π. Then, maximum tardiness is obtained according to

$$T_{max}(\pi, d) = \max_{j \in \mathbb{N}}\left\{T_j(\pi, d)\right\} \tag{13.2}$$

The classic scheduling problem involves finding π^* such that $T_{max}(\pi, d)$ is minimal, that is:

$$T_{max}\left(\pi^*, d\right) \le T_{max}(\pi, d), \quad \text{for any schedule } \pi. \tag{13.3}$$

Assume now the case of the single machine problem with maximum tardiness as objective. The only parameters we need are p_j and d_j, due to the processing times and due dates of work j, respectively. Suppose that only due dates can be modified, so for each job j we will have an interval $d_j \in \left[\underline{d}_j; \overline{d}_j\right]$. Therefore, the adjusted delivery date, $\hat{d}_j$, must be such that $\hat{d}_j \in \left[\underline{d}_j; \overline{d}_j\right]$ for each job j, producing a vector $\hat{d} = \left(\hat{d}_1, \hat{d}_2, \ldots, \hat{d}_n\right)$. Inverse scheduling amounts to minimize the Euclidean difference $\left\|\hat{d} - d\right\|$ (or any other norm) of d, making π optimal. Thus, the inverse scheduling problem is here

$$\min \left\| \hat{d} - d \right\| \tag{13.4}$$

$$s.t. \quad T_{max}\left(\pi, \hat{d}\right) \leq T_{max}\left(\sigma, \hat{d}\right), \tag{13.5}$$

$$\text{For any schedule } \sigma, \ \underline{d}_j \leq \hat{d}_j \leq \overline{d}_j, j \in N. \tag{13.6}$$

Other norms to measure the deviation $\hat{d} - d$ may yield different results (Ahuja and Orlin 2001).

Another problem arises when the goal is to achieve a fixed value of the objective function. Starting from an initial solution, the initial values of the parameter are adjusted so as to reach a given value of the objective function. This problem is known as the Reverse Scheduling problem, which in the case of minimizing maximum tardiness on a single machine is as follows. We seek to find $\hat{d}_j \in \left[\underline{d}_j; \overline{d}_j\right]$, such that the value T^* of the objective function T_{max} remains the same or gets improved.

There are other schemes that involve modifying the parameters, such that from an initial solution, adjust them so as to make the initial solution get a higher objective function value. This problem is called, as said, Reverse Scheduling Problem (Brucker and Shakhlevich 2009). For the previous case (a single machine with maximum tardiness), the Reverse Scheduling problem would look as follows:

$$\min \left\| \hat{d} - d \right\| \tag{13.7}$$

$$s.t. \quad T_{max}\left(\sigma, \hat{d}\right) \leq T^*, \tag{13.8}$$

$$\text{For any schedule } \sigma, \ \underline{d}_j \leq \hat{d}_j \leq \overline{d}_j, j \in N. \tag{13.9}$$

Here T^* is the given objective function value that the Schedule σ should improve by modifying the parameters of the problem. Other parameters are subject to choice so as to solve inverse and reverse scheduling problems, a sample of which can be found in Heuberger (2004).

13.4.2 The Tolerance Scheduling Problem

Rescheduling processes are triggered when the production process deviates from the established plan. These deviations derive from unexpected, and thus unplanned, events, so they were not considered in the design of the production schedule. However, since these events affect the performance of the schedule, corrective actions must be taken and decided according to rescheduling processes.

Rescheduling processes are event-driven. But the events that can trigger rescheduling processes must be distinguished from events that cannot start such processes. Furthermore, determining whether an event triggers or not, a rescheduling process is not enough since it is necessary to analyze also the magnitude of the event. The sensitivity of rescheduling processes to differences in the nature and magnitude of potential triggering events becomes an additional source of instability, making the production process subject to further uncertainty (Pinedo 2012; Framinan et al. 2014). To buffer the system against this, our proposal is to incorporate tolerances as to absorb disruptive events of low impact.

The Tolerance Scheduling problem can be conceived as a mathematical programming one (Rossit et al. 2019b). The decision-making process starts with the solution of a standard scheduling problem. The parameters have given values and the decision variables are related with the features of the production process. A solution is an optimal (or almost optimal) schedule. Then, we proceed to look for a range of tolerances for which the initial solution remains being acceptable. These tolerance ranges are computed assuming some shocks on the parameters of the initial problem, representing disruptive events that are not at the discretion of the scheduler. The issue here is to determine ranges of tolerance within which the produced goods are still considered appropriate, even if some degree of imperfection in the plan is allowed. Consider, for example, situations in which the actual processing times differ from the processing times specified by the original schedule. It is obvious that this affects the performance of the production process (e.g., worsening the makespan), calling for rescheduling the plan if the advantages of doing this exceed its costs.

To introduce the Tolerance Scheduling problem, consider the case previously presented by Brucker and Shakhlevich (2009). It is a simple single machine problem that can be solved in an inverse way, that is, by adjusting the parameters in such a way that a given schedule π becomes optimal. While in the problem of reverse scheduling, the idea is to make minor adjustments to the parameters with respect to initial reference values, in order to ensure that the π solution becomes good enough, in the Tolerance scheduling problem, a range of parameter values has to be found for which the objective function is acceptable and does not require being rescheduled.

Formally, given an optimal or near-optimal schedule π, $F(\pi) \approx T^*$, and the families of parameters d_j and p_j, we seek the maximal interval of variations for them, according to an *inertia factor*, δ, expressing the weight given to the stability of the system. A high δ indicates that the design favors high stability, meaning that few events can trigger reschedules. Then:

$$\max \left\| \hat{d} - d \right\| \tag{13.10}$$

$$s.t. \quad T_{max}\left(\pi, \hat{d}\right) \leq T_{max}\left(\sigma, \hat{d}\right) \cdot (1 + \delta), \tag{13.11}$$

$$T_{max}\left(\pi, \hat{d}\right) \leq T^* \cdot (1 + \delta), \tag{13.12}$$

$$\text{For any schedule } \sigma, \quad \underline{d}_j \leq \hat{d}_j \leq \overline{d}_j, \delta \geq 0, \; j \in N. \tag{13.13}$$

That is, the goal is to maximize the distance between the d parameters, while ensuring that schedule π improves over the original objective function up to an inertia factor $\delta \geq 0$. This provides a tool that not only detects possible rescheduling events but also determines whether or not to proceed with the rescheduling process. The choice of δ is not arbitrary: if the idea is to reschedule only at high levels of disruption, δ must be large. On the contrary, a low inertia system should be ready to react, requiring a lower δ.

This procedure is rather easy to automatize, providing another tool to be added to the DSS embedded in the CPPS, making the latter more prone to autonomous behavior. The value of δ should, in that case, be set at the design stage.

13.4.3 The Scheme of Smart Scheduling[1]

Smart Scheduling is introduced as a framework for scheduling in production planning by using the tools of Smart Manufacturing and Industry 4.0 environments. As shown in Fig. 13.4, this is handled mainly by a CPPS. The goal of Smart Scheduling is to automatize the solution to the scheduling problems in the integrated frame of CPPS.

We have to prove that the functionalities of a MSS (Manufacturing Scheduling System) (shown in Fig. 13.1), which is part of the MSS-scheduler ensemble in an autonomous CPPS, are captured in our approach. First, let us analyze the MSS in itself, then the scheduler and finally their combination. The functionalities provided by the MSS are easily integrated into a CPPS, since they can be connected to different business functions. CPPSs are by design computer (and physical) systems wired in such a way as to support all those functions (Monostori 2014). This amounts to say that CPPSs can replace and even improve over the MSS, being able to run quality analyses or failure diagnoses, increasing the global efficiency of the system.

This is not the case of the skills of the scheduler, which are not easy to integrate into the basic design of CPPSs (Lee et al. 2015). To extend the functionalities of CPPSs in this respect, we introduce the Smart Scheduling framework, which fundamentally addresses the Tolerance Scheduling Problem. Our proposal provides reliable optimization-based tools to assess the criticality of events, in terms of their nature and magnitude. Endowed with this ability, the CPPS can trigger reschedules only when the goals of the schedule become considerably affected, improving the resilience of the system and decreasing its sensitivity to the noise generated by the environment.

Smart Scheduling follows the logic of dynamic scheduling, as proposed in Fig. 13.5. In a first stage the problem, in either the standard or stochastic version,

[1]The majority of the analysis in this section is based upon the study by Rossit et al. (2019b).

is solved, yielding an initial schedule. Next, the tolerances are set to be used in solving the Tolerance Scheduling problem. In a second stage the production process is started, following the original schedule, until a disruptive event is detected. The event is analyzed to determine if it requires triggering a rescheduling procedure. If not, the production process continues. But if, by its very nature, the event belongs to the class of those that might trigger reschedules, the Tolerance Scheduling procedure is invoked. If the current schedule falls within range of the tolerance, the production process continues. Otherwise, a new schedule is generated to address the disruption caused by the event. Then, once the schedule is obtained applying the dynamic scheduling strategy, new tolerances are set for a future eventual solution of the Tolerance Scheduling

13.5 Examples

To show how Smart Scheduling works on the tolerance scheduling problem, we present a case study. Consider a dynamic scheduling problem in a single machine context with the goal of minimizing Total Weighted Tardiness (TWT). The resulting production process is subject, in the real world, to numerous disruptive events affecting the plan. In our case we will consider machinery breakage or failure, the usual instances studied in the literature (Zandieh and Gholami 2009; Ahmadi et al. 2016; Liu et al. 2017). We will draw data instances from the OR-library.

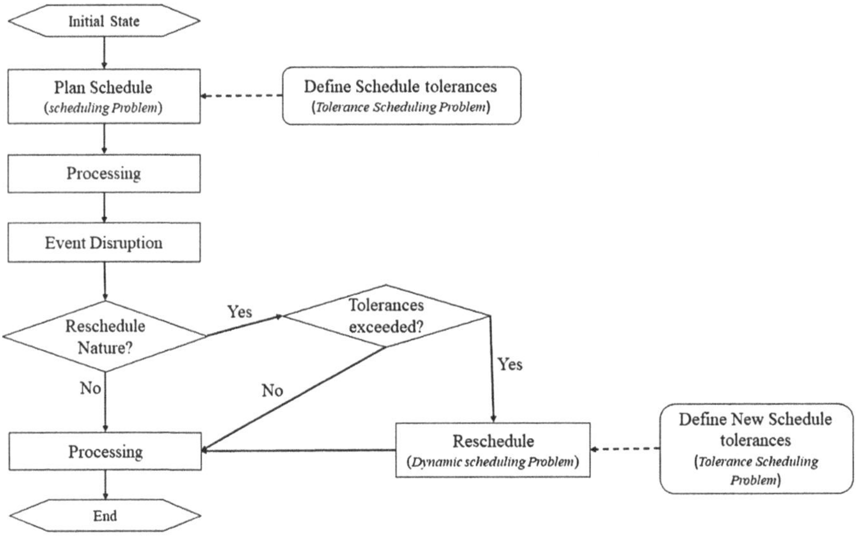

Fig. 13.5 Smart scheduling scheme

We will focus on the following issues: (1) obtain a quasi/optimal schedule, (2) defining tolerances, and (3) analyze events in the light of the schedule and the tolerances. (1) Has been widely studied in the literature and there exist a large variety of methods to solve the problem (Ruiz and Maroto 2005; Ruiz and Vázquez-Rodríguez 2010; Rossit et al. 2018), as well as a way of modeling the problem (Ivanov et al. 2012; Battaïa and Dolgui 2013; Pinedo 2012) and strategies to address it (Blazewicz et al. 2012; Sokolov et al. 2018; Dolgui et al. 2019). For our purposes we choose a heuristical method based on the ATC (Apparent Tardiness Cost) rule, which analyzes jobs according to two classical dispatch rules: WSPT (Weighted Shortest Processing Time) and Minimum Slack First. For details, see Sect. 14.2 in Pinedo (2012).

Then, (2) requires defining the tolerances for the schedule obtained with the ATC rule. In order to do this, we consider the model defined by (13.10)–(13.13) on the initial π. We have to define an inertia factor δ, which in turn yields the tolerance tol_π, defined as in (Rossit et al. 2019e):

$$tol_\pi = TWT \cdot \delta \tag{13.14}$$

The larger the values of δ, the larger will be the magnitude of events that the system is able to "absorb" without triggering a reschedule. On the contrary, lower values of δ make the system more reactive.

The full Smart Scheduling Scheme represented in Fig. 13.5 connects (1) and (2) distinguishing between events that require or not a reschedule. In our case, machine failures are the events that may or may not lead to reschedules. If it involves a small failure (e.g., in terms of the time required to be back in operation), its impact on the objective function can be assessed (via simulations).

We consider 40 cases from the OR-Library and generate different scenarios according to classical examples in the literature (Zandieh and Gholami 2009), with a different number of failures: 2 and 5. The repairing time of failures is represented by means of a uniform distribution over [30, 70], with an expected value of 50, similar to the processing times. Thus, each failure involves an extra operation, in average.

Each scenario is solved under different values of tolerances, according to Eq. (13.14). We considered different values of δ, namely 0%, 15%, 30%, 50%, 75%, and 100%. The results of the experiments are presented in Table 13.2.

Table 13.2 shows the average values of TWT for ten replicas in each case, as well as the corresponding standard deviations. Each row represents either two or five failures in the planning horizon. The first thing to notice is the impact of considering either two or five failures: the average TWT duplicates under no tolerance (0%), increasing from 3447.1 to 6646.8. With two failures and low tolerances, no changes in the average value of the objective function can be detected. For higher values (over 50%), the objective function gets impacted. On the other hand, for five failures, we can see that the tolerances at which the objective function changes are higher than those for two failures (100%). These variations in the objective function can be explained by the lower reactivity in the face of failures, leading to a higher tolerance to their impact and thus to less readjustments of the schedule.

Table 13.2 Experimental results

	Tolerance											
	0%		15%		30%		50%		75%		100%	
	Mean	SD	Mean	SD	Mean	SD	Mean	SD	Mean	SD	Mean	SD
Two failures	3447.1	717.3	3447.1	717.3	3447.1	717.3	3651.7	317.2	3651.7	317.2	3793.2	440.1
Five failures	6646.8	1253.7	6646.8	1253.7	6646.8	1253.7	6646.8	1253.7	6646.8	1253.7	6761.9	1035.4

Ref: *SD* standard deviation

13.6 Conclusions

In this chapter we presented the Tolerance Scheduling Problem, discussing its solution and its performance in a case study. We can see that tolerances allow modeling the capacity of the system to absorb deviations with respect to the initial schedule. We find that the larger that tolerance, the larger the deviation of the original plan that ensues.

Future work involves the analysis of this problem under the same objective function (TWT) in more complex production environments, like flow shop or job shop systems.

Acknowledgments This work was partially supported by the CYTED Ciencia y Tecnología para el Desarrollo project [P318RT0165].

References

Ahmadi, E., Zandieh, M., Farrokh, M., & Emami, S. M. (2016). A multi objective optimization approach for flexible job shop scheduling problem under random machine breakdown by evolutionary algorithms. *Computers & Operations Research, 73*, 56–66.

Ahuja, R. K., & Orlin, J. B. (2001). Inverse optimization. *Operations Research, 49*(5), 771–783.

Al-Hinai, N., & ElMekkawy, T. Y. (2011). Robust and stable flexible job shop scheduling with random machine breakdowns using a hybrid genetic algorithm. *International Journal of Production Economics, 132*(2), 279–291.

Almada-Lobo, F. (2016). The Industry 4.0 revolution and the future of manufacturing execution systems (MES). *Journal of Innovation Management, 3*(4), 16–21.

Arnaout, J. P. (2014). Rescheduling of parallel machines with stochastic processing and setup times. *Journal of Manufacturing Systems, 33*(3), 376–384.

Battaïa, O., & Dolgui, A. (2013). A taxonomy of line balancing problems and their solution approaches. *International Journal of Production Economics, 142*(2), 259–277.

Birge, J., Frenk, J. B. G., Mittenthal, J., & Kan, A. R. (1990). Single-machine scheduling subject to stochastic breakdowns. *Naval Research Logistics (NRL), 37*(5), 661–677.

Blazewicz, J., Ecker, K. H., Schmidt, G., & Weglarz, J. (2012). *Scheduling in computer and manufacturing systems*. Berlin: Springer Science & Business Media.

Bortolini, M., Ferrari, E., Gamberi, M., Pilati, F., & Faccio, M. (2017). Assembly system design in the Industry 4.0 era: A general framework. *IFAC-PapersOnLine, 50*(1), 5700–5705.

Brucker, P., & Shakhlevich, N. V. (2009). Inverse scheduling with maximum lateness objective. *Journal of Scheduling, 12*(5), 475–488.

Brucker, P., & Shakhlevich, N. V. (2011). Inverse scheduling: Two-machine flow-shop problem. *Journal of Scheduling, 14*(3), 239–256.

Dolgui, A., Ivanov, D., Sethi, S. P., & Sokolov, B. (2019). Scheduling in production, supply chain and Industry 4.0 systems by optimal control: Fundamentals, state-of-the-art and applications. *International Journal of Production Research, 57*(2), 411–432.

Dong, Y. H., & Jang, J. (2012). Production rescheduling for machine breakdown at a job shop. *International Journal of Production Research, 50*(10), 2681–2691.

Framinan, J. M., Leisten, R., & García, R. R. (2014). *Manufacturing scheduling systems. An integrated view on models, methods and tools*. New York: Springer.

Heuberger, C. (2004). Inverse combinatorial optimization: A survey on problems, methods, and results. *Journal of Combinatorial Optimization, 8*(3), 329–361.

Ivanov, D., & Das, A. (2020). Coronavirus (COVID-19 / SARS-CoV-2) and supply chain resilience: A research note. *International Journal of Integrated Supply Management*. in press.
Ivanov, D., & Dolgui, A. (2020). Viability of intertwined supply networks: Extending the supply chain resilience angles towards survivability. A position paper motivated by COVID-19 outbreak. *International Journal of Production Research*. in press.
Ivanov, D., Dolgui, A., & Sokolov, B. (2012). Applicability of optimal control theory to adaptive supply chain planning and scheduling. *Annual Reviews in Control, 36*(1), 73–84.
Ivanov, D., Dolgui, A., Sokolov, B., Werner, F., & Ivanova, M. (2016). A dynamic model and an algorithm for short-term supply chain scheduling in the smart factory industry 4.0. *International Journal of Production Research, 54*(2), 386–402.
Ivanov, D., Sethi, S., Dolgui, A., & Sokolov, B. (2018). A survey on control theory applications to operational systems, supply chain management, and Industry 4.0. *Annual Reviews in Control, 46*, 134–147.
Ivanov, D., Dolgui, A., & Sokolov, B. (2019). The impact of digital technology and Industry 4.0 on the ripple effect and supply chain risk analytics. *International Journal of Production Research, 57*(3), 829–846.
Katragjini, K., Vallada, E., & Ruiz, R. (2013). Flow shop rescheduling under different types of disruption. *International Journal of Production Research, 51*(3), 780–797.
Koulamas, C. (2005). Inverse scheduling with controllable job parameters. *International Journal of Services and Operations Management, 1*(1), 35–43.
Kuo, Y. H., & Kusiak, A. (2019). From data to big data in production research: The past and future trends. *International Journal of Production Research, 57*(15–16), 4828–4853.
Lee, E. A. (2008, May). Cyber physical systems: Design challenges. In *2008 11th IEEE International Symposium on Object Oriented Real-Time Distributed Computing (ISORC)*. IEEE (pp. 363–369).
Lee, J., Bagheri, B., & Kao, H. A. (2015). A cyber-physical systems architecture for industry 4.0-based manufacturing systems. *Manufacturing Letters, 3*, 18–23.
Lee, J., Jin, C., & Bagheri, B. (2017). Cyber physical systems for predictive production systems. *Production Engineering, 11*(2), 155–165.
Liu, F., Wang, S., Hong, Y., & Yue, X. (2017). On the robust and stable flowshop scheduling under stochastic and dynamic disruptions. *IEEE Transactions on Engineering Management, 64*(4), 539–553.
Monostori, L. (2014). Cyber-physical production systems: Roots, expectations and R&D challenges. *Procedia CIRP, 17*, 9–13.
Monostori, L., Kádár, B., Bauernhansl, T., Kondoh, S., Kumara, S., Reinhart, G., Sauer, O., et al. (2016). Cyber-physical systems in manufacturing. *CIRP Annals, 65*(2), 621–641.
Ouelhadj, D., & Petrovic, S. (2009). A survey of dynamic scheduling in manufacturing systems. *Journal of Scheduling, 12*(4), 417.
Peng, K., Pan, Q. K., Gao, L., Zhang, B., & Pang, X. (2018). An improved artificial bee colony algorithm for real-world hybrid flowshop rescheduling in steelmaking-refining-continuous casting process. *Computers & Industrial Engineering, 122*, 235–250.
Pinedo, M. (2012). *Scheduling*. New York: Springer.
Rahmani, D., & Heydari, M. (2014). Robust and stable flow shop scheduling with unexpected arrivals of new jobs and uncertain processing times. *Journal of Manufacturing Systems, 33*(1), 84–92.
Rossit, D., & Tohmé, F. (2018). Scheduling research contributions to smart manufacturing. *Manufacturing Letters, 15*, 111–114.
Rossit, D. A., Tohmé, F., & Frutos, M. (2018). The non-permutation flow-shop scheduling problem: A literature review. *Omega, 77*, 143–153.
Rossit, D. A., Tohmé, F., & Frutos, M. (2019a). An Industry 4.0 approach to assembly line resequencing. *The International Journal of Advanced Manufacturing Technology, 105*(9), 3619–3630.
Rossit, D. A., Tohmé, F., & Frutos, M. (2019b). Industry 4.0: Smart scheduling. *International Journal of Production Research, 57*(12), 3802–3813.

Rossit, D. A., Tohmé, F., & Frutos, M. (2019c). A data-driven scheduling approach to smart manufacturing. *Journal of Industrial Information Integration, 15*, 69–79.
Rossit, D. A., Tohmé, F., & Frutos, M. (2019d). Production planning and scheduling in Cyber-Physical Production Systems: A review. *International Journal of Computer Integrated Manufacturing, 32*, 385–395.
Rossit, D. A., Tohmé, F., & Frutos, M. (2019e). Designing a scheduling logic controller for Industry 4.0 environments. *IFAC-PapersOnLine, 52*(13), 2164–2169.
Ruiz, R., & Maroto, C. (2005). A comprehensive review and evaluation of permutation flowshop heuristics. *European Journal of Operational Research, 165*(2), 479–494.
Ruiz, R., & Vázquez-Rodríguez, J. A. (2010). The hybrid flow shop scheduling problem. *European Journal of Operational Research, 205*(1), 1–18.
Sokolov, B., Dolgui, A., & Ivanov, D. (2018). Optimal control algorithms and their analysis for short-term scheduling in manufacturing systems. *Algorithms, 11*(5), 57.
Vieira, G. E., Herrmann, J. W., & Lin, E. (2003). Rescheduling manufacturing systems: A framework of strategies, policies, and methods. *Journal of Scheduling, 6*(1), 39–62.
Zandieh, M., & Gholami, M. (2009). An immune algorithm for scheduling a hybrid flow shop with sequence-dependent setup times and machines with random breakdowns. *International Journal of Production Research, 47*(24), 6999–7027.
Zhang, J., Ding, G., Zou, Y., Qin, S., & Fu, J. (2019). Review of job shop scheduling research and its new perspectives under Industry 4.0. *Journal of Intelligent Manufacturing, 30*(4), 1809–1830.

The manufacturer's authorised representative in the EU is Springer Nature Customer Service Centre GmbH, Europaplatz 3, 69115 Heidelberg, Germany. If you have any concerns regarding our products, please contact ProductSafety@springernature.com

Printed and bound by CPI Group (UK) Ltd, Croydon, CR0 4YY
05/10/2026
02228124-0008